Nation, Ethnicity and Race on Russian Television

Russia, one of the most ethno-culturally diverse countries in the world, provides a rich case study on how globalization and associated international trends are disrupting and causing the radical rethinking of approaches to inter-ethnic cohesion. The book highlights the importance of television broadcasting in shaping national discourse and the place of ethno-cultural diversity within it. It argues that television's role here has been reinforced, rather than diminished, by the rise of new media technologies.

Through an analysis of a wide range of news and other television programmes, the book shows how the covert meanings of discourse on a particular issue can diverge from the overt significance attributed to it, just as the impact of that discourse may not conform with the original aims of the broadcasters. The book discusses the tension between the imperative to maintain security through centralized government and overall national cohesion that Russia shares with other European states, and the need to remain sensitive to, and to accommodate, the needs and perspectives of ethnic minorities and labour migrants. It compares the increasingly isolationist popular ethno-nationalism in Russia, which harks back to 'old-fashioned' values, with the similar rise of the Tea Party in the United States and the UK Independence Party in Britain.

Throughout, this extremely rich, well-argued book complicates and challenges received wisdom on Russia's recent descent into authoritarianism. It points to a regime struggling to negotiate the dilemmas it faces, given its Soviet legacy of ethnic particularism, weak civil society, large native Muslim population and overbearing, yet far from entirely effective, state control of the media.

Stephen Hutchings is Professor of Russian Studies at the University of Manchester, UK.

Vera Tolz is Sir William Mather Professor of Russian Studies at the University of Manchester, UK.

BASEES/Routledge Series on Russian and East European Studies

This series is published on behalf of BASEES (the British Association for Slavonic and East European Studies). The series comprises original, high-quality, research-level work by both new and established scholars on all aspects of Russian, Soviet, post-Soviet and East European Studies in humanities and social science subjects.

1 **Ukraine's Foreign and Security Policy, 1991–2000**
Roman Wolczuk

2 **Political Parties in the Russian Regions**
Derek S. Hutcheson

3 **Local Communities and Post-Communist Transformation**
Edited by Simon Smith

4 **Repression and Resistance in Communist Europe**
J.C. Sharman

5 **Political Elites and the New Russia**
Anton Steen

6 **Dostoevsky and the Idea of Russianness**
Sarah Hudspith

7 **Performing Russia – Folk Revival and Russian Identity**
Laura J. Olson

8 **Russian Transformations**
Edited by Leo McCann

9 **Soviet Music and Society under Lenin and Stalin**
The baton and sickle
Edited by Neil Edmunds

10 **State Building in Ukraine**
The Ukranian parliament, 1990–2003
Sarah Whitmore

11 **Defending Human Rights in Russia**
Sergei Kovalyov, dissident and Human Rights Commissioner, 1969–2003
Emma Gilligan

12 **Small-Town Russia**
Postcommunist livelihoods and identities: a portrait of the Intelligentsia in Achit, Bednodemyanovsk and Zubtsov, 1999–2000
Anne White

13 **Russian Society and the Orthodox Church**
Religion in Russia after Communism
Zoe Knox

14 **Russian Literary Culture in the Camera Age**
The word as image
Stephen Hutchings

15 **Between Stalin and Hitler**
Class war and race war on the Dvina, 1940–46
Geoffrey Swain

16 **Literature in Post-Communist Russia and Eastern Europe**
The Russian, Czech and Slovak fiction of the changes, 1988–98
Rajendra A. Chitnis

17 **The Legacy of Soviet Dissent**
Dissidents, democratisation and radical nationalism in Russia
Robert Horvath

18 **Russian and Soviet Film Adaptations of Literature, 1900–2001**
Screening the word
Edited by Stephen Hutchings and Anat Vernitski

19 **Russia as a Great Power**
Dimensions of security under Putin
Edited by Jakob Hedenskog, Vilhelm Konnander, Bertil Nygren, Ingmar Oldberg and Christer Pursiainen

20 Katyn and the Soviet Massacre of 1940
Truth, justice and memory
George Sanford

21 Conscience, Dissent and Reform in Soviet Russia
Philip Boobbyer

22 The Limits of Russian Democratisation
Emergency powers and states of emergency
Alexander N. Domrin

23 The Dilemmas of Destalinisation
A social and cultural history of reform in the Khrushchev era
Edited by Polly Jones

24 News Media and Power in Russia
Olessia Koltsova

25 Post-Soviet Civil Society
Democratization in Russia and the Baltic States
Anders Uhlin

26 The Collapse of Communist Power in Poland
Jacqueline Hayden

27 Television, Democracy and Elections in Russia
Sarah Oates

28 Russian Constitutionalism
Historical and contemporary development
Andrey N. Medushevsky

29 Late Stalinist Russia
Society between reconstruction and reinvention
Edited by Juliane Fürst

30 The Transformation of Urban Space in Post-Soviet Russia
Konstantin Axenov, Isolde Brade and Evgenij Bondarchuk

31 Western Intellectuals and the Soviet Union, 1920–40
From Red Square to the Left Bank
Ludmila Stern

32 The Germans of the Soviet Union
Irina Mukhina

33 Re-constructing the Post-Soviet Industrial Region
The Donbas in transition
Edited by Adam Swain

34 Chechnya – Russia's "War on Terror"
John Russell

35 The New Right in the New Europe
Czech transformation and right-wing politics, 1989–2006
Seán Hanley

36 Democracy and Myth in Russia and Eastern Europe
Edited by Alexander Wöll and Harald Wydra

37 Energy Dependency, Politics and Corruption in the Former Soviet Union
Russia's power, Oligarchs' profits and Ukraine's missing energy policy, 1995–2006
Margarita M. Balmaceda

38 Peopling the Russian Periphery
Borderland colonization in Eurasian history
Edited by Nicholas B. Breyfogle, Abby Schrader and Willard Sunderland

39 Russian Legal Culture Before and After Communism
Criminal justice, politics and the public sphere
Frances Nethercott

40 Political and Social Thought in Post-Communist Russia
Axel Kaehne

41 The Demise of the Soviet Communist Party
Atsushi Ogushi

42 Russian Policy towards China and Japan
The El'tsin and Putin periods
Natasha Kuhrt

43 Soviet Karelia
Politics, planning and terror in Stalin's Russia, 1920–1939
Nick Baron

44 Reinventing Poland
Economic and political transformation and evolving national identity
Edited by Martin Myant and Terry Cox

45 The Russian Revolution in Retreat, 1920–24
Soviet workers and the new communist elite
Simon Pirani

46 Democratisation and Gender in Contemporary Russia
Suvi Salmenniemi

47 Narrating Post/Communism
Colonial discourse and Europe's borderline civilization
Nataša Kovačević

48 Globalization and the State in Central and Eastern Europe
The politics of foreign direct investment
Jan Drahokoupil

49 Local Politics and Democratisation in Russia
Cameron Ross

50 The Emancipation of the Serfs in Russia
Peace arbitrators and the development of civil society
Roxanne Easley

51 Federalism and Local Politics in Russia
Edited by Cameron Ross and Adrian Campbell

52 Transitional Justice in Eastern Europe and the former Soviet Union
Reckoning with the communist past
Edited by Lavinia Stan

53 The Post-Soviet Russian Media
Conflicting signals
Edited by Birgit Beumers, Stephen Hutchings and Natalia Rulyova

54 Minority Rights in Central and Eastern Europe
Edited by Bernd Rechel

55 Television and Culture in Putin's Russia
Remote control
Stephen Hutchings and Natalia Rulyova

56 The Making of Modern Lithuania
Tomas Balkelis

57 Soviet State and Society Under Nikita Khrushchev
Melanie Ilic and Jeremy Smith

58 Communism, Nationalism and Ethnicity in Poland, 1944–1950
Michael Fleming

59 Democratic Elections in Poland, 1991–2007
Frances Millard

60 Critical Theory in Russia and the West
Alastair Renfrew and Galin Tihanov

61 Promoting Democracy and Human Rights in Russia
European organization and Russia's socialization
Sinikukka Saari

62 The Myth of the Russian Intelligentsia
Old intellectuals in the new Russia
Inna Kochetkova

63 Russia's Federal Relations
Putin's reforms and management of the regions
Elena A. Chebankova

64 Constitutional Bargaining in Russia, 1990–93
Institutions and uncertainty
Edward Morgan-Jones

65 Building Big Business in Russia
The impact of informal corporate governance practices
Yuko Adachi

66 Russia and Islam
State, society and radicalism
Roland Dannreuther and Luke March

67 Celebrity and Glamour in Contemporary Russia
Shocking chic
Edited by Helena Goscilo and Vlad Strukov

68 The Socialist Alternative to Bolshevik Russia
The Socialist Revolutionary Party, 1917–1939
Elizabeth White

69 Learning to Labour in Post-Soviet Russia
Vocational youth in transition
Charles Walker

70 Television and Presidential Power in Putin's Russia
Tina Burrett

71 Political Theory and Community Building in Post-Soviet Russia
Edited by Oleg Kharkhordin and Risto Alapuro

72 Disease, Health Care and Government in Late Imperial Russia
Life and death on the Volga, 1823–1914
Charlotte E. Henze

73 Khrushchev in the Kremlin
Policy and government in the Soviet Union, 1953–1964
Edited by Melanie Ilic and Jeremy Smith

92 The Politics of HIV/AIDS in Russia
Ulla Pape

93 The Capitalist Transformation of State Socialism
The making and breaking of State Socialist society, and what followed
David Lane

94 Disability in Eastern Europe and the Former Soviet Union
History, policy and everyday life
Edited by Michael Rasell and Elena Iarskaia-Smirnova

95 The Making and Breaking of Soviet Lithuania
Memory and modernity in the wake of war
Violeta Davoliūtė

96 Ideologies of Eastness in Central and Eastern Europe
Tomasz Zarycki

97 Cinema, State Socialism and Society in the Soviet Union and Eastern Europe, 1917–1989
Re-visions
Edited by Sanja Bahun and John Haynes

98 Ethnic Relations in Post-Soviet Russia
Russians and non-Russians in the North Caucasus
Andrew Foxall

99 Eastern Europe and the Challenges of Modernity, 1800–2000
Stefano Bianchini

100 Nation, Ethnicity and Race on Russian Television
Mediating post-Soviet difference
Stephen Hutchings and Vera Tolz

101 Russia–China Relations in the Post-Crisis International Order
Marcin Kaczmarski

Nation, Ethnicity and Race on Russian Television

Mediating post-Soviet difference

Stephen Hutchings and Vera Tolz

Routledge
Taylor & Francis Group
LONDON AND NEW YORK

First published 2015
by Routledge

2 Park Square, Milton Park, Abingdon, Oxfordshire OX14 4RN
711 Third Avenue, New York, NY 10017

Routledge is an imprint of the Taylor & Francis Group, an informa business

First issued in paperback 2017

British Library Cataloguing in Publication Data
A catalogue record for this book is available from the British Library

Library of Congress Cataloging in Publication Data
A catalog record for this book has been requested

ISBN: 978-1-138-85328-7 (hbk)
ISBN: 978-0-8153-6232-6 (pbk)

Typeset in Times
by Sunrise Setting Ltd, Paignton, UK

Contents

Figures

Acknowledgements

The authors of this book have benefited immensely from the support and advice of a number of institutions and individuals and we would like to express our gratitude to them. We thank the University of Manchester for guaranteeing a stimulating and supportive working environment, and the UK's Arts and Humanities Research Council for funding the three-year project (2010–13) of which the book represents the primary outcome. Above all, the AHRC award (grant number AH/HO18964/1) made it possible for us to employ two outstanding postdoctoral researchers, Dr Sue-Ann Harding and Dr Elisabeth Schimpfoessl. They both contributed enormously to the research on which the book is based, establishing and managing the archive of television recordings that provided the bulk of our material, interviewing journalists and generating the statistics that form the basis of Chapter 2. Dr Ilya Yablokov was also a key member of the project team, conducting interviews, monitoring developments on Dozhd' TV and providing a wealth of vital insight and information. We are grateful to Timur Badmatsyrenov for compiling the archive of *Vesti-Buriatiia* news recordings which we analyse in Chapter 6.

Some of the material we present has already been published in a different form and we are grateful to the journals in question for giving us their permission to adapt and develop it for the purposes of this book. A different version of Chapter 4 appeared in *Slavic Review* 71/4, 2012, 873–99. Elements from Chapter 5 were published in *Russian Review*, 72/3, 2013, 470–91, and *Nationalities Papers*, 41/6, 2013, 1083–99.

We owe a large debt of gratitude to Peter Sowden, who commissioned our book for Routledge, and on whose professionalism, good sense and unfailing enthusiasm we have come to rely. As always, there are a large number of colleagues and other individuals whose judicious comments and constructive criticisms have significantly enriched and strengthened the arguments we make. They include Peter Rutland, Hilary Pilkington, Richard Sakwa, Lara Ryazanova-Clarke, Kenzie Burchell, Ellen Mickiewicz, Elena Prokhorova, Luke March, Tony Anemone, Viktor Shnirel'man, Helge Blakkisrud, Pål Kolstø, Marlene Laruelle, Vlad Strukov, Martin Dewhirst after Strukov and Helena Goscilo. Finally, we would like to thank our families for their unfailing support, encouragement and understanding throughout the project. And, as always, any flaws or weaknesses that remain are entirely our own responsibility.

Note on transliteration and translation

In transliterating from Russian into English, we have used the Library of Congress System throughout, except where custom has persistently favoured an alternative spelling. Unless stated otherwise, all translations from Russian are our own.

Introduction

A clash of two Russias, a tale of two cities

Two Russias

This book is ultimately concerned with the conditions and prospects for statehood in Russia. It intervenes in the debate around that issue by providing a country-specific case study of a phenomenon whose wider import transcends state boundaries: the fate of complex multicultural policies in the age of globalization, and the role of the broadcast media in shaping it. Broadly speaking, this is a study of the contemporary media's role in fostering or hindering community cohesion. Specifically, the book explores Russian state-aligned television's approaches to ethnicity, race and nationhood, contributing to our grasp of a highly topical issue – the nature and effectiveness of the tools used by the leaders of the largest transition state, still at an early stage of its formation, to forge a sense of belonging among its citizens. The book argues that the importance of both elite and popular understandings of nationhood to Russian political culture and policy-making has been seriously underestimated in the scholarly literature, and that milestone events such as the Ukraine conflict of 2014 cannot be fully understood without reference to it. For the first time in post-communist studies we attempt to bring together two theoretical strands: one addressing the nature of Soviet Russian imperialism and nation-building, and the second treating Russian television news as a consensus-management tool. To introduce the rationale for our dual focus, and to set the scene for the book, we begin *in medias res*: with a brief extract from a television talk show discussion, and a single, ephemeral news report.

First, the talk show.

Following the now infamous Pussy Riot episode in February 2012, Channel 1's *V kontekste* (*In Context*) discussion programme, hosted by the leading television personality Maksim Shevchenko, featured a debate about the incident in its edition of 15 March 2012.[1] The guests included a representative of the Orthodox Church, Maksim Kozlov; a leader of a moderate human rights group, Ol'ga Kostina; an art critic, Marat Gel'man; and a liberal campaigner, Kseniia Sobchak. In an exchange between Kozlov – who, in raising the stakes over Orthodoxy's supremacy in Russia, applies the same sardonic label to the women of Pussy Riot as that applied earlier by Shevchenko himself – and Sobchak, who invokes the waning mantra of official multi-confessionalism, they state:

> Kozlov: We have two cultures. One is connected to the great Russian culture … And there is another one, which for a long time will have these 'maidens' (*devits*) as its symbol. This is a clash of civilizations (*tsivilizatsionnyi konflikt*) between traditional Russian culture and this alluvial, film-like layer (*s etim nanosnym plenochnym sloem*).
>
> Sobchak: I think that today a conflict is taking place – here you are right. But, in my opinion … the powers that be are advancing Orthodoxy as their core ideology in what is a multi-confessional country … [2]

Huntington's 'civilizational clash' rhetoric, along with the related – but not identical – term 'conflict of cultures', has, as we shall later find, been used in reference to the rise of tension between Russians and North Caucasians (as well as Central Asians).[3] It is this which provides the most obvious hint at the links between our dual concerns. For although Kozlov is referring to Russia's 'civilizational conflict' with the deleterious western values adopted by counter-cultural forces like Pussy Riot, he clearly perceives Russia as a community bound together by 'traditional Russian culture'. His view therefore has implications for inter-ethnic relations within Russia. It is this that alerts Sobchak, who identifies an unresolved tension within the Kremlin's entire nation-building discourse: between the vision of Russia as a multi-ethnic, multi-confessional society of equal citizens and that of Russia as a national state in which Russian culture, language and Orthodoxy have primacy.

On the face of it, while the Pussy Riot incident became a leitmotif for the opposition protests which followed the controversial 2011 parliamentary elections and Vladimir Putin's re-election as president in March 2012, and which seemed momentarily to mark the dramatic and unexpected birth of a 'Russian Spring', it appears to have little relevance to the issue of ethnic cohesion in Russia. Indeed, conventional wisdom, and the weight of scholarly discourse, would have it that state television coverage of inter-ethnic relations is at most a sideshow in the drama at centre stage in Russia: that which pits Vladimir Putin's semi-authoritarian and increasingly paranoid regime against the forces for democratic change emboldened by the mobilizing potential offered by new social media platforms such as Twitter.[4] An indication of this conventional wisdom at work can be found in the contrast between the feverish excitement awakened in BBC journalists by the post-election protests in late 2011 and in 2012 – and, indeed, by precursors to those protests, including the outcry over the November 2010 re-sentencing of the dissident businessman Khodorkovskii – and the BBC's deafening silence about the terrifying race riots which occurred in Moscow only one month after the Khodorkovskii trial.[5]

Why, one might wonder, should things be otherwise? The parlous and inflammatory state of inter-ethnic relations in Russia is arguably a secondary issue, unlikely to be resolved until the primary tumour ravaging the body of Russian society is excised. Moreover, state television appears on the surface to offer the least promising tool of excision. Ever subject to the baleful influence of the Kremlin, and seemingly run by its acolytes, it is now struggling to keep pace with the paradigm shifts in media technology and practice, particularly inasmuch as their main beneficiaries are precisely those younger, more educated, metropolitan Russians who lead the political protest movement. It is, however, precisely these assumptions that we seek to challenge.

For the brief and inconsequential exchange between Kozlov and Sobchak is but one thread in a tapestry of connections, covert and overt, that tie the subject of inter-ethnic relations to that of Russia's political future. On one hand Kozlov and Sobchak are, mutual antagonism notwithstanding, both striving to attain the common ground of a broad societal consensus which will enable them to express their disagreement, and which is the precondition for civil society. On the other, our analysis indicates that it is issues bearing an ethnic tinge – in particular the widespread concern over corruption and the abuse of power – which provide the territory over which the battle to formulate, and to claim ownership of, that common ground takes place.

The central importance, and elusiveness, of the consensual idiom required for dialogic exchange, and thus for the Habermasian rational public discourse without which Russia will never achieve a stable civil society, is broached later in the *V kontekste* discussion. In a significant contribution, Kozlov articulates a meta-description of the course of the entire conversation. Spoken from the Orthodox position, his willingness to subject himself to a debate with those calling for broader freedoms draws both the discussion itself and the demand for democracy that his opponents advance within it inside the bounds of the consensual realm:

> This action was a conscious provocation … against the slowly maturing consensus in our society. Against the fact that, whilst we are maximally different, we can reach agreement with one another if we gather together the foundations of some shared concepts.[6]

The action he refers to is, of course, not an ethnic conflict, but what he and significant sections of Russian society perceived to be a provocation against traditional Russian values (the shocking 'punk prayer' performed in Moscow's Cathedral of Christ the Saviour). What he demonstrates all too starkly is the paradox that to include and unify is necessarily to exclude and divide: a community built around shared, consensual values must possess clearly defined boundaries and, thus, rules of entry. For Slavoj Žižek, democracy's very claim to universality is based on an original, now suppressed, 'founding act' of expulsion. It does not require a leap of faith to recognize that among the most clearly enunciated rules of admission and exclusion are those relating to the markers of ethnicity. Thus, Žižek goes on to trace the original act of exclusion in the past to what he terms the 'social symptom' in the present, 'the part which, although inherent to the existing universal order, has no "proper place" within it (say, illegal immigrants)'.[7]

Indeed, the paradox of consensus overlaps with the second way in which inter-ethnic relations insert themselves at the core of the struggle for Russia's future: before joining that battle, it is vital to establish the identity of the state on behalf of which it is being waged. Sobchak is in no doubt that the Russia whose future she is fighting for is characterized by an inclusive multi-confessionalism (and, by extension, multi-ethnicity). Kozlov, by contrast, perceives his nation in an altogether more exclusionary manner. His 'false' Russia is primarily intended to invoke deleterious western influence, but his reference to traditional Russian cultural values extends to the non-Orthodox peoples of the North Caucasus. In this context, indeed, Kozlov's divisive rhetoric is not restricted to the Orthodox Church. Aleksei Naval'nyi, the de facto figurehead of the democratic protest movement, is himself not averse to marching under such banners as 'Stop Feeding the Caucasus' and 'Russia for the Russians', a contradiction to which we return in a later chapter. Somewhat inconveniently, democratic transformation and ethnic inclusivity do not always coincide; to what extent the former is dependent on the latter is a debatable point well beyond Russia, of course, but it is a debate with particularly acute consequences for that country.

Nonetheless, the very fact that Kozlov and Sobchak presented their contrasting visions within a discussion show broadcast on Russia's leading state television channel is indicative of a level of dialogic exchange which, although tame by western standards, belies the broadcaster's image as a mere purveyor of Kremlin propaganda. Such licence is not often accorded to political issues on Channel 1 (though until the Ukraine crisis it was more frequent than some western commentators might imagine), and the fact that deep discord over the significance of events which bring ethnic and cultural difference into sharp relief is one of a number of exceptions brings us to the third factor linking the topic of our book to issues of civil society and Russia's political future. For, as we shall find, the terrain of inter-ethnic difference has the capacity to blur into and contaminate adjacent, but separate, fields. This is true in the 'positive' sense of fostering pluralism within public discourse (the Kremlin's own ambivalence over inter-ethnic cohesion policy facilitates the airing of disagreements of the sort demonstrated by the *V kontekste* clash). But it also has a profoundly 'negative' dimension, as indicated by the ethnicizing of consensus-building. The question of which of the two tendencies will eventually prevail is key to Russia's future.

It is also apparent – and the *V kontekste* exchange suggests as much – that television will have more of a say in determining the answer to the question than might be supposed. This, then, takes us to the core of the second challenge to percieved wisdom that we are posing. Here too there are three ways in which to interrogate the received view. First, although television is losing ground to other, freer and more flexible media in terms of its relevance to Russian civil society, it stubbornly remains the information source of choice for the majority of Russians,[8] just as Putin's presidency retains more than respectable levels of support beyond the rarefied world of the metropolitan intelligentsia.[9] Transforming Russia will be impossible without changing the outlook of the many millions of ordinary

Russians who continue to watch, enjoy and generally trust Channel 1 and other state broadcasters.

Second, as the *V kontekste* clash also illustrates, it is on television rather than via the Internet or new social media forms that the encounter between official and sub-official thinking of the sort represented by Kozlov, Kostina and, indeed, the host Shevchenko and the counter-cultural views of Sobchak, Gel'man and others is most likely to occur. Without such dialogues, any genuine transfer of power is likely to involve violence. Television's vital mediatory capacity extends beyond its function as a site for dialogue. It is precisely because it appeals to (and, in Russia, is oriented towards) a mass audience that it also serves as the point at which official diversity management policy must engage with the xenophobic fears and prejudices of a general populace resentful at an influx of post-Soviet migrant workers over which the Kremlin struggles to assert control. Again, it is difficult to underestimate the associated responsibilities that state-aligned broadcasting acquires.

The fact that national television's function in broadcasting to mass audiences is now threatened by new media narrowcasting modes (including satellite and cable television) adds to its responsibilities. As Evgenii Morozov has argued, far from offering a basis for national, democratic renewal, web-based communication and social media platforms have a tendency to fragment societies into closed, self-affirming communities of the like-minded.[10] For all its lack of agility, prohibitive expense and propensity to orient itself towards state orthodoxies, television remains one of the most effective tools for maintaining the vast, imagined community that is the modern nation. Moreover, it does so not merely by stating the fact of its existence, but by performing it through the very genres it adopts (in his own, limited way, Kozlov is both naming the 'maturing consensus' to which he refers and enacting it by striving to draw hostile interlocutors onto common ground). If officially sanctioned discourse fails to develop this role considerably further than the limited manner in which it is fulfilled in *V kontekste* (we shall refer more than once to the importance of television's 'performative' function), one Russia may literally become two and then multiple Russias, even as the prospects for national democratic renewal recede.

Two cities

Some of the preceding discussion would not be out of place in a wider European context. In order to signal the position we will be adopting in relation to this overlap, and to present the remaining issues which our book addresses, we turn to a second introductory example – this time taken from Channel 1's sister broadcaster. On 19 September 2010, as the repercussions of France's expulsion of Roma migrants who had travelled from Romania resonated across Europe's airwaves, the Russian state broadcaster Channel Rossiia's *Vesti* news bulletin ran a report on the crisis, describing how one Moscow-based Roma encampment was responding by co-writing a protest letter to French president, Nicolas Sarkozy. Shot from inside the encampment, it commences with a snapshot of a contented, legally compliant community at ease with its non-Roma neighbours, and enjoying a life superior to that of its European equivalents or its less fortunate domestic cousins:

> Rustam lives in Moscow … and runs a music school for gypsy children. He is sure that he owes his success to his qualifications: 'What can they do, these uneducated gypsies? Go and earn money on the streets! But if they know English, French, mathematics, they can easily find work'. Viacheslav Valentinovich is their neighbour: 'As you see, we get along well … Everything is fine, unlike in Romania'.

But as the 'baron' of the encampment phones his anxious relatives in Paris, he comments revealingly: 'If we can't give some sort of support to gypsies in France, then we could probably also be told that people like us are not needed here.' This seemingly off-the-cuff remark offsets the lyrical musings with which the journalist concludes: 'How can one not love gypsies if all the world's best songs are gypsy songs?', he asks with a rhetorical flourish, accompanied by the briefest snatch of a traditional gypsy ballad.

The Roma story translates the European issue into local idiom but also provides an interpretative key, ensuring that it is read according to the code implicit in the principle of 'inter-ethnic strife' (*mezhnatsional'naia rozn'*), a Soviet notion that applies to Russia 'virtually' (as something that happens elsewhere, or threatens to occur in the future), and which is outlawed under Russian legislation.[11] In an elision reflecting Russia's confused policy surrounding race, ethnicity and nationality, *mezhnatsional'naia rozn'* translates as 'inter-nationality strife', though it covers what in English is normally termed 'ethnic conflict'. *Vesti*'s implicit invocation of the concept is intended both to indicate the relative harmony between Russian gypsies and non-gypsies and to warn that those relations might deteriorate.

The offbeat story of the Moscow baron encapsulates several of our book's central concerns. First, it confirms that Russia is one of many European countries sharing the migration-related concerns generated by the new wave of globalization. (It is vital to point out, however, that – since Russia's significant gypsy population has been established for centuries – Rossiia's attempt to link Moscow to Paris, and to locate its report in the context of global migration trends, reflects a misleading elision, to which we shall return.) With the Soviet Union receding into history, and despite the heavy hand of Putin, even news agendas and media formats operate in a globalized environment and bear more than a passing resemblance to those found elsewhere; fly-on-the-wall reportage featuring ordinary people at the centre of events is reassuringly familiar, for example. Television's power to engage audiences with the embodied, lived experiences of those like themselves is no more lost on Russian newsmakers than it is on their western counterparts.

Yet Russia sees itself as different. The implication of the Rossiia reporter's privileged access to the baron's home, and of the concerned phone call to Paris, is that Russian inter-ethnic relations are more harmonious than those responsible for the plight of the Roma in France. Reality tells a different story; migration flows to Russia are second only to those to the United States and anti-migrant and anti-minority attitudes, underpinned by a post-imperial angst and attendant suspicion of 'the ethnic other', are no less pronounced than in other European countries with colonial legacies: the baron's less-than-subtle warning hints at as much.[12]

Our gloss on *mezhnatsional'naia rozn'* confirms that approaches to inter-ethnic cohesion in Russia continue to reflect the lingering influence of Soviet policies on nationality, representing the most ambitious attempt ever to transform a sprawling centripetal empire into a multicultural, federal state. In this far-reaching Soviet experiment, affirmative actions in support of minority cultures went hand in hand with repressions, and both policy lines at times implicitly reflected the perceptions of fixed, unchanging ethnic characteristics transmitted from generation to generation. It is hardly surprising, therefore, that the inept stereotyping of the reporter described above, and his reluctance to sacrifice the evocative connotations of the term 'gypsy' (*tsygan*) for the politically correct 'Roma' appellation common throughout Europe, should betray an unfamiliarity with the careful language of 'respect for minorities' now well established in the official discourses and quality press of west European nations – though, as hysteria in the UK in early 2014 about a feared EU-generated wave of Roma migration from Bulgaria and Romania testifies, this is far from fully embedded in the consciousness of those nations' popular media.

The most glaring consequence of Russia's communist past, however, is the continuing absence of western-style parliamentary democracy and a strong 'fourth estate', and thus, as we suggested earlier, of a public sphere united around consensual values. Under Putin in particular, and contrary to the relative licence accorded to programmes like *V kontekste*, news bulletins such as *Vesti* have become constrained in what they can report and how they may report it.[13] Significantly, of the 29 other stories that *Vesti* ran in the autumn of 2010 covering the Roma crisis, only one dealt with problems in the relationship between Russia and its large Roma community, despite genuine concerns about such problems among ordinary Russian citizens.[14]

Our Moscow–Paris conversation reveals a further legacy of the Soviet era: the stubbornly persistent tendency of the state-aligned media to construct Russian identity in relation to a western, as well as an 'ethnic', 'other'.[15] The predomination of reports on discrimination endured by Roma minorities elsewhere highlight a transparent effort to exploit European 'misfortunes' to Russian advantage, a strategy in relation to the representation of gypsies deriving from the late imperial period in Russia.

What is most revealing about *Vesti*'s instrumental approach to the Roma crisis is what it conceals. The narrative progression of the report generates a three-way contrast pitting a well-integrated Russian Roma community against: (i) its European counterparts; (ii) images of an alienated domestic underclass plagued by criminality and poverty; and (iii) other, unnamed minority communities lacking the gypsies' romantic qualities. It is unclear whether the baron 'let slip' his comment about the domestic implications of the crisis (one of the perils of *vox pop* is that ordinary people do not always adhere to their 'script'), or whether he was primed to do so. (The airing of occasional warnings that currently stable domestic situations might deteriorate is characteristic of official discourse.) The reporter clearly aims to juxtapose the European crisis with a somewhat more 'harmonious' situation in Russia. Yet behind this one-dimensionality is a thinly disguised polemic including the notion that disharmony reigns supreme in Russia too – for some of Europe's most explosive conflicts are those involving ethnic Russians and North Caucasians or Central Asian migrants. Gypsies thus provide a foil for what, as we shall show, was one of the perennial *non-dits* of Russian state broadcasting, but which, as we also indicate, has emerged from the shadows with a vengeance following the 2012 presidential elections. Moreover, the reporter's own harmonious presence in the community instigates a productive interplay between sameness and inclusion (an ethnic Russian implicitly addressing a viewing community of ethnic Russians, he is welcomed by his hosts) and exclusion and difference (he is reporting on the community *because* it is ethnically 'other') which serves as a model for exploring inter-ethnic relations at the national level.

There is, then, a disjunction between what the report 'says' (that Europe has a problem with migrants), what it 'implies' (that Russia boasts comparatively cohesive inter-ethnic relations) and what it 'does': both overtly (enacting the 'coming together' of communities through its narrative structures and modes of address) and inadvertently (pre-empting perceptions of a rather different reality, and in so doing conferring credence on that reality). The implications of the disjunction for national cohesion and the condition of Russian public discourse are only too apparent.

Concentrated here, then, are some key questions motivating the chapters to follow, all of which ultimately serve to illuminate the relationship between Russian television's diversity management function and the prospects for civic renewal that is our overarching concern: (i) how does Russia's status as a post-imperial European state shape its broadcasters' representations of problems arising from accelerating population movements, and from transnational flows of media formats and content? (ii) How does Russia's Soviet legacy differentiate its approach to such problems from those of its European neighbours? (iii) How do the latter two factors combine to detach referential ('saying'), associative ('implying') and performative ('doing') meanings?

Contemporary Russia is, in fact, a new state, struggling to stabilize and unify a plurality of identities in flux following the sudden disintegration of a vast, multi-ethnic empire and to formulate policies capable of dealing with the combustible aftermath of that event. That it is doing so at a time when many European states are facing doubts about the efficacy of multiculturalist policies in ameliorating the consequences of the demise of their own empires only adds to the complexity of the situation. The broader European context is in fact central to the rationale for this study, which we now present, before positioning it in relation to the existing body of relevant scholarship.

Rationale and the state of the field

The case for the study that we are undertaking is, as the preceding discussion has intimated, as multi-faceted as it is compelling. It has a number of dimensions to it:

1 Russia, one of the most ethno-culturally diverse countries in the world, provides a distinctive angle as to how globalization is disrupting and causing the radical rethinking of approaches to inter-ethnic cohesion.

2 Russia's authoritarian, centripetal state, weak civil society and high vulnerability to a variety of extreme, fundamentalist ideologies lends it particular importance in this context, since it tests to the limits the ability of the state, and of the nation-building led by public service broadcasters (PSB), to withstand the pressures that they currently face across the European continent.

3 Prominent among those pressures are global terrorism and an increasingly sophisticated network of international crime syndicates. These factors have contributed to the steady 'securitization' of advanced European democracies, to the point that some have suggested that the 'state of exception' justifying the incremental erosion of individual freedoms is inherent to the inner logic of democracy itself,[16] or that 'minorities', far from retreating into their communities – and aided by global communications – increasingly assert themselves as transnational 'majorities'.[17] In Russia as elsewhere, the media play a crucial role in fostering the climate of fear that feeds the voracious security culture. This makes it all the more necessary to delineate 'democratic' and 'authoritarian' mediations of national and international security issues.

4 The influx of migrants from predominantly Muslim countries as a particularly controversial contributor to securitization has coincided with the emergence of a covert 'new racism', a prejudice based on a pseudo-rational hostility to alien customs and beliefs rather than a blind hatred of biological difference: both more liable to infiltrate mainstream public opinion and no less damaging to the fabric of society.[18] As the European country with the largest domestic Muslim population and the largest number of migrants from predominantly Muslim societies (mostly of the former Soviet Central Asia and Transcaucasia), Russia is highly prone to these trends. Islam has a different status in Russia than in western Europe, since in the former it is legally recognized as one of the country's 'traditional religions', alongside Orthodox Christianity, Judaism and Buddhism. The Muscovite state had begun incorporating large regions with a predominantly Muslim population as early as the sixteenth century. Due to the nature of Russian imperial expansion, these regions are located on the state's periphery rather than dispersed throughout the 'host' community (the primary economic migration of Muslims to the Russian heartlands is thus of a peculiar, 'internal' variety). The ways in which this complex set of similarities with and differences from other European societies play out in the broadcast media have import well beyond Russia.

5 The free flow of capital unleashed by 1980s deregulation culminated in the 2008 global financial crisis which, along with mass immigration, has provided part of the backdrop for a revival of neo-conservative traditionalism founded on a retrenchment into a more inward-looking, protectionist nationalism and an embrace of the 'old-fashioned' values of family and church (cf. the American 'Tea Party' and the UK Independence Party phenomenon in Britain). An exploration of the Russian variant on this transnational phenomenon is as likely to enhance our understanding of the latter as of the former.

6 The co-existence of this isolationist popular nationalism with the neo-imperial nationalism of Putin's regime, strikingly manifested in the popularly supported annexation of the Crimea in 2014 and reflected in the long-term

Eurasian Union project about which Russia's citizens tend to be far less enthusiastic, reflects several contradictory faces of Russian nationhood. An understanding of how these contradictions are mediated is crucial for assessing Russia's current and future political trajectories.

7 Finally, the future of the Putin regime, of Russia's fledgling civil society and of the Russian Federation itself depends on how the challenge of achieving society's inter-ethnic and inter-faith cohesion is met. The country's political culture is shaped by a public discourse contaminated with the threat of secession of the Islamic North Caucasus, growing conflicts between non-Russian 'migrants'[19] and their 'host' communities and the dual toxins of nationalist extremism and a now rampant anti-western phobia. The responsibility of journalists to mitigate such dangers, hampered as they are by an embattled, repressive state, is onerous. The price of failure could, as we began to see in 2014, be paid by the rest of Europe.

In light of its varied and far-reaching significance, it is therefore surprising that Russian television's approach to inter-ethnic cohesion remains a blind spot in the scholarly literature, especially since there is an abundance of research relating to western Europe.[20] The media's influence on the playing out of 'crises of multiculturalism'[21] is even greater in post-communist countries than in western Europe, given the residual control that the state maintains here. Yet the majority of relevant publications are non-scholarly reports by various human rights agencies.[22] The body of scholarly works on media coverage of ethnicity-related matters in Russia and other post-communist countries remains small and, rather than offering a comprehensive and systematic analysis of the subject, usually focuses either on specific incidents or on single issues (such as representations of migrants or Roma communities). Moreover, for logistical reasons which render the intensive study of television problematic in other contexts, these works also tend largely to rely on evidence from print media and the internet.[23]

The dearth of antecedents to our study is explained partly by the specificities of the disciplinary intersection at which it is located: Russian Studies is traditionally conservative in its attitude to emerging fields such as media studies – a legacy of its historical concern with the seemingly weightier ethical and ideological issues raised by the twentieth century's most influential totalitarian regime and its aftermath. Even now, when the Russian media are subjected to analysis, the focus is typically on matters pertaining to press freedom, state manipulation and electoral politics.[24] When covering Russia, west European media outlets are more comfortable highlighting issues of press freedom and political repression than migration-induced tensions in multicultural policies, since the latter do not fit familiar stereotypes of a Russia entirely foreign to domestic (i.e. west European) preoccupations. Yet the way in which these issues unfold could have consequences for Russia's political future far greater than the fate of this or that celebrated pro-democracy 'dissident'. Indeed, while many observers of the post-Soviet world (and other regions) make the implicit and normative assumption that political trajectories in these regions must invariably culminate in western-style democracy, we might caution that there is little historical or contemporary evidence to support such a view. Indeed, a central part of our purpose is to challenge it.

Aims and objectives

In addressing the blind spots outlined above, we focus on Russian national television's mediations of the shifts and contradictions in government policy on ethnic diversity and the currents of populist xenophobia and nationalist extremism which infiltrate it from below, assessing how issues of inter-ethnic cohesion are affecting the balance, the emotional texture and the lexicon of the public discourse that is the corollary of these clashes. Even Russia's highly regulated media system (about which we will say more in Chapter 1) must accommodate a circulation of meanings emanating from official, sub-official and unofficial sources. Kremlin pronouncements at once internalize and moderate the often more extreme positions of Duma deputies, or those of parallel sources of legitimacy such as the Orthodox Church.[25] They also draw on the influences of ideologically convenient academic and pseudo-academic writings and of popular and subcultural discourses whose connotative inflections they fail to anticipate. We illuminate the nature of that circulation as it relates to issues of inter-community cohesion.

Key to the dynamics of the circulation is the relationship between approaches to issues within news bulletins and how they are exacerbated or mitigated in the non-news programming whose porosity provides sub-official discourses a point of access to the official sphere. Yet despite their predominance in the Russian viewer's diet, genres such as the talk show, fictional drama and comedy series rarely feature in existing news and current affairs-oriented studies of the post-Soviet media.[26] Just as important to the task of capturing the multi-directional mediatory trajectories of ethnicity discourses, and just as marginal within the scholarly literature, is the interpretation of state policies adopted by regional broadcasters.[27] We will make good a portion of both gaps.

We remain attuned throughout to the dual legacy of Russian imperialism and Soviet communism whose outlines we present in more detail in our first chapter. We will also maintain an implicitly comparative perspective, attempting wherever possible to locate Russian television's struggle to mediate the competing forces surrounding issues of ethnicity and nationalism within European media culture. We estimate the extent to which it deviates from trends prevalent in western broadcasting on one hand, while asking whether the Russian case challenges our assumptions about the very normative status of those trends on the other. We pay particular attention to the complex re-articulation of race, ethnicity, religion and culture which has occurred in the wake of globalization, and to the crisis in liberal values and the waning credibility of multiculturalism played out in media systems across the continent.

Our study also opens out in the direction of contemporary Russian society more generally. With or without Putin, Russia must confront two monumental challenges: the achievement of stable relations between communities, some of which are characterized by mutual perceptions of radical 'otherness', and the development of a vibrant civil society in which the media facilitates the expression of ideological difference. The shared legacy of communism and the now dramatically changing dynamic of Russia's relationship with its neighbouring states should ensure that our findings illuminate issues of ethnic tension in Europe's other transition societies.

The final context which we address lies beyond academic study altogether. For if, by analysing what in Russia is something of an 'outlier', we uncover fresh lenses through which to view other European broadcasters' role in fostering/hindering community cohesion, we will contribute, albeit modestly, to shaping how they see their own responsibilities and identify their own blind spots.

Theory and method

At the heart of the book lies a variant of critical discourse analysis which adopts the Bakhtinian principle that audio-visual communication, like language, is socially situated and shot through with the aftertaste of the ideological struggles which motivate it. It is also informed by a belief in the importance of the historical precursors which leave their indelible mark on its rhetorical structures. As qualitative research, the analysis locates the meanings it mines in the discursive shifts and nuances pervading individual texts, rather than in the broad aggregate patterns recovered by data-led quantitative approaches. It privileges significant difference over recurring norm, though the distinction is far from absolute. Battles over the representation of inter-ethnic cohesion and the inflammatory sentiments they expose are fought precisely on the territory of covert meaning, repressed historical precedent and lexical nuance.

The study draws on the concept of 'mediation', which Roger Silverstone describes as 'the fundamentally uneven, dialectical process in which institutionalized media are involved in the general circulation of symbols in social life'.[28] We remain cognizant of mediation's dialectical aspect and of the fact that, whatever manipulation it is subject to by the political elite, television operates squarely in the broader realm of culture. Even under Putin, the key to achieving the societal consensus capable of sustaining hegemonic pre-eminence over the plurality of competing opinions characterizing the Russian public sphere is appropriation of the definition of what passes for 'consensus'. In the process, culture becomes what Jesus Martin-Barbero describes as the 'strategic battlefield in the struggle to define the terms of the conflict', for 'there is no hegemony – nor counter hegemony – without cultural circulation … no imposition from above which does not imply an incorporation of what comes from below'.[29]

There is little question that the Putin regime, too, implicitly recognizes that a consensus as complex as that relating to inter-ethnic relations cannot be merely 'communicated', no matter how sophisticated the medium, and that, far from being tainted by 'degradation', it is national television's capacity to straddle the official realm and that of the unofficial discourses prevailing in culture at large which make it such a useful consensus-management tool.

But 'cultural circulation' cannot, by definition, be controlled. The tensions that this contradiction generates intersect with a related Gramscian paradox: for if hegemonic consensus relies on the circulation of meanings 'from above and from below', the dynamism of the process depends on the constant renewal of the antagonisms between those meanings. Thus, as Ernesto Laclau, whose account of populism informs several chapters, argues, populism is not a fixed ideological constellation 'but … proceeds by articulating fragmented and dislocated demands around a new core'.[30] He stresses the need for heterogeneity, but argues that it 'has as one of its defining features a dimension of deficient being or failed unicity'.[31] The demands must remain discrete, yet contain within them an unfulfilled urge towards unity. The populist impulse situates itself at the core of the creative tension pitting the need for the revitalizing dynamism of difference and antagonism against the urge for unity and subordination. For a semi-authoritarian government suffering a crisis of legitimacy, the imperative to navigate the treacherous waters of populism is particularly acute.

As broadcasters negotiate the paradoxes structuring media environments, a gap opens up in which their status as performers of communicative acts is decoupled from their function as purveyors of messages. This gap is best approached through J. L. Austin's notion of 'performativity' in speech acts, and in particular through his distinction between *locution* (the words spoken), *illocutionary force* (what the speaker is attempting to achieve) and *perlocutionary effect* (the consequence flowing from the utterance).[32] These acquire new dimensions when what is at stake is the mediated representation of a multi-ethnic nation. For representation always involves the performative claim to hegemony over the definition of the represented reality favoured by one discourse, against that of competing discourses; the trace of differences between them is never effaced from the object of representation. The battle for representational rights is waged at the borders between the discourses. In Bakhtin's words, 'meaning lives on the boundary between its own context and another, alien context'.[33]

The hegemonic power embodied in representation is associated with the drive to reorient the hybrid voices inhabiting all discourse towards what Bakhtin calls the unifying, centripetal discourse.[34] For dialogism is not restricted to hybridity *within* speech; even apparently single-voiced utterances are oriented towards an unnamed *external* collocutor. This has particular salience for a state broadcaster the subtext of whose utterances is directed at rebuffing hostile opponents.

The final element in our toolset is the genealogical prism through which we capture the reconfiguration and reinvention of the imperial and Soviet legacies within the current context. The genealogical prism also illuminates the chain of wider associations which media representations are likely to trigger among viewers who share a set of perceptions rooted in this legacy. Much of the historical background informing this prism is presented in our next chapter. But if, in operating the genealogical tool, we emphasize the historical contingency of ethnicity, we must simultaneously stress that notions of ethnicity, race and nation are always contingent in a more immediate sense, too; they are articulated for particular purposes by particular groups and applied to situations and events which may, in other contemporary contexts, lack those meanings. Here we draw on the work of Rogers Brubaker, whose argument that ethnicity, race and nationhood are 'perspectives on the world rather than entities in the world' complements Bakhtin's insistence on the contextual dialogism of all meaning. It is an argument that underlies every chapter in this book.

Brubaker's reminder that when politicians, journalists and members of society throughout the world talk about 'ethnic *groups*', they tend to impose the *category* of 'ethnicity' onto disparate people who in reality usually have little in common,[35] immediately comes to mind when we encounter in contemporary Russian public discourse the term 'ethnic diasporas' in reference to people of Caucasian or Central Asian origin who reside in Moscow or St Petersburg. But Brubaker's analysis of the variable and contingent nature of groupness enables us to go beyond the now obvious argument that ethnicity, race and nationhood are constructed, setting the agenda for exploration of *how* these categories are constructed, reified and naturalized.[36] Our book pursues this same agenda by analysing how Russian television's representations constitute reified ethnicity as a powerful social agent.

It should by now be self-evident that our own understanding of the terms 'ethnic', 'national' and 'racial' differs radically from that of the broadcasters we analyse. For reasons of convenience, however, this does not mean that whenever we import instances of their usage into our discourse we feel obliged to place every such instance in scare quotes. As Edward Said commented, 'labels purporting to name very large and complex realities are notoriously vague and at the same time unavoidable'.[37] We remain acutely aware of the problematic status of the terminology that Russian broadcasters are prone to use with neither discrimination nor consistency. Rather than cluttering the text with endless typographical reminders and other caveats (which, however, we do not altogether eschew when we deem them to be necessary), we hope the reader accepts that the need for that awareness can be taken as axiomatic.

Brubaker further warns us against treating ethnicity, race and nationalism as separate subfields of inquiry, and emphasizes instead their interconnectedness.[38] Again, this is relevant to the Russian case, where the word 'nation' (*natsiia*) is utilized not only to define the entire Russian Federation as the imagined community of all its citizens (*grazhdanskaia rossiiskaia natsiia*) but, in line with the Soviet approach, continues to be used interchangeably with the term 'ethnos'. In the latter usage both 'nation' and 'ethnos' describe another type of imagined collectivity – a sub-state community of people who claim common ancestry, specific cultural traditions and even supposedly common behavioural characteristics. Race in the rigidly biological sense is explicitly utilized by marginal activists.[39] In Kremlin-sponsored discourse, race is not explicitly evoked, yet it is nevertheless implicitly present. The genealogical prism, then, gives our apparatus the historical perspective vital for an understanding of how our topic is shaped by the unique Soviet legacy. The apparatus will be supplemented by other tools keyed to the particular issues under scrutiny.

In order to extend the findings derived from a preliminary statistical analysis of news recordings and to corroborate, contextualize and enrich the theoretically informed close readings which form the greater part of the book, we conducted interviews with 16 prominent television journalists. They work primarily for Channel 1 and Rossiia, but some also for NTV, two independent broadcasting companies and the main channel of the Siberian republic of Buriatiia (see below). Among them were leading television personalities such as Dmitrii Kiselev, deputy director of the All-Russian State Television and Radio Company (2008–12) and since August 2012 anchor of one of Russia's most widely viewed television news programmes, *Vesti nedeli*, as well as director of the newly established Russian information agency Russia Today as of December 2013; Arkadii Mamontov, the highly controversial journalist from the Rossiia channel; and a prominent presenter and political commentator from Channel 1, Maksim Shevchenko. They were happy to allow us to refer to them by name in this book. As one might expect of employees of Russian state television, they are loyal to the government and act as executors of state policies. Yet they are also bold characters whose semi-radical views play a crucial role in maintaining viewer ratings. It became clear during the course of our interviews that their personal views and interests have a significant influence on what they say and present on screen. They are, as we shall see, much more than merely the passive mouthpieces of their political masters.[40]

We scheduled the interviews for late 2012 and early 2013 to ensure that they took place after we had already conducted a preliminary analysis of our broadcasting sample. This allowed us to ask questions about the context for significant shifts which we had noted in the coverage of specific issues. We also asked journalists to comment on our interpretations of individual broadcasts. Our interviewees – who, despite the provocative nature of some of our questions, often spoke with virtually no inhibition – offered us important insights into the actual processes by which news and other television programmes are produced, including the important role played by self-censorship, which most Russian journalists exercise with great skill. In the absence of explicit written guidelines on how to cover sensitive issues such as those followed by the BBC, our interviewees' explanations of unwritten rules in the coverage of ethnicity-related issues proved useful for our analysis. The information we obtained through the interviews, and the findings we derived from it, are deployed in a supporting capacity and are dispersed throughout the book rather than presented in one discrete section.

The approaches we adopt reflect our respective disciplinary backgrounds, combining aspects of cultural theory with a strong commitment – to the extent that our evidence allows it – to the careful tracing of lines of empirical causation, and offering complementary synchronic and diachronic accounts of the intertextual affinities and contrasts and affinities in the media texts we analyse. Thus, we seek both to locate those texts within an associative network of parallel and paratextual cultural meanings and to explicate their genealogical influences and historical antecedents. At times the synthesis is managed within a single chapter; elsewhere it takes shape over several. The different critical idioms deployed require divergent lexicons and styles of writing, the tensions between which are hopefully productive and contribute to the scope and subtlety of our argument.

Purely qualitative research cannot adequately convey the accumulated meanings generated in a two-year recording period. Striking the right balance between close readings of small numbers of texts and engaging with the volume dimension is a challenge. We have addressed it through the very structure of our book, whose chapters alternate between: (i) broad-ranging analyses of themes drawing on multiple news bulletins; (ii) case studies based on the coverage of specific events (the 'Manezhnaia' disturbances in December 2010, annual celebrations of the Day of National Unity; the Pussy Riot scandal of 2012); and (iii) an 'intermediate' category centring on particular television genres (the talk show; the fictional drama). The events providing the focus for two case studies (Manezhnaia and Pussy Riot) are located at opposite ends of our corpus (2010 and 2012 respectively). The third (Day of National Unity) examines coverage from 2010, 2011 and 2012. These case studies thus facilitate the diachronic analysis of shifts in the tendencies that our thematic chapters uncover.

Two intertwined themes run throughout this book. First, we demonstrate the pivotal role of television in contributing (with varying degrees of success) to the transformation of particular occasions and incidents into 'media events'. The latter we define, after Couldry and Hepp, as 'situated, thickened, centering performances of mediated communication that are focused on a particular thematic core, cross different media products and reach a wide and diverse multiplicity of audiences and participants'.[41] We identify four sub-genres of media event, corresponding to four different sets of relationship between the phenomenon's component elements ('media' and 'event'): one more closely aligned with Dayan and Katz's original definition of the term as an occasion when media and state openly collaborate in constituting a carefully planned, ritualized event as an official occasion of nation-building value (the Day of National Unity celebrations);[42] a second reflecting Couldry and Hepp's recasting of the original definition, encompassing minor, unplanned disruptions to normality which media and state then covertly collude in inflating into major societal stand-offs (the Pussy Riot scandal); a third drawing on Couldry and Hepp's identification of the media event's inherently agonistic nature, incorporating unanticipated major disturbances in the mitigation of whose consequences the media are assigned the key role (the Manezhnaia riots); and a fourth, 'endogenous' type in which a media product itself generates significant societal conflict (the scandal over the 2010 reality television serial, *Shkola*). Together these different sub-genres provide the architecture for the argument we develop in the book.

The second theme is the role of television in generating different, and often conflicting, modes of combining national unity and ethno-cultural diversity – the two features which, according to Kremlin-sponsored discourse, in equal measure distinguish Russia as a distinct type of community. The 'media events' we discuss lay bare the contradictory nature of these combinations and reveal the inability of the state-aligned media to reflect this ideal systematically in their coverage. Demonstrating the instability and fluidity of different combinations of unity and diversity throughout the broadcasts we analyse, we also note a pronounced shift towards one particular combination during Putin's third presidency.

We preface our interpretative readings with some basic statistical analyses of our recorded data, enabling us to identify the key tendencies that emerge from it. Their purpose is not to generate our overarching arguments (these emerge from the subsequent chapters); the book does not centre on a comprehensive, multi-aspectual 'content analysis' informed by the full battery of statistical tools available to social scientists who research media trends. Instead, it offers an introduction to the corpus and a point of reference for the more focused analyses to follow. A complete account of its design will be provided in Chapter 2.

Sources

Located at the intersection of the official policy positions of the state and the beliefs and concerns of citizens, the news bulletin has a unique mediatory capacity and is our primary source of material. With its immersion in the routine patterns of the everyday, it also offers access to representation in its typifying dimension: the production of ethnicity discourses in their recurrent, 'representative' mode. Over a two-year period (September 2010 to August 2012) we recorded alternating three-month blocks of the main evening news bulletin of the two state channels: *Vremia* (Channel 1) and *Vesti* (Rossiia). Since both *Vremia* and *Vesti* provide a full archive of broadcasts dating back several years on their respective websites, we were able to access reports relevant to our topic which fell in between our three-month recording blocks or outside the corpus altogether. After our recording period finished, we continued to monitor Channel 1 and Rossiia's coverage through the channels' web archives until the summer of 2014. We also used websites to view reports relating to the events and issues in question on NTV, REN TV and Dozhd' TV.

If we imagine a spectrum running from the unofficial discourses generated by popular culture to the Kremlin's rehearsed policy statements, news bulletins, particularly those on Channel 1, tend to occupy a position considerably closer to the latter than the former. Other genres belong elsewhere on the spectrum. Russian talk shows and documentaries rarely stray too far from the official line, but their vernacular style of delivery, pseudo-dramatic modes of presentation and inclusion of figures from beyond the mainstream render them a filter for unorthodox viewpoints. Comedy formats and fictional dramas likewise serve as gateways to the popular mythologies, narratives and stereotypes which rarely find their way into news bulletins. With their location at the margins of that realm and away from the political spotlight, they also provide space in which mainstream positions can be interrogated or developed to their logical conclusion. We draw on all these forms.

While national television is our primary focus, the regional dimension cannot be altogether bracketed out. The way in which the Putin government's nation-building effort is interpreted at the edges of the Russian Federation will tell us much about its inner coherence and outward effectiveness. We devote one chapter to the coverage of *Vesti-Buriatiia* (Buriatiia News) broadcast by the

Buriatiian State Television and Radio Company based in Ulan-Ude, Buriatiia, one of Russia's autonomous republics, on the eastern shores of Lake Baikal in Siberia. Buriatiia has a large Buddhist population, reflecting its previous colonial status. It is an ethnically diverse republic which includes a majority Russian contingent and is overwhelmingly loyal to Moscow, but as we shall see, *Vesti-Buriatiia* expresses that loyalty in a quite distinctive manner. For this programme we obtained a news recording set, organized on the same principles as those for *Vremia* and *Vesti*.

Chapter outline

In putting the rationale we have outlined into practice, we divide the book into three parts. In the first, we analyse national news reports which tend (though not exclusively) to cleave closely to the stated government position, and through which broadcasters strive to manage issues – anticipated and unanticipated – arising from ethno-cultural difference. In Part II, we explore how difference plays out at the margins of the official sphere, focusing on: (i) non-news genres whose relative distance from the centre of power accords them greater licence to breach official mantras (fictional dramas; comedy series); (ii) regional television news bulletins which reflect the geo-political context in which they operate. Part III deals with the radical changes in coverage of issues of ethnicity during and after the transition to Putin's controversial third term as president.

Preceded by Chapter 1, which offers an overview of the historical background to our topic and of the current state of Russian television broadcasting, Part I opens with our second chapter, in which we present in numerical form the data derived from the full set of daily recordings. We identify the broad patterns and trends that these data bring to light, and thus the themes which guide us in subsequent chapters. Drawing on these themes, Chapter 3 analyses Russia's Day of National Unity (DNU), celebrated on 4 November, as a failed media event of the first, collaborative, kind. We examine broadcasters' approach towards DNU's interpretation of the 'Unity in Diversity' mantra and of the meaning of Russian nationhood, tracing the shifts that have occurred since DNU was introduced, and its inexorable decline into irrelevance. Chapter 4 situates the relationship between news broadcasting, multiculturalist nation-building and the media's management of growing inter-ethnic tension in the context of a media event of the second, agonistic, type: the Manezhnaia riots, which became a defining moment in the recent history of inter-ethnic relations in Russia.

In Part II, the emphasis shifts to the refraction of inter-ethnic cohesion issues and policies through perspectives emanating from culturally and geographically peripheral realms. Chapter 5 reads recent Russian fictional drama and comedy series in terms of their function as a gateway through which competing subcultural prejudices and counter-discriminatory discourses infiltrate the spaces of official culture; we treat the most controversial of these programmes, the 2010 Channel 1 serial *Shkola*, as a media event of the endogenous variety since it was itself the source of raging societal conflict. Chapter 6 considers the discursive strategies which *Vesti-Buriatiia* utilizes in order to address the republic of Buriatiia's borderland position and to negotiate its cultural and religious diversity.

Part III explores the febrile environment of Russian electoral politics, focusing on the lead up to, and aftermath of, the 2012 presidential election. Chapter 7 begins by noting how, at first glance, news coverage of the campaign downplayed the importance of the 'national question'. We consider why this should be the case, arguing that it was in fact far more important to key actors on both government and opposition sides than has been acknowledged. In Chapter 8 we demonstrate how a collusive media event co-ordinated by the Orthodox Church and the state media generated from the Pussy Riot affair a public scandal linked to the production of a highly exclusionary form of Russian-ness. Our concluding chapter explores the paradoxical relationship between an increasingly repressive Kremlin and federal broadcasters whose demonization of the dual figures of the migrant and the (radical) Muslim signalled a newfound capacity to drive rather than be driven by official rhetoric, and a retreat from the last vestiges of the diversity management function.

Notes

1 *V kontekste* (Channel 1), 15 March 2012. Available online at www.1tv.ru/sprojects_edition/si=5851&d=15.03.2012 (accessed 1 November 2012).
2 Ibid.
3 Aleksandr Malashenko, a specialist on Islam in Russia, notes that the Chechen wars of the 1990s were often inflated into a 'clash of civilizations', and that the concept persists in contemporary Russian media coverage of conflict in the North Caucasus. See 'Islam: The Way We See It', *Russia in Global Affairs*, 4, 2006. Available online at http://eng.globalaffairs.ru/number/n_7325 (accessed 15 November 2013).
4 In an insightful analysis of the state of Russian politics, Richard Sakwa focuses on the struggle between the elites surrounding Putin, making only passing reference to the ethnic question. See 'Putin Redux: Continuity and Change', *Open Democracy Russia: Post-Soviet World*, 14 September 2012. Available online at www.opendemocracy.net/od-russia/richard-sakwa/putin-redux-continuity-and-change (accessed 15 November 2013); see also Richard Sakwa, *The Crisis of Russian Democracy: The Dual State, Factionalism and the Medvedev Succession*, Cambridge: Cambridge University Press, 2011. Daniel Treisman's authoritative book *The Return: Russia's Journey from Gorbachev to Medvedev* (New York: Free Press, 2011) also pays very little attention to the role of 'the national question' in post-communist Russian politics. The following works on the developments in Russia in the aftermath of the 2011 parliamentary and the 2012 presidential elections ignore inter-ethnic relations altogether or refer to this issue only in passing: Lilia Shevtsova, 'Russia under Putin: Titanic Looking for Its Iceberg?', *Communist and Post-Communist Studies*, 45, 2012, 209–16; Denis Volkov, 'The Protesters and the Public', *Journal of Democracy*, 23/3, 2012, 55–62; Sharon Wolchik, 'Putinism under Siege: Can there Be a Color Revolution?', *Journal of Democracy*, 23/3, 2012, 63–70; Dmitri Trenin, 'No Return to the Past for Russia', *The International Spectator*, 47/3, 2012, 8–12; Lilia Shevtsova, 'Putinism under Siege. Implosion, Atrophy, or Revolution?', *Journal of Democracy*, 23/2, 2012, 19–32. Nicu Popescu, 'The Strange Alliance of Democrats and Nationalists', *Journal of Democracy*, 23/3, 2012, 46–54 limits the discussion of the 'national and ethnic factors' in post-election Russian politics to analysis of the new, anti-immigrant Russian nationalism.

5 Belatedly, some four days after the riots, BBC television news did feature a brief report of less than two minutes in duration: available online at www.bbc.co.uk/news/world-europe-12003385 (accessed 1 November 2013).
6 V kontekste (Channel 1), 15 March 2012.
7 Slavoj Žižek, *The Ticklish Subject*, London: Verso, 1999, p. 224.
8 Although the balance is shifting, in 2012, according to Levada Centre polling of March 2013, television was still considered by more than 57 per cent of the Russian population as the most trustworthy source of information, as opposed to only 11 per cent who gave their preference as the Internet. See 'Otkuda rossiiane uznaut novosti', 8 July 2013. Available online at www.levada.ru/08-07-2013/otkuda-rossiyane-uznayut-novosti (accessed 17 October 2013).
9 A Levada Centre poll taken over the period 18–21 January 2013 recorded a 62 per cent approval rating for Putin, still far higher than that enjoyed by most western leaders. See 'Russian Pollster Says Approval for Putin at 12-Year Low'. Available online at www.reuters.com/article/2013/01/24/us-russia-putin-approval-idUSBRE90N0IX20130124 (accessed 17 October 2013). Following the annexation of Crimea, Putin's ratings soared.
10 Evgenii Morozov, *The Net Delusion: The Dark Side of Internet Freedom*, New York: Public Affairs, 2011.
11 Article 29 of the Constitution of the Russian Federation; Part 1 of Article 282 of the Criminal Code of the Russian Federation, Criminal Code of the Russian Federation. Available online at www.ug-kodeks.ru/ (accessed 21 September 2012). See also a useful discussion by Vladimir Sokolov, 'K voprosu o razzhiganii natsional'noi rozni', available online at www.m3ra.ru/2010/12/01/inciting-ethnic-hatred/ (accessed 21 September 2012), and the USSR law 'Ob usilenii otvetstvennosi za posiagatel'svo na natsional'noe ravnopravie grazhdan', 2 April 1990. Available online at: www.bestpravo.ru/ussr/data01/tex10976.htm (accessed 21 September 2012).
12 A 2013 report issued jointly by the SOVA Centre for Information and Analysis in Moscow and the St Petersburg-based Anti-Discrimination Centre identifies continuing widespread anti-migrant sentiment and expresses 'significant concern about the failure of the Russian Government to counter racist and xenophobic statements in the media'. 'Implementation of the International Convention on the Elimination of All Forms of Racial Discrimination by the Russian Federation', report available online at www.2.ohchr.org/English/bodies/cerd/docs/ngos/SOVA_ADCM-IPHR_FIDH_RussianFederation_CERD82.pdf (accessed 22 March 2013).
13 For an account of the process by which Putin's government established effective control over Russian television, see Tina Burrett, *Television and Presidential Power in Putin's Russia*, London: Routledge, 2010.
14 A 2008 opinion poll found that the highest levels of hostility in Russia were focused at gypsies (40 per cent) and Chechens (36 per cent), with other key racialized groups including Chinese (16 per cent), Tajiks and Azerbaidzhanis (both 15 per cent), Jews (12 per cent) and Africans (11 per cent). See Levada Centre, *Nationwide Survey*, Moscow: Levada Centre Publications, 2008.
15 Vera Tolz, *Russia: Inventing the Nation*, London: Arnold/Hodder, 2001.
16 Giorgio Agamben, *State of Exception*, Chicago: University of Chicago Press, 2008.
17 Arjun Appadurai, *Fear of Small Numbers: An Essay on the Geography of Anger*, Durham: University of Duke Press, 2006.
18 For an early definitive account of new racism, see Martin Barker, *New Racism: Conservatives and the Ideology of the Tribe*, New York: Junction Books, 1981.
19 In Chapter 9, we discuss the laxity with which this term is used in the Russian public discourse.
20 Antonis Ellinas, *The Media and the Far Right in Western Europe: Playing the Nationalist Card*, Cambridge: Cambridge University Press, 2010; Chris Flood, Stephen Hutchings, Galina Miazhevich and Henri Nickels, *Islam, Security and Television News: A Comparative Study*, Basingstoke: Palgrave Macmillan, 2012; Simon Cottle,

Ethnic Minorities and the Media: Changing Cultural Boundaries, Buckingham: Open University Press, 2000; Ella Shohat and Robert Stam, *Unthinking Eurocentrism: Multiculturalism and the Media*, London: Routledge, 1994.

21 The term features in Alan Lentin and Gavan Titley, *The Crises of Multiculturalism: Racism in a Neoliberal Age*, New York: Zed Books, 2011, which devotes a whole chapter to 'the mediation of the crisis [of multiculturalism]'.

22 Various references to media coverage of Roma can be found on the website of the European Roma Rights Centre. Available online at www.errc.org (accessed 15 November 2013).

23 Galina Zvereva, 'Nationalisticheskii diskurs i setevaia kul'tura', *Pro et Kontra*, 2/29, 2005. Available online at http://uisrussia.msu.ru/docs/nov/pec/2005/2/ProEtContra_2005_2_03.pdf (accessed 21 December 2011); Viktor Shnirel'man, *Porog tolerantnosti'. Ideologiia i praktika novogo rasizma*, in two volumes, Moscow: Novoe literaturnoe obozrenie, 2011, pp. 277–84; Mikhail Kroz, 'Ksenofobiia v rossiiskikh SMI kak proiavlenie ekstremizma', in Marlène Laruelle (ed.) *Russkii natsionalizm. Sotsial'nyi i kul'turnyi kontekst*, Moscow: Novoe literaturnoe obozrenie, 2008, pp. 425–44; Vladimir Malakhov, *Ponaekhali tut...Ocherki o natsionalizme, rasizme i kul'turnom pliuralizme,* Moscow: Novoe literaturnoe obozrenie, 2007; Natalia Rulyova and Taras Zagibalov, 'Blogging the Other: Construction of National Identities in the Blogosphere', *Europe-Asia Studies* 64/8, 2012, 1524–45; Andrew Foxall, 'Discourses of Demonisation: Chechens, Russians, and the Stavropol' Riots of 2007', *Geopolitics,* 15/4, 2010, 684–704; Caress Schenk: 'Nationalism in the Russian Media: Content Analysis of Newspaper Coverage Surrounding Conflict in Stavropol, 24 May–7 June 2007', *Nationalities Papers: The Journal of Nationalism and Ethnicity*, 40/5, 2012, 783–805; Karen Erjavec, 'Media Representation of the Discrimination against the Roma in Eastern Europe: The Case of Slovenia', *Discourse Society*, 12/6, 2001, 699–727; Magdalena Ratajczak, 'Representation and Visibility: Roma in the Media', *Global Media Journal*, German edition, 1/2, 2011. Available online at www.globalmediajournal.de/current-issue/ (accessed 4 January 2012).

24 In addition to Burrett, *Television and Presidential Power*, among the most prominent examples are Ellen Mickiewicz, *Changing Channels: Television and the Struggle for Power*, Durham: Duke University Press, 1999; ibid., *Television, Power, and the Public in Russia*, Cambridge: Cambridge University Press, 2008; Sarah Oates, *Television, Democracy and Elections in Russia*, London: Routledge, 2008; Ivan Zassoursky, *Media and Power in Post-Soviet Russia,* New York: M.E. Sharpe, 2004.

25 Vera Tolz, 'The Radical Right in Post-Communist Russian Politics', in Peter H. Merkel and Leonard Weinberg (eds.), *The Revival of Right-Wing Extremism in the Nineties*, Portland: Frank Cass, 1997, pp. 177–202.

26 For exceptions, see Stephen Hutchings and Natalia Rulyova, *Television and Culture in Putin's Russia: Remote Control*, London: Routledge, 2009; Elena Prokhorova, 'Can the Meeting Place Be Changed? Crime and Identity Discourse in Russian TV Series of the 1990s', *Slavic Review*, 62/3, 2003, 512–24; David MacFadyen, *Russian Television Today: Primetime Drama and Comedy*, London: Routledge, 2007; Vera Zvereva, '"Zakon i kulak": Rossiiskie militseiskie serialy', *Novoe literaturnoe obozrenie*, 78, 2006: 305–25.

27 However, see Hutchings and Rulyova, *Television and Culture in Putin's Russia*, pp. 176–95.

28 Roger Silverstone, 'Proper Distance: Toward an Ethics for Cyberspace', in Gunnar Liestøl, Andrew Morrison and Terje Rasmussen (eds.), *Digital Media Revisited: Theoretical and Conceptual Innovations in Digital Domains*, Cambridge: MIT Press, 2003, pp. 469–90.

29 Jesus Martin-Barbero, *Communication, Culture and Hegemony: From the Media to Mediations*, London: Sage, 1993, p. 99.

30 Ernesto Laclau, *On Populist Reason*, London: Verso, 2005, pp. 76–7.

31 Ibid., p. 225.
32 J. L. Austin, *How to Do Things With Words*, Cambridge: Harvard University Press, 1962.
33 Mikhail Bakhtin, *The Dialogic Imagination: Four Essays*, Austin: University of Texas Press, 1981, p. 284.
34 Valentin Voloshinov, *Marxism and the Philosophy of Language*, Cambridge: Harvard University Press, 1986, pp. 115–124.
35 Rogers Brubaker, 'Ethnicity Without Groups', *Archives Européennes de Sociologie,* XLIII, 2, 2002, 163–189. See also his book *Ethnicity Without Groups*, Cambridge: Harvard University Press, 2004.
36 Brubaker, 'Ethnicity Without Groups'.
37 Edward Said, *Covering Islam: How the Media and the Experts Determine How We See the Rest of the World*, London: Vintage, 1997, p. 9.
38 Brubaker, 'Ethnicity Without Groups'.
39 See, for instance, Andreas Umland (ed.), *Theorizing Post-Soviet Extreme Right: Comparative Political, Historical and Sociological Approaches*, Armonk: M.E. Sharp, 2008.
40 Elisabeth Schimpfossl and Ilya Yablokov, 'Coercion or Conformism? Perceptions of Censorship and Self-Censorship among Russian Television Reporters in the 2010s', *Demokratizatsia: The Journal of Post-Soviet Democratization*, 22/2, 2014, 295–311.
41 Andreas Hepp and Nick Couldry, 'Introduction. Media Events in Globalized Media Cultures', in Nick Couldry, Andreas Hepp and Friedrich Krotz (eds.), *Media Events in a Global Age*, London: Routledge, 2010, p. 12.
42 Daniel Dayan and Elihu Katz, *Media Events: The Live Broadcasting of History*, Cambridge, Harvard University Press, 1992, p. 16.

1 Television and nationhood

The broader context

In this chapter we set out the context, both historical and contemporary, for our subsequent analysis of Russian television broadcasts. The disciplines which this book brings together – empire and nationalism studies and media studies – are vibrant fields of scholarly inquiry. It is, therefore, necessary to outline where we stand on the key debates, particularly on the role of the imperial and Soviet legacies in contemporary politics of belonging in Russia and the specific role television plays in fostering or hindering community cohesion world-wide. We reject the view which claims that Russian developments are unique; yet we acknowledge that – as is the case with any country – in Russia global processes take specific forms, depending on local circumstances. Throughout the book we aim to capture the interaction between the global and the local, and this chapter links the Russian case to existing analysis of the relationship between television and nationhood in other parts of the world.

Identity, empire and the 'national question'

When watching Russian television today one is struck by the frequency with which broadcasters evoke the category of ethnicity in discussing current affairs. As we will see, it is only in a minority of cases that what in Russia is called *mezhetnicheskie otnosheniia* (inter-ethnic relations) constitutes the main subject of news. But ethnicity, even if implicitly, often frames the discussion of social and political events. Such an ethnicized view in itself is not specific to Russia. The notion of a world divided into discrete ethno-cultural communities, often assumed to be homogenous, dates back to nineteenth-century Romantic nationalism, which shaped the way Europeans came to see the world for years to come. It is not only in Russia that this legacy persists. In fact, its underlying assumptions can inform the arguments of both promoters and critics of Europe's policies of multiculturalism.[1]

In the Russian case, the ethnicized vision of the world has been significantly strengthened by Soviet policies, which defined nationality (*natsional'nost'*) not as citizenship but as ethnicity (ethno-cultural identity), which was institutionalized at the sub-state level and turned into a key marker of social accounting.[2] It is no coincidence that Rogers Brubaker's argument that ethnicity, race and nationhood are 'perspectives on the world rather than entities in the world' – an argument which informs our understanding of ethnicity here – was, in part, based on his study of the Soviet experience.[3] The Soviet government's top-down policies, which imposed ethno-national labels on people who did not by and large think of themselves in ethnic and national terms, rendered the constructed nature of ethnicity and nationhood in particularly vivid colours.

Russia's historical legacy is above all linked to the imperial nature of the state. Its beginnings are usually dated to the sixteenth century. Soviet leaders depicted the USSR as anti-colonial, claiming equality of all its nationalities. Yet retrospectively, the Soviet state is also widely perceived as an empire in which the peripheries were subordinate to and exploited by the centre.[4] Today the description of historical Russia as an empire often has a pejorative meaning, but in Russia itself the imperial past is not infrequently evoked with pride.[5]

The post-1991 Russian Federation has retained the Soviet approach of institutionalized and territorialized ethnicity. It contains 21 ethnic republics and four further autonomous ethnic districts named after a particular nationality and represented as this nationality's homeland. In most cases a sense of 'groupness' and its imaginings in national terms are a result of Soviet policies.[6] The continuing existence of separate administrative units for non-Russian nationalities allows Russian politicians to boast about the country's strong support for minority cultures. The Russian–Chechen war waged by both El'tsin's and Putin's governments, not to mention the recent annexation of Crimea, reminded the world about the colonial origins of Russia's relationship with its peripheries.

Still, imperial imaginings alone do not tell us the full story of Russia's construction as a community, either in the past or post-1991. In her penetrating analysis of how 'the congruence of nation with state has historically been resisted' by the Russians, Nancy Condee concludes that the 'Russia of Tsarist, Soviet and even post-Soviet times as an "imagined community" is not a nation'.[7] She further argues that observers' presumption that 'imagined community' is a 'nation' for the Russians can only hinder our 'understanding of Russia's discursive formations', and she urges scholars to employ analytical strategies which would help to capture Russia's 'imperial particularity'. Condee is right in her critique of the deployment of the term 'nation' as 'the default category' for defining collectivity in the modern world.[8] Yet her broad argument about the Russian imagined community not being a nation glosses over the complexity of long-term pursuits to forge collective identities in Russia.

Condee's conclusion implies that 'Russia' is always a clearly defined category. She explains that she uses it as a shortcut for the Tsarist empire, the USSR and post-Soviet Russia, all of which, she contends, are not nation-states but empires.[9] This (at first glance obvious) definition of what Russia is overstates its congruence with the empire, and obscures the fact that since the nineteenth century members of the Russian elites have been attempting to separate the Russian nation from the empire, defining this nation in multiple ways in terms of its geography (which rarely coincided with the entire imperial state) and membership.[10] Vigorous attempts to distinguish 'Russia' from the entire state, now in its Soviet incarnation, were resumed by national(ist) intellectuals in the 1960s. The geographical space of this imagined Russian nation and its membership were again contested.[11] The identification of 'Russia' with the Russian Soviet Federative Socialist Republic (RSFSR) was one such definition, yet it was far from dominant; still, in his often instrumental struggle against the 'imperial' Soviet centre, El'tsin benefited from this tradition of separating 'Russia' from the larger state, exploiting it during the 1990 and 1991 elections.[12]

Since the demise of the USSR, vigorous debates have raged about what a nation is in the Russian context and how the collectivity called the Russian Federation (the former RSFSR) should be defined.[13] The fact that most Russian actors speak about contemporary Russia in terms of 'nationhood' should be taken seriously when studying the discursive formation of the imagined community. After all, Aleksandr Prokhanov's insistence that he is an imperialist is primarily intended to shock, as are most other aspects of this writer's public performances. In turn, the term 'liberal empire' – proposed in 2003 by a leading member of El'tsin's first administration, Anatolii Chubais – failed to catch on.[14] Significantly, in 2014, in contrast with the western media's utilization of the Russian imperialism frame in coverage of Russia's annexation of the Crimea, Channel 1 and Rossiia represented the Crimea as a core of the Russian nation, where the ethnic Russian spirit (*russkii dukh*) was stronger than in the Russian Federation itself.[15]

By refusing to recognize the Russians' own attempts to define themselves in national terms – which today, if not historically, is the dominant way of constituting the Russian imagined community discursively – we imply that there exists a single normative experience of nation-building. Instead we should recall John Hall's observation that the umbrella term 'nationalism' describes highly varied processes, which are shaped by their specific temporal and geographical settings.[16] Furthermore, by rejecting the significance of the current Russian discourses of nationhood, we unwittingly reify the category of the nation, implying that it is an entity defined by a set of objective characteristics. As we argued in the Introduction, we do not share this view. When describing today's Russia as an empire we should also recall Michael Hechter's conclusion that only *post factum* can it be determined whether nation- or empire-building has occurred.[17] The membership of communities defined as national is fully equalitarian and non-hierarchical only in imagination, not in reality. The legacy of empire, with its tradition of Orientalizing the ethno-cultural 'other', is alive in the west too. Public fears about the threat to European values emanating from migrants from countries which used to be colonial domains of Europe testify to that.[18]

The construction of a compound identity for the peoples of Russia, with the help of an authoritative discourse of nation-building, has been part of Vladimir Putin's broad agenda of strengthening the Russian state in the new millennium. It is significant, not least in terms of policy implications, that – in contrast to the Soviet period – in today's official discourse the compound identity is now imagined as national, rather than supranational or imperial. The discourse emphasizes that the process of globalization has not made the nation-state a remnant of the past.[19] It is too early to dismiss the current community-building project as a failure. Yet it is important to analyse the effectiveness with which it is being implemented, because the territorial unity of the Russian state in part depends on it. Whether the Russian Federation is able to remain within its current borders is of importance not only to Russia, but to the world at large.

Given the especially contradictory nature of the identity-forging strategies of the Soviet regime, the power of the imperial legacy and the recent beginnings of the current community-building project, it comes as no surprise that it is surrounded by acute uncertainties. The official discourse of the nation emanating from the Kremlin since 2000 has been eclectic and – similarly to the situation across the entire post-Soviet region – marked by a strong tension between civic and ethnic definitions of nationhood and the resulting high degree of terminological laxity and conceptual ambiguity. On the one hand, the regime has put major efforts, at least at the discursive level, into promoting a vision of Russia as a multi-ethnic, civic community (*mnogonatsional'naia rossiiskaia grazhdanskaia natsiia*), which is claimed to derive its power from Russia's historical experiences of multi-ethnicity and multi-confessionalism. On the other, the primacy of the Russian language and culture, including Orthodox Christianity, as the binding force of the state-framed national community is also highlighted.[20]

What further complicates the situation is that the discourse of the pan-Russian civic community is itself contradictory. One variant of this discourse leaves the Soviet legacy of managing multi-ethnicity uninterrogated, or even sees it in a positive light. Another variant is critical of the Soviet Union's 'multi-ethnic regime', which legally and institutionally endorsed multiple ethnic categories, thereby encouraging public manifestations of ethnic particularism. This model of a civic national community, while accepting people with diverse ethnocultural backgrounds as citizens, sees the Soviet institutionalization of ethnicity as overtly detrimental to community cohesion. Many politicians in Moscow are concerned about the legacy of the Soviet Union's 'multi-ethnic regime', but they are also aware that policies which openly reject the Soviet model of institutionalized ethnicity will not be accepted by the elites of Russia's ethnic autonomies, which have existed since Soviet times.[21]

The political leadership does not have a monopoly on the discourse of the nation, in whose invention intellectuals invariably play a leading role. Intellectual discourses are highly heterogeneous, embracing ideas ranging from definitions of Russia as inseparable from its [former] empire to Russia as a community of ethnic Russians where minorities have no place.[22] The concept of a democratic political nation of Russia's citizens has been consistently promoted only by a few public figures.[23] Politicians recognize the role of intellectual discourses in nation-building and they instrumentally utilize particular intellectual constructs, depending on the requirement of the moment.[24]

The contradictions are not only within the Kremlin-sponsored discourse itself; there is also a mismatch, which is not unique to Russia, between the official rhetoric of multi-ethnic harmony and societal xenophobia. The latter seems to have grown since the mid-2000s.[25] Riots in the Karelian town of Kondopoga in 2006 set the pattern for future similar incidents in other Russia's cities. Expressions of public discontent tend to follow an ordinary fight that leads to a fatality: the involvement of representatives of non-Russian minorities triggers an interpretation of the events through the lens of ethnic conflict and the ineffectiveness and perceived corruption of the authorities handling the case leads to mass demonstrations or riots that feature extreme nationalist slogans and demands to expel members of minorities, irrespective of their citizenship status. The Kondopoga scenario was replayed in the centre of Moscow in December 2010 and in the Ural town of Sagra in July 2011. In July 2013 alone, three similar clashes took place in various parts of Russia.[26]

Such developments, which the state-aligned media cannot ignore, contradict the optimistic interpretation of inter-ethnic relations in Russia which Putin articulated during his presidential election campaign in 2012, when he contrasted 'the crisis of multiculturalism' in the west with Russia's history of peaceful co-existence of different nationalities.[27] In fact, many parallels can be drawn between post-imperial Russia and post-imperial Europe. Political actors and members of society at large both in Russia and in the wider world often assume that ethnic origin determines a person's cultural belonging, which in turn impacts on his or her public behaviour, and so arguments about the cultural incompatibility of various groups of people are employed in the discussion of social, political and economic issues.[28] In Russia such perceptions and practices are particularly strong, in part due to the influence of the following four aspects of its historical legacy:

1 The first aspect is the contradictory relationship between the Soviet regime's aim to construct an overarching, ostensibly civic, pan-Soviet identity among citizens and the simultaneous cultivation of people's separate ethno-cultural identities at the sub-state level, which we mentioned earlier. In Terry Martin's words, post-1917 Bolshevik leaders decided to confront 'the rising tide of nationalism' by promoting the national consciousness of the state's ethnic minorities 'and by establishing for them [at the sub-state level] many of the characteristic institutional forms of the nation-state'. [29] Significantly, in the centralized Soviet state – which was called misleadingly, yet purposefully, a federation – the borders of its administrative units were drawn along ethnic lines. Soviet leaders believed that the eventual fusion (*sliianie*) of the state's numerous nationalities could be achieved only when the ethno-national demands of the minorities had been satisfied; yet, in reality, many affirmative actions aimed at minorities undermined and at times took precedence over the pan-Soviet community-building.

Two major consequences of this legacy are felt today. One is the fear felt among elites and within wider society that the USSR's disintegration, to which the Soviet policies on nationalities are believed to have contributed, could be replayed in the Russian Federation.[30] The war in Chechnya seems to have confirmed to the current leadership that Soviet-style ethnic federalism continues to pose a threat to Russia's unity. In reality, Russia's ethnic autonomies never enjoyed the same rights as the USSR's Union republics. National institutions of the former were less developed, were economically even more dependent on Moscow and had a far more Russified population. Furthermore, whereas in the USSR ethnic Russians constituted around half of the population, in the Russian Federation they enjoy a majority of 82 per cent. The lesser ability of Russia's ethnic republics to act as independent national homelands of their titular nationalities, as compared to the USSR's Union republics, can be interpreted in different ways: it can be seen as indicating that the danger of disintegration is lower in Russia compared to the Soviet Union, or alternatively it can be assumed that if Russia's disintegration were to occur the process would be violent, in contrast to the peaceful developments in 1991, as the potential lines of disintegration would be more vigorously contested. The Putin administration has chosen to highlight the second interpretation, stressing Russia's insecurity and thereby justifying the need for authoritarian policies.[31]

A further consequence of Russia's administrative structure, rooted in the Soviet past, is the continuing tendency, particularly among the ethnic Russian majority, to associate representatives of non-Russian minorities solely with ethnic autonomies within Russia, rather than seeing them as equal citizens of the state as a whole. This tendency shapes perceptions of individuals particularly from the North Caucasian republics, when they reside outside these autonomies in the cities of central Russia, as 'visitors' (*priezzhie*) or 'migrants' (*migranty*), even though they are *bona fide* citizens of the Russian Federation.

2 Historically, the ideological underpinnings of state policies also strongly emphasized the separateness of people of different ethnicities and nationalities. From the 1930s onwards, the official understanding of the nation had been that of a primordial entity whose members had a common ancestry.[32] In the course of the implementation of state policies – both those supportive and repressive of minorities – immutable characteristics were at times assigned to every member of a national group, to one of which every citizen of the USSR was perceived as belonging. Scholars disagree on whether, despite the Soviet regime's official rejection of the racist ideology of Nazi Germany, the Soviet policies on nationality had racist undertones.[33] They do agree, however, that the highly ambivalent Soviet discourse of belonging offered a specific interpretation of the tension, common across Europe, between exclusive and inclusive understandings of membership in a state-framed community and between belief in the societal power to change human behaviour and perceptions that certain 'deviances' were so deeply rooted that their bearers could only be purged from society.

Today, both the primordial view of the nation as an ethnocultural community with common ancestry and the tendency to perceive people as carriers of immutable ethnocultural characteristics which mark certain political discourses elsewhere in the world are present in Russia in a particularly strong form, and the conflicting Soviet perceptions of ethnicity and nationality are supplemented by west European and pre-revolutionary Russian theories of race. We discuss this issue in detail in Chapter 4.

3 Historical legacy also impacts on the situation of the ethnic Russian majority. Current ethno-nationalist leaders talk incessantly about historically rooted discrimination against Russians in their own state, extrapolating the situation of the 1920s – when the official discourse branded the Russians as the oppressor nation – to the entire Soviet period, and claim continued discrimination even today. In reality, the 1930s Soviet propaganda presented the Russians as first among equals in the family of Soviet nationalities, triggering among non-Russians the opposite perception of Russia's privileged position in the Soviet state.[34]

The fact that Russians were encouraged to identify above all with the USSR as a whole, rather than to perceive the RSFSR as their primary national homeland, was widely seen as another sign of privilege. Yet Russian nationalists insist that this peculiarity turned ethnic Russians into stateless people and that this state of affairs persists in contemporary Russia. According to this line of argument, it is discriminatory against ethnic Russians that the state today is defined by the civic term *rossiiskaia*, and, while containing ethnic republics (that is, national homelands) for various minorities, it still lacks any clearly designated territory which the ethnic Russian majority can regard as truly its own.[35]

Such arguments significantly heighten the tension between civic and ethnic aspects of nationhood that exists in most nation-building projects. Because of the salience of these arguments, leading politicians increasingly feel compelled to evoke the special status of the ethnic Russian majority, despite the fact that it is not stipulated by the constitution and its official acknowledgement has the potential to be damaging to the country's ethnic cohesion.[36]

4 Finally, most discourses of nation in today's Russia, both official and oppositional, are marked by a reluctance to reflect critically on the country's imperial and colonial legacy, with its highly unequal power relations between the centre and the periphery. The post-colonial perspective has little public resonance.[37] Instead, among the most dominant perspectives on Russian past are multiple variants of the ideology of Eurasianism, first articulated by a group of Russian émigrés in the 1920s. As we discuss in Chapter 6, this ideology mythologizes Russia's experiences of multi-ethnic statehood, rather than critically assessing them.[38] The lack of reflection on the imperial and colonial origins of Russia's ethnic diversity both undermines the potential appeal of the concept of a civic Russian nation among leaders of Russia's ethnic autonomies, encouraging its perception as Russification in disguise,[39] and perpetuates the relentless 'othering' of certain minorities by the ethnic Russian majority.

Television: Roles and responsibilities

Historically, the media have been at the heart of every nation-building project as they disseminate particular imaginings of the community, of its supposedly shared values and its constitutive 'others'.[40] By selecting certain issues to be covered as news and by framing news reports in one way or another, the media also significantly contribute to building community consensus around particular perceptions, accommodating a plurality of voices.[41] Since the 1960s, television has become the main source of news for Europeans.[42] Today, because of the spread of the 'narrowcasting' modes favoured by many new technologies, television's unique capacity to 'broadcast' to an entire 'imagined community' paradoxically acquires still greater value. With its capacity to unite householders around the television set, the medium would seem to have been designed for nation-building purposes, for imagining the nation as an extended family. This value relates not only to domestic community-building, but also to the foreign policy ambitions of the political elites, as (national) television is watched widely by an ever-growing émigré population abroad to which national governments attempt to reach out. It is, therefore, unsurprising that in its attempts to establish a strong vertical line of power (*vertikal' vlasti*), Putin's leadership from the outset decided to align the main television channels closely with the Kremlin.[43]

With its audio-visual mode, television favours spontaneous forms of representation which operate at a subliminal level and which enable it to exploit the techniques of 'banal nationalism' (the trivial, barely perceptible markers of nationhood).[44] The unifying effect of such techniques is the corollary of their power to alienate and exclude, and the impact of this process on societal perceptions of ethno-cultural diversity can hardly be overstated. Television's heightened iconicity and ability to address mass audiences in a common lexicon also confers a sense of representational authenticity, both in the sense of reproducing reality in credible, photographic form and that of conveying what is 'representative' of it. It also elides the distinctions between issues of race as reflected in visual attributes, and those of ethnicity conveyable in verbal narratives about cultural specificities. By the same token, broadcasters have the potential to shape their representations through their own practices. By selecting a journalist from a minority community to report an ethnic cohesion issue, by preferring one lexical term over another to describe it or by choosing to combine a verbal narrative with a particular visual image, a newsmaker is able to inflect the way that the issue is portrayed.

The rise of television coincided with the decline of the great European empires, and with the flows of post-imperial migration which so changed the 'look' of European nations in the latter half of the twentieth century. In Europe, television was for much of that period synonymous with public service broadcasters (PSBs), the most notable example of which remains the BBC. Committed to standards of impartiality, free speech and objectivity, PSBs nonetheless also sought to embody specific 'national values' which tautologically (and paradoxically) included respect precisely for democracy, free speech and an independent press.[45]

The hidden contradiction at the heart of the PSB mission is occasionally exposed, including in the context of coverage of inter-ethnic cohesion issues. In 2006, for instance, the BBC provided covertly filmed evidence supporting the prosecution of Nick Griffin, leader of the neo-fascist British National Party, for inciting anti-Muslim violence; when Griffin was exonerated, the BBC was left stranded between its adherence to multicultural tolerance and its commitment to free speech and an independent judiciary.[46] More recently, in 2013 a BBC1 interview with Tommy Robinson, then leader of the English Defence League, prompted audience complaints that the interviewer had been 'too aggressive' in his attempts to reassure viewers by presenting a coruscating exposure of racist prejudice, after the corporation had previously drawn criticism for giving airtime (in the interests of free speech) to Robinson, whose far-right group had mounted a number of demonstrations in the wake of a British soldier's murder in London by two Islamist fanatics.[47] The contradictions are exacerbated by the fact that PSBs must track, without fully endorsing, cohesion policies pursued by their host states, and by the fact that these policies – and the public consensuses that they reflect – shift such that there is frequently a time lag before PSBs 'catch up'. For example, the BBC's continuing preference for 'reassuring' white male newsreaders and reporters during the 1970s, a period in which, influenced by the US civil rights movement, state policy was shifting towards diversity, brought temporary opprobrium upon the broadcaster.

'Public consensus' is itself something of a myth, and societies are characterized by multiple splits – between rival contingents within ruling elites, between elites and the populace and within public groups. Even as a culture of diversity was sweeping European establishments, currents of subcultural hostility to ethnic difference and anxiety over the dilution of national identities were given sporadic, but vivid, expression by 'rogue' politicians. When the new impetus given to migration flows by financial deregulation coincided with the end of the Cold War and the emergence of Islam as the preferred anti-hegemonic ideology of the dispossessed, those currents acquired new strength – at the very point at which anti-discrimination policies were bearing fruit, but the reaction against a failing multiculturalist model was strengthening.[48]

The first decade of the twenty-first century was a time in which new media communication platforms achieved pre-eminence. Thus, the securitization of the European public sphere has been underpinned by the ease with which transnational tropes such as the Danish cartoons of Prophet Mohammed, the French *hidjab* controversy and the Swiss minarets conflict circulate via the globalized communication system;[49] it was only in 2013 that the debate over whether Muslim medical staff should be permitted to wear the *niqab* while treating patients in UK hospitals entered the British media, having hit the French headlines in 2007. Such themes enter the news agenda and structure news output, as the three-way clash between securitization, Europe's commitment to secular states and its post-Enlightenment traditions of tolerance for cultural difference is played out in living rooms across the continent.

The advent of new media forms has bolstered both Islamist fundamentalism and the extreme right-wing groupings for which Muslims are now the primary target. Far from fostering global homogeneity, digital technology often has the opposite effect: closeting users in self-reaffirming virtual communities detached from the public sphere. PSBs bear particular responsibilities here: that of countering the prejudices peddled by such communities, even as they gain currency beyond their bounds, and that of continuing to provide the main mechanism for binding nations together as new media influences fracture them down multiple axes (gender, class, ethnicity, age).

Television and diversity in Russia

The very term 'public service broadcasting' is difficult to apply to a context in which 'public service' as we understand it does not exist and where, even in its commercial variant, broadcasters operate under the watchful eye of – and sometimes under direct instruction from – a semi-authoritarian state keen to deploy television to implement its own agenda. This situation reflects the emergence of Russian television from a Soviet system in which the media were controlled fully by the party state, and in which they were seen as a tool of policy rather than as the space of the public sphere.[50]

Russia has three main national television broadcasters: Channel 1, Rossiia and NTV (*Nezavisimoe Televidenie*, or Independent Television). In 2012, Dmitri Medvedev initiated the launch of a new national broadcaster, OTR (*Obshchestvennoe Televidenie Rossii*), as a more open platform for the airing of national issues of importance, but it has so far failed to gain a substantial viewership; it is at best ignored and at worst regarded as a sham. Technically only part-owned by the state, Channel 1 follows the Kremlin's line particularly closely. It is also external-facing, as it broadcasts to communities of so-called Russian compatriots (*sootechestvenniki*) outside Russia's state borders, many of whom do not have Russian citizenship but speak the Russian language. This trans-border reach of Channel 1 adds a crucial foreign policy dimension to its remit, which accounts for the extra caution with which it treats sensitive issues.[51]

Nonetheless, and despite their symbiotic relationship with the Kremlin, the large industrial interests holding the controlling stake cannot disregard issues of commercial viability (the channel relies on advertising revenue). This enables it to deliver the high-quality output which means that its political messages are delivered in a polished manner. But that revenue is in part dependent on high viewing figures, and therefore on responding to popular preferences. This tension means that Channel 1's propaganda function is encumbered by its need to consolidate its viewing audience.

Rossiia, which replaced Soviet Channel 2, is fully state-owned. Curiously, the financial constraints under which it operates mean that it plays second string to Channel 1 as regards its information management function. It is therefore accorded less attention from its political overseers, paradoxically leading to a wider range of political perspectives than might be expected. Rossiia has been assigned the task of integrating local interests with the national perspective, thereby playing a particularly important role in promoting a national cohesion agenda.

NTV was Russia's first independent television channel, and built its reputation via a path-breaking critique of the first Chechen war. It was taken over by Gazprom, the state energy giant, in 2001 and has now fallen into line with other broadcasters in maintaining an editorial stance close to the government, though the residue of its earlier bold informality remains evident in much of its output.[52]

NTV relies more heavily on audience ratings than Channel 1 and experimental informality is frequently deployed in support of an interpretation of the official line that spills over into the kind of populist nationalism which threatens the government's authority, even as it is appropriated for legitimating purposes – a paradox of particular relevance to this book.

REN TV, though small, is important, since during the period analysed in this book it remained the only private channel with an independent voice and national reach.[53] Owned until 2005 by Anatoly Chubais, in December of that year it was taken over by a consortium with ties to the Kremlin. Its audience consists primarily of the metropolitan intelligentsia and it has a market share of 5 per cent. It is part of a wider media environment which is more varied than one might think from the impression given by national television on its own.

We should also mention RT (Russia Today), Russia's relatively successful international broadcaster, targeted at audiences across the world and intended as a rival to the likes of CNN and the BBC World Service. In a highly significant development to which we will return, December 2013 saw the dissolution of Russia's respected news agency, RIA-Novosti, and its replacement with a new organization bearing the name *Rossiia Segodnia* (Russia Today) and overtly presented as a tool with which to improve Russia's image in the world.

Central to the wider environment is the internet, still relatively uncontrolled in Russia, along with other new technologies and media platforms.[54] The opposition protests in 2011 and 2012 owed much to the 'crowdsourcing' function of social media like Facebook and Twitter. The penetration that new media are achieving in Russia's large cities is allied with the phenomenon of global media convergence. No Russian channel is without multiple online audience discussion forums, whose content, though moderated, is often critical of the government.[55] More important are new, web-based television channels which bypass state control mechanisms and can operate on relatively modest resources. The most significant of these in Russia is Dozhd' TV (Rain Television), established in 2010, which targets a young liberal audience but has grown steadily in influence, having reached an agreement with the BBC enabling it to broadcast BBC news bulletins and eventually acquiring a presence within Russia's cable television network. (At the time of writing this is under threat, owing to political pressure applied to companies distributing Dozhd' following the channel's ill-advised poll seeking views on whether the Soviet Union should have ceded Leningrad to the Nazis in order to avoid the terrible suffering of the Leningrad Siege).

New media developments do not necessarily represent a mortal threat to Kremlin supremacy; for example, the YouTube video clips sponsored by government sources are of infinitely superior quality to the vast majority of home-produced clips.[56] Nor, crucially, does such threat necessarily come from the liberal wing of Russian politics (witness the mushrooming of various shades of nationalist extremist websites).[57]

Television's space within the Russian media environment is, then, more complex than its ownership and control structures would seem to allow. It is true that the news agenda is shaped quite actively by the Kremlin; our own fieldwork demonstrates that broadcasting executives have weekly meetings with their political masters. But, in the absence of a guiding ideology, there is uncertainty surrounding television's mode of operation. The guidance lacks consistency, especially insofar as it relates to inter-ethnic relations, creating the possibility that the instructions might be misconstrued.[58] Paradoxically, because broadcasters like the BBC share a set of societal values with mainstream opinion, their editorial independence is no more likely to produce radical deviations from the government line than Russian state television's obedience to a 'master', the underlying principles of whose policies it struggles fully to internalize; we identify several such deviations in subsequent chapters.

Deviations of a similar kind were, however, occasionally observable even in Soviet television, which played an important part in community-building, cleaving closely to the policies of nationality designed to maintain the integrity of a vast multi-ethnic state. During the 1970s, it responded to the clamour for the promotion of regional (non-Russian) cultures as each constitutive republic acquired the right to broadcast for a limited period in its own designated national language. This, however, led to concerns over growth in 'localism'.[59] Indeed, television's unique power to unify and centralize is offset by its equal capacity to give voice to those living at the peripheries. For this reason, much of the Soviet period was dominated by centrally produced programming intended to reinforce the image of a harmonious blend of multi-ethnic peoples, free to develop their own cultural traditions but committed to the integrity of the Soviet Union. Elements of the Soviet strategies have been retained today. We discuss in this book how current television's imaginings of multi-ethnic harmony still occasionally utilize the Soviet 'friendship of the peoples' metaphor. Our example of a 'peripheral voice', that of the main news programme of the Republic of Buriatiia's State Television and Radio Company, has its origins in the Soviet era.

But the differences with the Soviet past seem greater than the similarities. At Moscow State University's Faculty of Journalism the training is now of a practical rather than an ideological orientation, and reflects the influences of contemporary western practices and pedagogical trends. The Dean, Elena Vartanova, has implicitly acknowledged that the nature of some aspects of its new curriculum point to a new culture of media freedom.[60] Because they now lack an all-embracing Marxist–Leninist framework, Russian journalists are also more liable to project the grassroots opinions and prejudices of their non-journalist peers, particularly as formal editorial guidelines of the kind published by the BBC have been notable by their absence. It was only in February 2013, in response to an alarming rise in street-level xenophobia, that the Presidential Council on Inter-ethnic Relations – which was established in the aftermath of the 2010 Manezhnaia riots in Moscow – proposed setting up an agency to provide guidance on media coverage of inter-ethnic relations.[61] At the time of writing these guidelines have not yet been made public.

Russia's current community-building project is still a work in progress. Its overall vision is outlined by the ruling elite only in general terms as a continuation of the long-term tradition of unity in diversity across much of Eurasia, combined with a central role played by ethnic Russians in state-building. Within this project old approaches and strategies of community-building are not reproduced but reinvented, and the fact that, for the first time in Russian history, the community is envisaged by political leaders as national rather than supranational or imperial is a particularly important innovation. At present, television remains a key tool with which the elites can build community consensus, and Putin's government has appreciated its importance. Despite the Kremlin's control over the main television channels, broadcasters operate in a political and discursive environment which is significantly different from its Soviet predecessor. The current media are inevitably open to infiltration by ideas and forms formerly deemed 'alien' and there is a much greater need to respond to grassroots voices external to approved discourse. Most importantly, in the absence of a single ideological framework, the relationship between state and broadcaster is inevitably uncertain. In an interview with our team, leading TV personality Dmitrii Kiselev argued that the task of Russian broadcasters was far more challenging than that of their British counterparts, precisely because of the newness of Russia's current nation-building stage.[62] In what might be seen as an odd paradox, the next chapters, which consider how Russian broadcasters respond to the multiple challenges that they face, broadly endorse Kiselev's assertion, but depart from very different premises. We now begin to construct those premises from an overview of our corpus of recordings.

Notes

1 Seyla Benhabib, *The Claims of Culture: Equality and Diversity in the Global Era*, Princeton: Princeton University Press, 2002.

2 Terry Martin, *The Affirmative Action Empire. Nations and Nationalism in the Soviet Union, 1923-1939,* Ithaca: Cornell University Press, 2001; Francine Hirsch, *Empire of Nations. Ethnographic Knowledge and the Making of the Soviet Union*, Ithaca: Cornell University Press, 2005.

3 Rogers Brubaker, 'Ethnicity Without Groups', *Archives Européennes de Sociologie* XLIII.2, 2002, 163–89; and *Nationalism Reframed: Nationhood and the National Question in the New Europe*, Cambridge: Cambridge University Press, 1997, pp. 23–54.

4 Ronald G. Suny, 'The Empire Strikes Out: Imperial Russia, "National Identity", and Theories of Empire', in Suny and Terry Martin (eds.), *A State of Nations. Empire and Nation-Making in the Age of Lenin and Stalin*, New York: Oxford University Press, 2001, p. 27; Adeeb Khalid, 'The Soviet Union as an Imperial Formation: A View from Central Asia', in Ann Stoler, Carole McGranahan and Peter Perdue (eds.), *Imperial Formations*, Santa Fe: SAR Press, 2007, pp. 123–51.

5 Vera Tolz, 'Forging the Nation: National Identity and Nation Building in Post-Communist Russia', *Europe-Asia Studies*, 50/6, 1998, 996–99.

6 Ronald G. Suny, *The Revenge of the Past. Nationalism, Revolution, and the Collapse of the Soviet Union*, Stanford: Stanford University Press, 1993.

7 Nancy Condee, *The Imperial Trace. Recent Russian Cinema*, Oxford: Oxford University Press, 2009, p. 10.

8 Ibid., pp. 10–12.

9 Ibid.

10 Aleksei Miller, 'Istoriia poniatiia natsiia v Rossii', in D. Sdvizhkov and I. Schirle (eds.), *Poniatiia o Rossii*, vol. 2, Moscow: Novoe literaturnoe obozrenie, 2012, pp. 7–49. See also Miller, 'The Empire and the Nation in the Imagination of Russian Nationalism', in

Miller and Alfred J. Rieber (eds.), *Imperial Rule*, Budapest: Central European University Press, 2004, pp. 9–26; and Olga Maiorova, *From the Shadow of Empire: Defining the Russian Nation through Cultural Mythology, 1855–1870*, Madison: The University of Wisconsin Press, 2010.

11 Yitzhak Brudny, *Reinventing Russia: Russian Nationalism and the Soviet State, 1953–1991*, Cambridge: Harvard University Press, 1998.

12 Ibid., pp. 243–50.

13 Vera Tolz, 'Forging the Nation' and 'Conflicting "Homeland Myths" and Nation-State Building in Post-Communist Russia', *Slavic Review*, 57/2, 1996, 267–94; Oxana Shevel, 'Russian Nation-building from Yel'tsin to Medevdev: Ethnic, Civic or Purposefully Ambiguous?', *Europe-Asia Studies*, 63/2, 2011, 179–202.

14 Anatolii Chubais, 'Missiia Rossii v XXI veke', *Nezavisimaia gazeta*, 1 October 2003. Available online at www.ng.ru/ideas/2003-10-01/1_mission.html (accessed 1 January 2014).

15 *Vremia*, 16 March 2014. Available online at www.1tv.ru/news/world/254236 (accessed 1 August 2014); *Vesti*, 17 March 2014.

16 John A. Hall, 'Nationalisms: Classified and Explained', *Daedalus*, 122/3, 1993, 1–28.

17 Michael Hechter, *Internal Colonialism: The Celtic Fringe in British National Development, 1536–1966,* Berkley: University of California Press, 1975, pp. 60–4. See also Mark R. Beissinger, 'The Persisting Ambiguity of Empire', *Post-Soviet Affairs*, 11/2, 1995, 149–84.

18 Rinella Cere and Rosalind Brunt (eds.), *Postcolonial Media Culture in Britain*, Basingstoke: Palgrave Macmillan, 2011; Alana Lentin, 'Imagining the West, Perceiving Race: Social Sciences and Political Imagination' in Chiara Bottici and Benoit Challand (eds.), *The Politics of Imagination*, Abingdon: Birkbeck Law Press, 2012, pp. 109–23.

19 See Putin's addresses to the Federal Assembly, where contemporary Russia is defined as a nation (*rossiiskaia natsiia*): May 2004. Available online at http://archive.kremlin.ru/appears/2004/05/26/0003_type63372type63374type82634_71501.shtml; May 2007, available online at http://archive.kremlin.ru/appears/2004/05/26/0003_type63372type63374type82634_71501.shtml; Medvedev's address in November 2008, available online at http://archive.kremlin.ru/appears/2008/11/05/1349_type63372type63374type63381type82634_208749.shtml (all accessed 1 September 2013). See also speeches by Vladislav Surkov, an ideologist of Putin's regime until 2012, in which he talks about contemporary Russia as a nation (*natsiia*): 'Natsionalizatsiia budushchego', *Ekspert* 43, 20 November 2006; and his speech at a United Russia meeting of 7 February 2006, 'Suverenitet – eto politicheskii sinonim konkurentnosposobnosti', *Moskovskie novosti*, 10 March 2006.

20 Oxana Shevel, 'Russian Nation-building from Yel'tsin to Medevdev'; Marlène Laruelle (ed.), *Russian Nationalism and the National Reassertion of Russia*, London: Routledge, 2009; Alfred Evans, 'Putin's Legacy and Russia's Identity', *Europe-Asia Studies*, 60/6, 2008, 899–912.

21 For an analysis of the effects of 'regimes of ethnicity' on perceptions of nationhood, including the Soviet model, see Sener Akturk, *Regimes of Ethnicity and Nationhood in Germany, Russia and Turkey*, Cambridge: Cambridge University Press, 2012, pp. 197–257.

22 Tolz, 'Conflicting "Homeland Myths"'; Vladimir Malakhov: '"Natsional'naia politika" kak fenomen politicieskoi rechi', in Malakhov, *Ponaekhali tut: Ocherki o natsionalizme, rasizme i kul'turnom pliuralizme*, Moscow: Novoe literaturnoe obozrenie, 2007, pp. 47–65.

23 Valerii Tishkov, 'Samoopredelenie rossiiskoi natsii', *Mezhdunarodnye protsessy*, 3/2, 2005, 17–27.

24 Malakhov, *Ponaekhli tut*, pp. 59–63.

25 Marlène Laruelle, *In the Name of the Nation. Nationalism and Politics in Contemporary Russia*, Basingstoke: Palgrave Macmillan, 2009, pp. 35–48.

26 Emil Pain, 'From Protests to Pogroms', *openDemocracy,* 27 August 2013. Available online at www.opendemocracy.net/od-russia/emil-pain/from-protests-to-pogroms (accessed 15 December 2013).

27 Vladimir Putin, 'Rossiia: National'nyi vopros', *Nezavisimaia gazeta*, 23 January 2012.

28 See, for instance, Greg Philo, Emma Briant and Pauline Donald, *Bad News for Refugees*, London: Pluto Press, 2013; Elisabeth Poole, *Reporting Islam: Media Representations of British Muslims*, London: I.B. Tauris, 2002; Unni Wikan, *Generous Betrayal: Politics of Culture in the New Europe*, Chicago: Chicago University Press, 2002; Alan Lentin and Gavan Titley, *The Crises of Multiculturalism: Racism in a Neoliberal Age*, New York: Zed Books, 2011; Viktor Shnirel'man, *'Porog Tolerantnosti': Ideologiia i praktika novogo rasizma*, in 2 volumes, Moscow: Novoe literaturnoe obozrenie, 2011.

29 Martin, *The Affirmative Action Empire,* pp. 1–27; Brubaker, *Nationalism Reframed*, pp. 23–54.

30 See Chapter 7 on how this fear was exploited during Putin's election campaign in 2012.

31 Dmitrii Medvedev, 'Sokhranit' effektivnoe gosudarstvo v sushchestvuiushchikh granitsakh', *Ekspert*, 13, 2005, 70–6.

32 Martin, *The Affirmative Action Empire*, pp. 311–93.

33 Eric D. Weitz, Francine Hirsch, Amir Weiner and Alaina Lemon, 'Discussion of Eric D. Weitz's "Racial Politics without the Concept of Race: Reevaluating Ethnic and National Purges"', *Slavic Review*, 61/1, 2002, 1–65.

34 David Brandenberger, *National Bolshevism: Stalinist Mass Culture and the Formation of Modern Russian National Identity, 1931–1956,* Cambridge: Harvard University Press, 2002.

35 Such a position is often reiterated by the high-profile right-wing parliamentarian, Vladimir Zhirinovskii.

36 See Chapter 9.

37 On the lack of engagement with post-colonialism in Russia, see: 'Postsovetskoe soznanie i Postcolonial Studies', *Novoe literaturnoe obozrenie*, 94, 2008. Available online at http://nlobooks.ru/sites/default/files/old/nlobooks.ru/rus/magazines/nlo/196/1208/index.html (accessed 1 July 2013).

38 Marlène Laruelle, *Russian Eurasianism: The Ideology of Empire,* Baltimore: Johns Hopkins University Press, 2008; Andreas Umland, *The Nature of Russian "Neo-Eurasianism": Approaches to Aleksandr Dugin's Post-Soviet Movement of Radical Anti-Americanism*, Armonk: M. E. Sharpe, 2009.

39 Tolz, 'Forging the Nation', 1006–7, 1013–14.

40 Benedict Anderson, *Imagined Communities: Reflection on the Origins and Spread of Nationalism*, London: Verso, 1991; John Postill, *Media and Nation Building. How the Iban Became Malaysian*, New York: Berghahn, 2006.

41 Maxwell McCombs, 'Building Consensus: The News Media's Agenda-Setting Roles', *Political Communication*, 14/4, 1997, 433–43.

42 European Commission, Standard Eurobarometer 78, 2012, 'Media Use in the European Union: Report'. Available online at http://ec.europa.eu/public_opinion/archives/eb/eb78/eb78_media_en.pdf.

43 Tina Burrett, *Television and Presidential Power in Putin's Russia*, London: Routledge, 2011.

44 Michael Billig, *Banal Nationalism*, London: Sage, 1995.

45 The BBC outlines its commitment to its core value of 'due impartiality' in its Editorial Guidelines, available on its website: www.bbc.co.uk/editorialguidelines/ (accessed 15 November 2013). For an analysis of the tension at the heart of those values, see Chris Flood, Stephen Hutchings, Galina Miazhevich and Henri Nickels, 'Between Impartiality and Ideology: The BBC's Paradoxical Remit and the Case of Islam-related Television News', *Journalism Studies*, 12/2, 2011, 221–38.

46 Chris Flood, Stephen Hutchings, Galina Miazhevich and Henri Nickels, *Islam, Security and Television News*, Basingstoke: Palgrave Macmillan, 2012, pp. 144–6 and 183–4.
47 'BBC Interview with EDL Leader, Tommy Robinson "Too Aggressive"', *Huffington Post*, 19 June 2013. Available online at www.huffingtonpost.co.uk/2013/06/19/bbc-interview-with-edl-leader-aggressive_n_3466543.html (accessed 15 December 2013).
48 Flood *et al.*, *Islam, Security and Television News*, pp. 72, 92, 170, 244.
49 For an analysis of the circulation of these transnational Islamophobic tropes see Lentin and Titley, *The Crises of Multiculturalism*, pp. 123–60.
50 Kristin Roth-Ey, *Moscow Prime Time: How the Soviet Union Built the Media Empire That Lost the Cultural Cold War*, Ithaca: Cornell University Press, 2011.
51 Interview with a Channel 1 journalist, 29 January 2013.
52 Burrett, *Television and Presidential Power in Putin's Russia*, pp. 77–80, 192–3.
53 REN TV's weekly news programme *Nedelia* often contained coverage which was critical of the Russian government. In August 2014, *Nedelia* was closed down. See 'REN TV zakrylo programmu "Nedelia" s Mariannoi Maksimovskoi', *Vedomosti*, 1 August 2014. Available online at www.vedomosti.ru/companies/news/29787971/ren-tv-zakrylo-programmu-nedelya-s-mariannoj-maksimovskoj (accessed 8 August 2014).
54 Vlad Strukov, 'Russia's Internet Media Policies: Open Space and Ideological Closure', in Birgit Beumers, Stephen Hutchings and Natalia Rulyova (eds.), *The Post-Soviet Russian Media: Conflicting Signals*, London: Routledge, 2009, pp. 208–23.
55 For examples of such criticism, see Stephen Hutchings and Natalia Rulyova, *Television and Culture in Putin's Russia: Remote Control*, London: Routledge, 2009, pp. 37–9.
56 For a detailed analysis of two such clips, see Beumers *et al.*, *The Post-Soviet Media,* pp. 14–17.
57 For the rise of Russian ultranationalism on the web, see Dennis Zuev, 'The Russian Ultranationalist Movement on the Internet: Actors, Communities and Organization of Joint Actions', *Post-Soviet Affairs*, 27/2, 2011, 121–57.
58 Interviews with a Channel 1 journalist, 29 January 2013, and an NTV journalist, 31 January 2013.
59 Ellen Mickiewicz, *Split Signals: Television and Politics in the Soviet Union*, New York: Oxford University Press, 1988, p. 5.
60 Elena Vartanova, 'Producing Journalists at University: A Professor's Opinion', *Moscow Times*, 23 March 2011. Available online at www.themoscowtimes.com/business/business_for_business/article/producing-journalists-at-university-a-professors-opinion/433541.html (accessed 29 March 2013).
61 Alena Sivcheva and Vladimir Dergachev, 'Mezhnatsional'nyi mir v Rossii obespechit novoe agentsvo', *Izvestiia*, 6 February 2013.
62 Interview with Kiselev, 27 March 2013.

Part I

Managing difference

2 Mapping an uncertain terrain

An overview of the corpus

Framework, method, sources

In this chapter we present an overview of the body of news recordings that we collected over a two-year period from September 2010 to August 2012. From those recordings we generated a simple content analysis which enabled us to gain a grasp of the broad patterns and trends in coverage of inter-ethnic relations in Russia, to map out the contested and shifting terrain that we will be exploring in more detail throughout the rest of the book and to identify some of the key themes and issues that inform the arguments that we advance. We should stress, however, that the quantitative data we analyse here is a point of departure for these arguments, not their underpinning base; our approach is grounded in the principles of qualitative research and the data provides a stimulus for, or at times a corroboration of, points that hopefully are independently sustainable. We explicate many of the patterns revealed by the data with illustrative references to individual stories and reports, some of which we pick up in subsequent chapters. Here, then, the individual examples serve as straightforward instantiations of larger phenomena; elsewhere in the book, the relationship between particular instance and general principle is often rather more complex.

The period to which the recordings belong encompassed some important changes in Russia's political landscape. The winter of 2011–12 saw the first major street protests experienced in Russia for nearly two decades, following the 2011 elections to the State Duma, the Russian parliament (the elections had become mired in suspicions of falsification and malpractice). Despite the unprecedented scale of the protests, Vladimir Putin returned to the presidency in March 2012. Putin's perceived manipulation of the constitution to permit him to run for a third term led to further mass demonstrations on the streets of Russia's cities. The period leading up to Putin's re-election witnessed the scandalous arrests of members of the punk collective Pussy Riot, who staged a protest performance on 21 February 2012 in Moscow's Cathedral of Christ the Saviour. The upheaval coincided with and, at certain key junctures – a primary focus for the chapters to follow – intersected with deteriorating inter-ethnic relations throughout Russia. It was also immediately preceded by major Islamist-inspired suicide bombings on Moscow's metro system in March 2010 and at Moscow's Domodedovo airport in January 2011, when the separatist insurgency that Russia had been fighting in its North Caucasian periphery dealt devastating blows to the (post-)imperial heartland. The state-aligned broadcast media bore responsibility for some of that deterioration, yet frequently resorted to suppressing the controversial topics associated with it in order to avoid fuelling the fires of conflict. Our daily news recordings captured some of the major milestones in this contradictory process, notably the media's confused reaction to the racially motivated riots in Moscow's Manezhnaia Square in December 2010.

In depicting the interpretative framework that news broadcasters applied to events to which they ascribed, whether implicitly or explicitly, an inter-ethnic dimension, we developed a coding system based on seven key thematic categories: 'ethnic cohesion' (reports foregrounding evidence of or celebrating inter-ethnic harmony); 'the Russian Orthodox Church' (coverage of the Church as a key tool for binding together the peoples of Russia's multi-ethnic, multi-faith state); 'other religions' (reporting, both positive and negative, on issues involving other religions, including Islam); 'migration' (stories centring on issues raised by population movements within and beyond the Russian Federation); 'inter-ethnic conflict' (stories detailing clashes, often violent, between individuals and groups, to which ethnic motivations are attributed); 'separatist violence in the North Caucasus' (coverage of terrorist and other assaults on Russian interests and personnel launched by armed opponents of Russia's rule in the troubled autonomous regions of Chechnia, Dagestan and Ingushetiia); and 'other/miscellaneous' (reports with an inter-ethnic aspect, but belonging to none of the above categories).

When coding items we used the content of the headlines under which they were introduced to guide us, but watched each item carefully before assigning it to a category. Overall, in most cases, we worked on the principle of thematic preponderance; thus, an item which dealt with issues other than ethnicity would only be coded if the invocation (implicit or explicit) of ethnicity outweighed that of other factors. This approach was not always applied to reports in the category 'separatist violence in the North Caucasus'. In their coverage of this topic, state-aligned broadcasters often denied religion- or ethnicity-related factors, using the alleged efficiency and effectiveness of the Special Forces and the government as the most common frame. Our decision to incorporate such reports into our dataset is a response to the widespread tendency among the Russian public to ethnicize developments in the North Caucasus. Furthermore, ethnic and religious factors were at times visually underscored in the coverage, even if they were not verbally acknowledged. Thus, reports about violence in the North Caucasus give us an example of how state-aligned television confronts interpretations which are undesirable from the leadership's point of view, yet are widespread in society and promoted by those media outlets which the government cannot control (such as the internet).

Finally, items which dealt with more than one of our chosen categories would be assigned to the one which predominated, ensuring that no item was coded more than once. Since the coding was done overwhelmingly by one person, we did not carry out systematic inter-coder reliability tests, but we discussed the coding system as a team and at length and, when particular items seemed typologically problematic, we resolved the issue together. We catalogued every news item in every news bulletin we watched over the two-year period, noting for each item, whether ethnicity-related or not, the length of time allotted to it within the bulletin and its position in the running order. This enabled us to gauge both the frequency (number of items) and the intensity (amount of time allotted) of the coverage, and to gain a sense of the saliency (aggregate running order position) of our topic within the overall Russian news agenda.[1] As we watched the daily news broadcasts, we also collected transcripts of relevant news items from each channel's website (audio-visual recordings of all of the reports were also available on websites, though not in the form of continuous news bulletins). We used these printed transcripts as a tool in the qualitative analyses at the centre of subsequent chapters, but only in conjunction with our own audio-visual recordings. Finally, we selectively applied to those transcripts NVivo qualitative data analysis tools in order to gauge the frequency with which certain key terms recurred in our material, and to identify the contexts in which they were used.

We developed our coding typology in two stages, applying both deductive, or a priori, and inductive, or 'grounded' approaches. Thus, as a first step, we selected the two primary categories dominating contemporary discourse on ethnicity-related topics throughout the world: 'migration' and 'inter-ethnic conflict'. We supplemented these with two categories based on our prior knowledge of the specific situation in Russia:[2] 'ethnic cohesion' and 'separatist violence'. We then watched selected news programmes for a month and, following an inductive processing of that material, identified two further categories: 'the Russian Orthodox Church' and 'other religions'. A very small number of news items – which were categorized 'other/miscellaneous' – revealed no clear patterns and were not included in the interpretation of our data.

There are a number of clarifications we should make about the typology. The first is that the categories included news about inter-ethnic relations in countries other than Russia; as we shall see, these stories fulfil a vital function for news broadcasters in providing telling points of contrast with, and similarity to, domestic events. Second, the categories are shaped both by our own understanding of the terms we selected to name them and by what the broadcasters themselves believe those terms to mean. Thus, in a Russian context, 'inter-national' (*mezhnatsional'nyi*) often encompasses what we would define as 'inter-ethnic'; the latter term (*mezhetnicheskii*) is at times used by the Russian broadcasters interchangeably with what we might interpret as 'inter-racial'. (At the same time, one should bear in mind that the very definition of 'ethnicity' is notoriously difficult to pin down.) Moreover, the radical contingency of the term in Brubaker's account means that some of the events we include in our typology have been overtly assigned an ethnic dimension by the broadcasters, when we might have elected not to accord them this designation; conversely, the ethnicity-related aspect which we perceived in other events is occluded by the broadcasters. Likewise, under 'migration' we include material which relates to the internal movement of Russian citizens from the North Caucasus to cities such as Moscow and which is covered by the specifically Russian usage of the term (*vnutrenniaia migratsiia*), as well as the flow to such cities of foreign citizens from Central Asia and elsewhere, which conforms with the term in its generally accepted sense.

As we noted in the Introduction, rather than clutter our text with systematically inserted scare quotes designed to designate the critical distance we wish to assert from the many questionable uses of 'migration' and 'ethnicity' we find in our material, we have, except on those occasions where a particular usage merits special comment, tended to treat the terms in their undifferentiated, multi-faceted guise. Similarly, our statistical data relates primarily to coverage of 'migration' and 'ethnicity' in the sense that those terms are deployed in the Russian media. We do this both as a matter of pragmatic convenience (we lack the space to engage in elaborate, contorted discussions of the contingency of these terms at their every point of mention or appearance in a statistical graph) and because we are interested in building an inclusive picture of the rich variety of ethnicity-related meanings, legitimate and illegitimate, accorded them by Russian television news, not in establishing restrictive, parsimonious definitions reflecting our own theoretical positions. We trust that the qualifications we provide here are sufficient to assure readers that we are, throughout, aware of the fraught and problematic nature of discourses of ethnicity in any context. Indeed, this becomes one of the major points of focus in the arguments we develop.

It follows from the preceding discussion that the coding we undertook entailed a complex navigation between broadcaster-generated and researcher-generated meanings. In countering any suggestion of inconsistency on our part, we venture the following points. First, and most importantly, we focus primarily on media *representations*, not on empirically verifiable truths. There is a deep ambiguity lurking at the heart of representation itself, for in claiming to depict what is *representative* of reality, it invariably re-constructs or *re-presents* that reality. It is this duality that our navigation process is intended to track. Second, the data we produced from the coding process is self-contained and does not form part of a larger comparative project; our concern is with the complexities and contradictions of television news coverage in Russia, and those selective cross-national comparisons we do make are based on qualitative observation and not quantitative data, thereby minimizing our vulnerability to methodological inconsistency. Finally, and to reiterate, the quantitative data analysis does not actively determine our key arguments; it serves merely as a useful means of establishing the parameters of the territory on which they are to be made. Readers will judge for themselves the strength of those arguments, and thus the extent to which the data analysis helped us in laying the ground for them.

The sources for our analysis are, as we established in the Introduction, derived from daily recordings of two flagship news programmes: *Vremia* (Channel 1) and *Vesti* (Rossiia).[3] (When researching subsequent chapters we supplemented these sources with material from other channels, including NTV, REN TV and Dozhd'.) *Vremia* is broadcast every weeknight for 30 minutes and on Saturday for 15 minutes, without any commercial breaks. The hour-long Sunday edition, *Voskresnoe Vremia*, is billed as an information-analytical programme and typically features longer, magazine-style stories from the previous week. All news items are tightly scripted and read with little deviation by one of the programme's veteran anchors.

Vesti typically runs for 40–45 minutes from Mondays to Saturdays, with one commercial break, after which regional channels broadcast 10–15 minutes of local news. The Sunday edition of the programme, *Vesti nedeli*, runs for an hour. Like *Vremia's* Sunday news programme, *Vesti nedeli* functions as a magazine-style news round up. However, in contrast to *Vremia's* more cautious approach based on summarizing the highlights of the previous week's news, *Vesti nedeli* positions itself as a briefing programme designed to shape the news agenda for the week ahead.

By studying our topic over a two-year period (our recordings ran from 1 September 2010 to 30 May 2012), we were able to acquire an acute sensitivity to the cumulative weight of ethnicity-related reporting in Russia and to the patterns of distribution and the rhythms of emphasis that it displays. Indeed, most content analyses centring on television, including those of the eminent Glasgow Media Group and of the BBC itself, are based on far smaller samples (sometimes of as little as a few weeks).[4] Nonetheless, the sheer volume of material to be processed ruled out the possibility of a continuous analysis covering the whole two years. Instead, we recorded and analysed the material in equally spaced blocks. Thus, three months of recording were followed by a three-month break in recording, producing four recording periods which contained a total of 9,352 items viewed, of which 654 were coded.[5] However, in order to guard against omissions and arbitrariness in our analysis, we continued monitoring and watching ethnicity-related news in between our recording blocks, relying on the two channels' websites. The Manezhnaia riots and the months leading up to the presidential elections in March 2012, for instance, occurred outside of our recording periods, but we examine coverage of both in detail in Chapters 4 and 7.[6] While we cannot trace the peaks and falls in coverage in a continuous line from September 2010 to August 2012, our blocks enable us to comment on broad changes in emphasis over the entire period.

In presenting our content analysis, we begin by assessing the overall presence of ethnicity-related news on state-aligned national television. We then look in detail at coverage within each coding category, beginning with those relating loosely to the positive promotion of the nation-building agenda ('ethnic cohesion', 'Russian Orthodox Church', 'other religions'). We next discuss the reporting of migration issues – in many ways the fulcrum of the analysis – as we begin to focus on how news events liable to provoke national discontent are handled within the nation-building framework. Finally, we analyse items assigned to the categories dealing with events in which ethnically motivated discontent explodes into interpersonal and inter-group strife ('inter-ethnic conflict' and 'separatist violence').

Inter-ethnic relations in the news

The overriding impression produced by the data we generated is that the importance of inter-ethnic relations to the government's political agenda and to popular opinion was not properly reflected in the patterns of news coverage. Stories coded as relevant made up only a small portion of the total news coverage on the two channels – between 6 and 8 per cent respectively, as Figure 2.1 shows – both in terms of frequency (number) and intensity (time). These figures remain relatively constant within each recording period (see Figures 2.2 and 2.3).

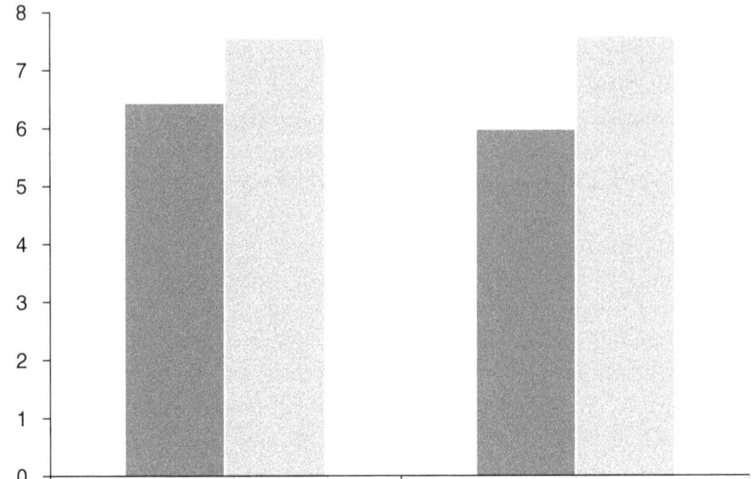

	Frequency	Intensity
Vremia	6.44	5.99
Vesti	7.56	7.6

Figure 2.1 Frequency and intensity of ethnicity-related news as a percentage of the overall news content.

Of all the stories on inter-ethnic relations, a significant portion is accounted for by coverage of issues which relate to ethnicity in other nations (particularly migration and violent conflict) and which implicitly alleviate any impression that Russia is unusually plagued by inter-ethnic tensions (Figures 2.4 and 2.5). This is the reason why international stories feature as prominently on the domestically oriented *Vesti* as they do on *Vremia*.

In the context of the barrage of confused and conflicting messages that national television was compelled to release in reaction to unanticipated crises such as that sparked by the Manezhnaia riots, and of the alarming number of ethnically motivated hate crimes on the streets of Russia's cities as measured by the SOVA centre, the paltry airtime usually received by domestic inter-ethnic relations indicates the extent to which the Kremlin had been struggling with the contradictions of its own nation-building policy.[7]

Within this overall picture, however, the topic of separatist violence in the North Caucasus demonstrated a relatively high degree of salience, at least on *Vremia*, which follows the Kremlin's line more closely than *Vesti*, and which in this case aimed to reaffirm this line in relation to a particularly sensitive problem. As we see from Figure 2.6 (depicting the number of items per coding category for each channel), over 40 North Caucasus-related stories featured among the first three items within the running order of *Vremia* bulletins during the recording period, with all other categories on both channels attracting fewer than 25 items in the top three.

	Sep 2010	Oct 2010	Nov 2010	March 2011	April 2011	May 2011	Sep 2011	Oct 2011	Nov 2011	March 2012	April 2012	May 2012
Frequency	10%	6%	5%	4%	8%	9%	5%	4%	7%	8%	6%	4%
Intensity	9%	5%	5%	4%	7%	8%	4%	4%	7%	9%	6%	4%

Figure 2.2 Frequency and intensity of ethnicity-related news by recording periods as a percentage of the overall news content: *Vremia*.

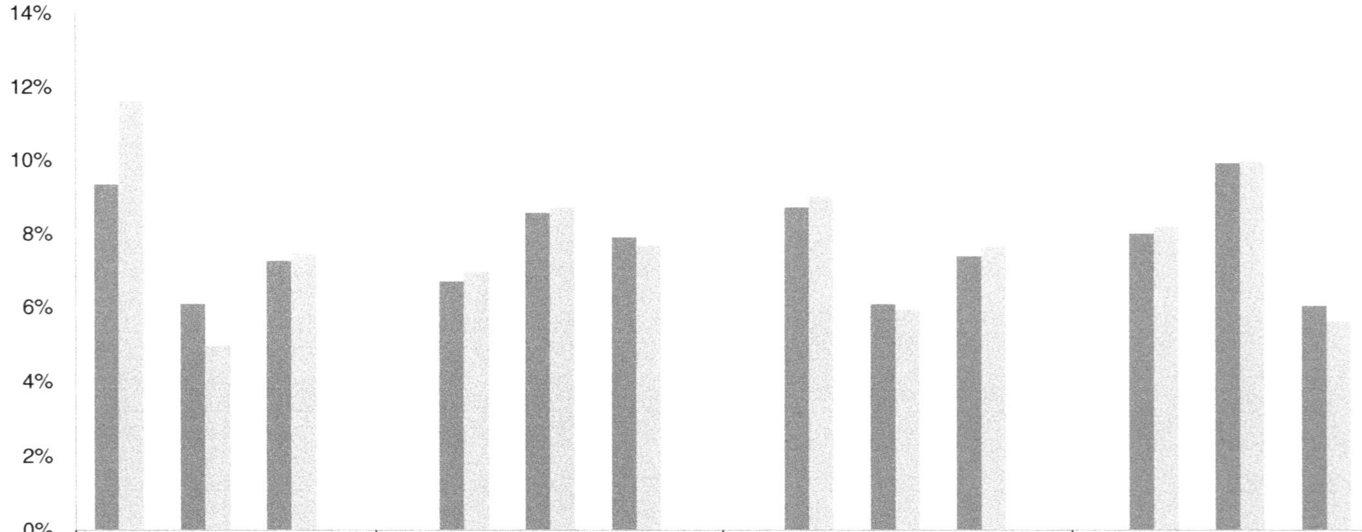

	Sep 2010	Oct 2010	Nov 2010	March 2011	April 2011	May 2011	Sep 2011	Oct 2011	Nov 2011	March 2012	April 2012	May 2012
Frequency	9%	6%	7%	7%	9%	8%	9%	6%	7%	8%	10%	6%
Intensity	12%	5%	7%	7%	9%	8%	9%	6%	8%	8%	10%	6%

Figure 2.3 Frequency and intensity of ethnicity-related news by recording periods as a percentage of the overall news content: *Vesti*.

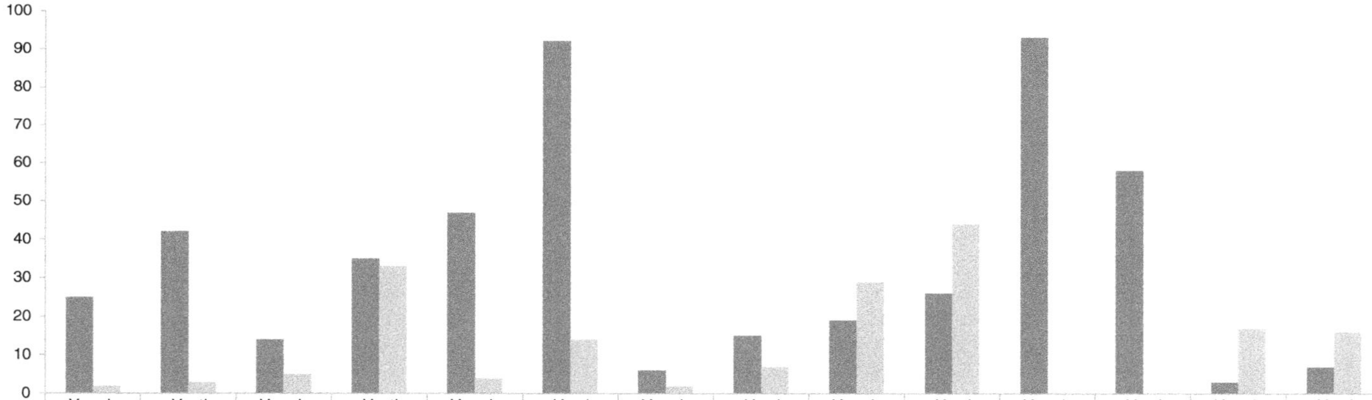

	Ethnic Cohesion		Migration		ROC		Other Religions		Violent Conflict		Separatist Violence in NC		Other	
	Vremia	Vesti	Vremia	Vesti	Vremia	Vesti	Vremia	Vesti	Vremia	Vesti	Vremia	Vesti	Vremia	Vesti
Inside RF	25	42	14	35	47	92	6	15	19	26	93	58	3	7
Outside RF	2	3	5	33	4	14	2	7	29	44			17	16

Figure 2.4 Frequency of ethnicity-related news inside and outside the Russian Federation: *Vremia and Vesti.*

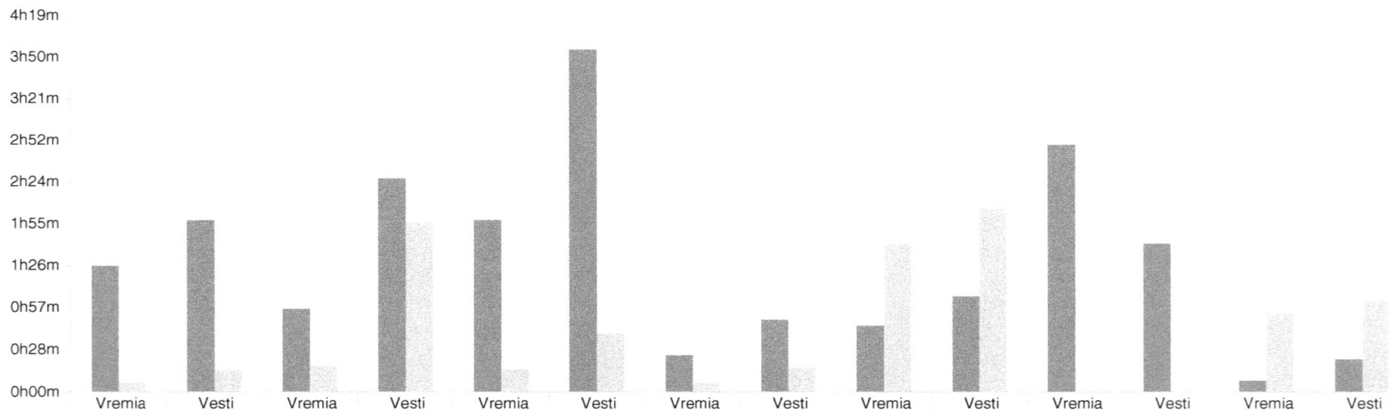

	Ethnic Cohesion		Migration		ROC		Other Religions		Violent Conflict		Separatist Violence in NC		Other	
	Vremia	Vesti	Vremia	Vesti	Vremia	Vesti	Vremia	Vesti	Vremia	Vesti	Vremia	Vesti	Vremia	Vesti
Inside RF	1h26m	1h57m	0h56m	2h26m	1h57m	3h55m	0h25m	0h49m	0h45m	1h05m	2h49m	1h41m	0h07m	0h22m
Outside RF	0h06m	0h14m	0h17m	1h56m	0h15m	0h40m	0h06m	0h16m	1h41m	2h06m			0h53m	1h02m

Figure 2.5 Intensity of ethnicity-related news inside and outside the Russian Federation: *Vremia and Vesti.*

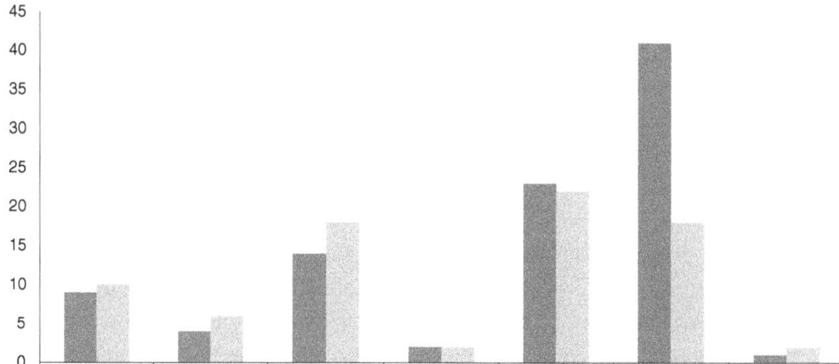

	Ethnic Cohesion	Migration	ROC	Other Religions	Inter-Ethnic Conflict	Separatist Violence In NC	Other
Vremia	9	4	14	2	23	41	1
Vesti	10	6	18	2	22	18	2

Figure 2.6 Salience of ethnicity-related news: *Vremia and Vesti.*

To explore the tensions a little further, on the one hand, the Kremlin was quite consistent throughout most of our recording period in promoting an image of multi-ethnic harmony and in underscoring ethnic diversity as the country's strength. These assertions were, not unsurprisingly, highlighted in news bulletins. For example, a *Vremia* report of July 2011 quoted Putin as stating that 'tolerance is in our blood' and stressing Russia's historical uniqueness with regard to ethnic relations: 'Nowhere on the planet, not on any territory, live more than 180 nationalities and ethnicities. Moreover, these peoples have not migrated from other places … but have lived on the territory of this country since time immemorial.'[8]

On the other hand, the confidence exuded by these claims is not borne out by the fact that the level of news coverage of inter-ethnic relations actually dropped at politically sensitive moments. At the lowest point on *Vremia* in terms of intensity it accounted for only 4.15 per cent; this was in May 2012, at the time of Putin's inauguration as president. Already prior to this, during the entire presidential election campaign, the media had largely refrained from reporting on inter-ethnic relations. During this period, most of the relevant coverage related to the Russian Orthodox Church, as will be discussed. As our interviewees noted, during certain periods reporters received instructions not to report on issues of a potentially inflammatory nature, including, specifically, inter-ethnic relations.[9]

The underreporting of ethnic issues is in part connected to unresolved tensions deriving from the Russian Federation's status as a multi-ethnic, multi-faith state. (The controversial implications of these contradictions are discussed at length in the Introduction and throughout the book.) Russian nationalists traditionally see ethnic Russians as marginalized by the state and other nationalities as favoured, but our word frequency analysis of the term 'Russian' (*russkii*) indicates that the state-aligned media are far from neglecting things Russian, the Russian language or Russian culture.[10] In fact, as the context of those usages confirms, the Russian language, Russian culture and Russian Orthodoxy have been allocated an unambiguous role as the key binding forces in the Federation, and the role of the state as a key factor in creating the pan-Russian (*rossiiskaia*) national community has remained without serious challenge throughout the Putin period. In his interview with us, Channel 1 presenter Maksim Shevchenko acknowledged his own responsibility to contribute to resolving the duality at the heart of the Federation as follows:

> We have more peoples than the whole of Europe put together. But in contrast to Europe, which has not seen a unified state since the Roman Empire, Russia has always been such. This is why our task is to figure out how to preserve it, how to establish a united political nation and at the same time preserve the diversity of ethnicities in Russia and give them the opportunity to develop within the country. I think that the American model might be suitable, but not entirely. America was set up from below, from the communities up to the state. We need to proceed the other way around. First there is the state, and now we need to figure out how to incorporate the communities within it.[11]

We begin our more detailed examination of our corpus of recordings by focusing on the coding category designed to capture those reports most actively and deliberately deployed in support of the ambitious mission that Shevchenko describes: ‘ethnic cohesion’ and national unity.

Ethnic cohesion

In terms of both intensity and frequency, and as we see from Figures 2.4 and 2.5, ‘ethnic cohesion’ amounted to a relatively modest portion of all ethnicity-coded news across both channels. In percentage terms, ‘ethnic cohesion’ accounts for approximately 12 per cent of the intensity of news coverage relating to our topic area for both *Vremia* and *Vesti* (see Figures 2.7 and 2.8).

This is, of course, lower than the mean across all seven categories, but still high when one considers the difficulties that stories in this category raise in terms of their newsworthiness and potential interest to viewers (in the post-Soviet, semi-commercialized news environment, Russia’s state-aligned broadcasters cannot afford to entirely ignore these issues). All of the events that we included in the category amounted to regularized, pre-planned, state-initiated activities such as festivals and anniversaries, none of which offered anything much in the way of spontaneous narrative content. The most significant of these events was the Day of National Unity (a new public holiday introduced in 2005), to which we subsequently devote an entire chapter of this book. Other reports related largely to traditional regional and local festivities. These stories highlighted thriving minority cultures and harmonious ethnic relations. The folk-cultural approach characterizing them was reminiscent of the Soviet presentation of ethnic diversity and harmony. While this might resonate nostalgically with viewers old enough to recall the Soviet period, the same is not true of the younger audience demographic that Channel 1 in particular has periodically hankered after.

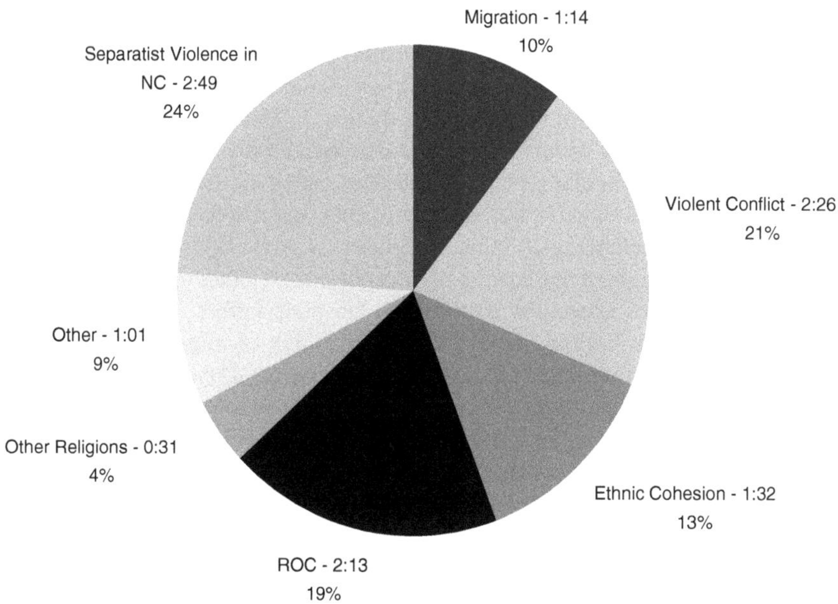

Figure 2.7 Intensity of each category as a percentage of all ethnicity-related news: *Vremia.*

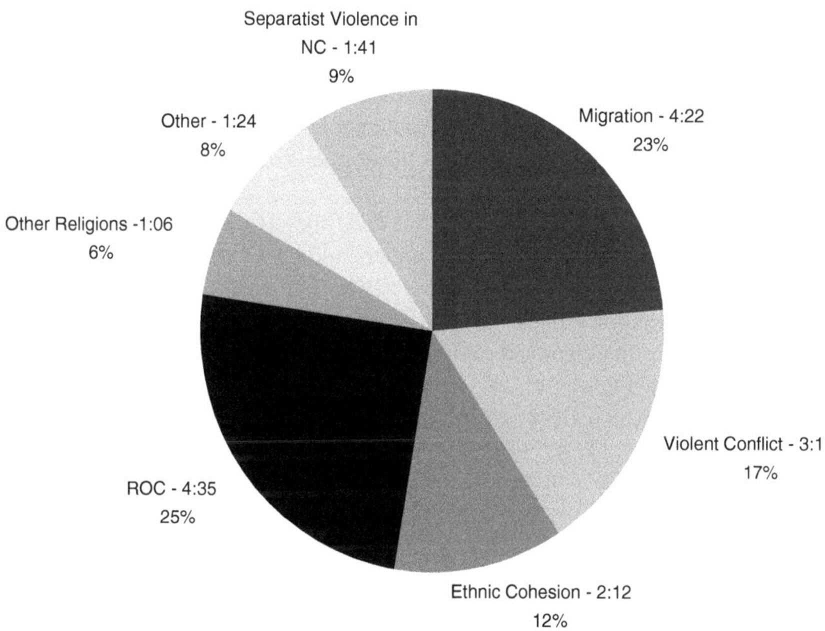

Figure 2.8 Intensity of each category as a percentage of all ethnicity-related news: *Vesti.*

The only negative news item in the 'ethnic cohesion' category covered a 2011 meeting of the Federation Council in which President Medvedev stated that the 'inflation of inter-ethnic conflict and religious dissension during the upcoming election campaign [would] be punishable by law'.[12] This measure had received consistent legitimation within media discourse from earlier points in our recording period, through regular reports on deteriorating ethnic relations in the west. Their key message – that, in Europe, ethnic cohesion is doomed – was present in many reports belonging to other categories as well. Within all categories, these stories highlighted the *lack* of ethnic cohesion. Among them was a report on Angela Merkel's speech of 2010 on the 'absolute failure' of multiculturalism, presented as a 'failing battle'[13] against an influx of migrants who do not speak German and have failed to integrate. Such events – David Cameron's later speech on the same topic received similarly prominent attention from *Vremia* and *Vesti* – provided the Russian authorities with cover not only for announcements like Medvedev's, but also for the harsh and uncompromising anti-migration and anti-Islamic rhetoric which later took hold during Putin's third presidency, as will be described in Chapter 9. During our recording period in 2010 and 2011, media coverage of domestic community cohesion problems was generally much more cautious.

Dmitrii Kiselev, Rossiia's deputy director and *Vesti nedeli*'s presenter, was one of several prominent broadcasters who heavily promoted the idea of the failure of multiculturalism and western European liberal values. In his interview with us, he explained that 'tolerance' is a purely western concept which is not suitable for Russian society:

> Tolerance is not our term. That is why we don't use it often. Tolerance is a very cold word for a Russian. It means that one needs to endure others. The Russian term instead is love. We should not endure each other, but love each other… We are all brothers. This is the correct Russian notion.[14]

Kiselev was particularly passionate in his hostility toward the west, and despite his stance as a Kremlin loyalist, his understanding of tolerance was at odds with that adopted by Putin and Medvedev, who often used the term in its more positive, European meaning. However, Kiselev's interpretation is more closely allied to the traditional Russian discourse, which juxtaposes (hollow) toleration with (genuine) recognition.[15] His emphasis on the contrast between positive Russian values, which foster ethnic harmony, and hypocritical western ones, which inevitably lead to failure, will recur in several of the categories discussed below.

Russian Orthodox Church

The contrast between Russian and western values was reinforced in coverage of the Russian Orthodox Church. It became particularly sharp toward the end of our recording period, during the presidential election campaign and in the unfolding case against Pussy Riot. As we discuss in Chapter 8, in that period, leading television journalists discursively transformed Orthodox Christianity from merely an important national value into the very foundation of the Russian state, which had historically protected it from harmful foreign influences. The Church's centrality to the state-sponsored nation-building project was reflected in the number of *Vesti* reports on Orthodox Christianity – more, in fact, than on any other category (see Figure 2.4). Furthermore, the de facto superior status of the Church compared to other 'traditional religions' of Russia was confirmed by the fact that both *Vremia* and *Vesti*'s coverage of Orthodoxy ran four times longer than the amount of time devoted to all other religions combined (see Figure 2.5). In the case of *Vremia*, which aims to act as the primary news source for Russian-speaking communities outside Russia, the prioritization of the Orthodox Church also indicates its importance as a national soft-power agent. In our recording period, the Russian government became especially keen to increase the effectiveness of its cultural diplomacy, in particular targeting Russian speakers across the world. The Church has been allotted a major role in these efforts.[16]

The times during which the coverage of Orthodox Christianity peaked during our recording period further point to the special relationship that the Church – and Patriarch Kirill personally – enjoys with the state and the state-aligned media. There were two peak months in terms of both frequency and intensity of the relevant coverage: November 2011 and April 2012 (see Figures 2.9 and 2.10).

Both peaks fall at the time when the alliance between the Church and the state became even stronger than before, following the announcement of Putin's decision to run for a third presidential term in September 2011. The Church's overt support for Putin provoked criticism from the opposition and alternative media, which began featuring damaging revelations about the lavish lifestyle of the Patriarch and other hierarchs and examples of questionable activities through which the Church had attempted to increase its material wealth.[17] In response the Church mounted a well-organized public relations campaign, in which it was closely assisted by state-aligned television.

	Sep 2010	Oct 2010	Nov 2010	Mar 2011	April 2011	May 2011	Sept 2011	Oct 2011	Nov 2011	Mar 2012	April 2012	May 2012
Vremia	3	3	5	5	4	4	2	2	10	0	10	3
Vesti	1	7	10	4	9	9	6	5	14	12	23	6

Figure 2.9 Frequency of Russian Orthodox Church-coded stories over the total recording period: *Vremia and Vesti.*

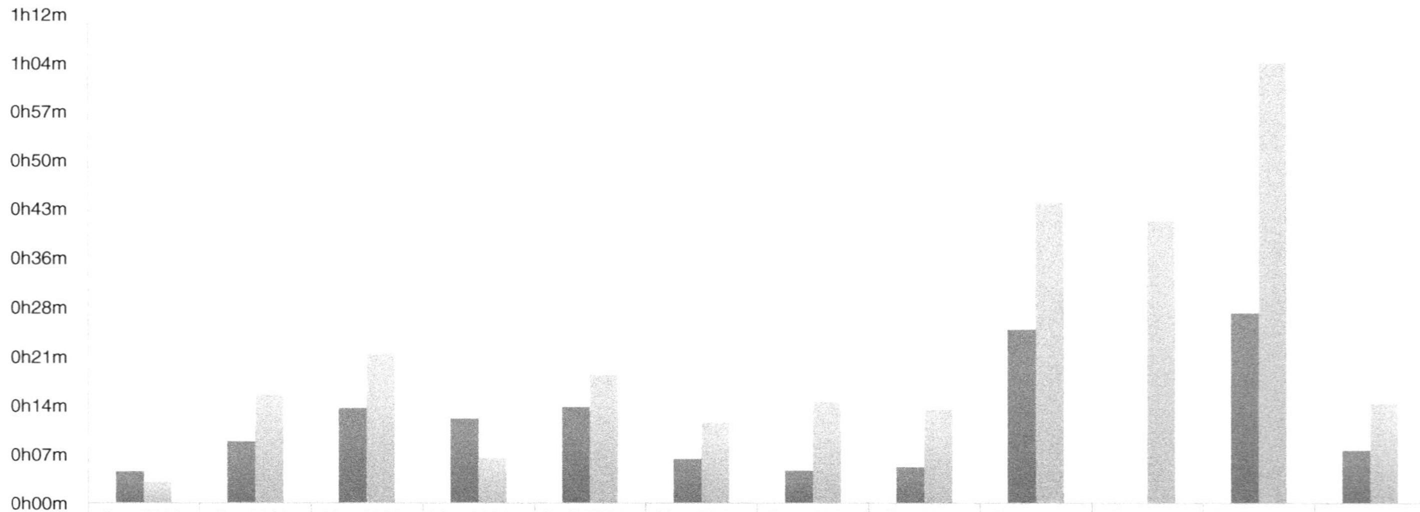

	Sep 2010	Oct 2010	Nov 2010	Mar 2011	April 2011	May 2011	Sept 2011	Oct 2011	Nov 2011	Mar 2012	April 2012	May 2012
Vremia	0h04m	0h09m	0h14m	0h12m	0h14m	0h06m	0h04m	0h05m	0h25m	0h00m	0h28m	0h07m
Vesti	0h03m	0h16m	0h22m	0h06m	0h19m	0h12m	0h15m	0h13m	0h44m	0h41m	1h04m	0h14m

Figure 2.10 Intensity of Russian Orthodox Church-coded stories over the total recording period: *Vremia and Vesti.*

In the first step in this campaign, a revered relic was brought to Russia from Mount Athos – the cincture of the Theotokos (or the Virgin Mary's belt). Its display in Moscow and a number of other cities attracted numerous visitors. The relic's journey across Russia was systematically televised, and relevant reports accounted for the November 2011 rise in the coverage of Orthodoxy-related issues.[18]

The second peak was even more striking, as in April 2012 the Russian Orthodox Church accounted for more than half of all our coded *Vesti* reports. There were three reasons for this increase. One was particularly heavy coverage on both channels of the celebration of Easter, considered in the Orthodox tradition to be the most important holiday. Whereas in 2011 this extended only to the Easter weekend, in 2012 it stretched over most of Passion Week. This expansion provided an indication of the further elevation of the Church's status in the context of Putin's re-election. Second was Pussy Riot's alleged desecration of an Orthodox cathedral, which triggered an intensification in the coverage of Church activities, with reports featuring the reaction of the clergy and ordinary believers to the event. Notably, whereas *Vesti* began its reporting on the case in March, further increasing coverage in April, *Vremia* delayed its first report on Pussy Riot to 19 April.[19]

The final reason for the rise in coverage of the Church in April 2012 was another major public relations initiative organized by Patriarch Kirill. This was the so-called prayer vigil 'of Orthodox believers in defence of their faith', held in Moscow and across the country on 22 April. With the help of the state, thousands of people from around Russia were brought to Moscow to pray with the Patriarch for the end of what he dramatically described as a 'war' against Orthodox Christianity, allegedly triggered by the Pussy Riot performance. As expected, *Vesti* and *Vremia* covered the event at length.[20] This coverage actively promoted an image of Russia as primarily the homeland of ethnic Russians, completely marginalizing the alternative, state-sponsored, vision of a multi-confessional and multi-ethnic society. To some extent this marginalization could be observed throughout the entire recording period, as the minimal attention accorded to other religions attests. In terms of frequency and intensity, and in relation to events inside the Russian Federation, 'other religions' was the weakest category across both channels (see Figures 2.4 and 2.5).

Other religions

Under the category of 'other religions' we expected above all to see stories about Islam, Buddhism and Judaism, which, like Orthodoxy, enjoy official status among Russia's 'traditional religions'. Yet Buddhism had no presence at all on the federal news, and Judaism had virtually none; the only relevant report related to the celebration of the New Year in Israel in September 2011.[21]

Islam was less peripheral to the news agenda. In the official discourse, Russia's multicultural nature is at times described with reference to centuries of peaceful co-existence between Orthodoxy and Islam. This line was particularly strongly endorsed in the coverage of the celebrations of Muslim religious holidays in Moscow. Reporting on one such celebration in September 2011, *Vesti nedeli* gave a brief history of the life of 'the Muslim community' in Moscow, stressing its beginnings in the fourteenth century and noting that an estimated 20 million Muslims live in Russia today.[22]

Nonetheless, overall coverage of Islam was still limited, particularly on *Vremia* (six stories). On *Vesti* there were 21 stories, many of which were about celebrations of religious holidays in Russia's predominantly Muslim regions of Tatarstan and the North Caucasus. As with Orthodox Christianity, the display of relics revered by believers was a familiar theme. In September 2011, for instance, *Vesti* reported on the arrival in the capital of Chechnya, Groznyi, of a bowl allegedly belonging to the Prophet Mohammed.[23] These parallels helped to project an image of the harmonious co-existence of Orthodoxy and Islam in Russia. The message of harmony, in accordance with the Eurasianist outlook reflecting the official rhetoric, was further reinforced by repeated characterization of the form of Islam said to be 'historically traditional' to Russia as 'moderate and peaceful'. According to one *Vesti* report, spreading Islamic education drawing on the indigenous traditions of the religion was the best way 'ideologically to counter radicalism and extremism' penetrating Russia from abroad.[24]

As we will see below, with the exception of major terrorist events in the Russian heartlands, Islam was rarely evoked in the reporting of violence in the North Caucasus and, if references to it were made, this was usually in order to deny the relevance of religious factors to the events. In the recording period, inter-confessional disharmony was stressed mainly in relation to western Europe, usually in the context of stories we placed in the category 'migration'. These pointed to growing societal Islamophobia in response to western governments' policies on multiculturalism, which were invariably described as failures.[25] As we will see in Chapters 8 and 9, the situation changed significantly after the end of our recording period, and during the first 18 months of Putin's new presidency both channels began featuring negative depictions of Islam itself when covering events in western Europe. Blurring the boundary between radical Islamism and Islam, *Vesti* also began including this negative view of Islam in its interpretation of developments in Russia.

Christian denominations other than Russian Orthodoxy were barely mentioned by state-aligned broadcasters. That all four reports on Catholicism referred to events outside Russia symbolically reinforced its status as a religion alien to Russian culture. Protestant communities (which, with 300,000 members, are twice as numerous as Catholics) were also ignored. Although the Moscow Patriarchate has historically been far less antagonistic to Protestantism than to the Vatican, the Church has been negatively disposed toward the country's largest Protestant community – the Baptists – both in the past and to this day. Together with representatives of other Protestant groups – Pentecostals and Adventists – Baptists were fiercely persecuted in the Soviet period, especially for their refusal to enter military service; discrimination against them continues today.[26]

Other religious groups, such as Jehovah's Witnesses, the Church of Jesus Christ of Latter-Day Saints (commonly known as Mormons) and the Church of Scientology, often branded negatively as 'sects' or 'cults', received minimal airtime on *Vesti*. They were portrayed as alien and destructive to Russian culture and traditions.[27] In view of strong disapproval of their activities by the Orthodox Church, which objected to their very presence in Russia, such coverage comes as no surprise. Jehovah's Witnesses were actually banned from practising their religion in the Soviet period; it was legalized only in 1991.[28] Under the conditions of full religious freedom in the first half of the 1990s, the well-funded foreign missionaries representing these groups actively proselytized in Russia. It was in response to the perceived threat to its dominant position in the country from such proselytizing activities that the Russian Orthodox Church pushed through a controversial law on religion in 1997. This law singled out Orthodoxy, Islam, Buddhism and Judaism as religions 'traditional to Russia' and therefore worthy of state support, and at the same time imposed strict limitations on the activities of other religious groups.[29]

In short, Russia's religious diversity, which Kremlin-sponsored discourse at times evokes, found little reflection in our corpus of recordings. The promotion of harmonious multi-confessionality was secondary to the need to highlight the central importance of Orthodox Christianity to Russia, past and present. Indeed, the very term 'multi-confessional' (*mnogokonfessional'nyi*) was used only six times during the entire recording period, suggesting a subtle divergence between official Kremlin rhetoric and that of state-aligned television news, as well as contradictions in the Kremlin's own position on religion-related issues.

Migration

Similar divergences and inconsistencies were identified in relation to migration. In academic literature, definitions of migration are complex and often contradictory. As Bridget Anderson and Scott Blinder note, there is no consensus on a single definition of 'migrants', who can be defined by foreign birth and citizenship as well as by their temporary or long-term geographical mobility across and within national boundaries.[30] The confusion increases in the context of mass media representations and in the discourses of politicians, who regularly politicize migration-related issues. Media outlets in many European countries have been criticized for their discriminatory treatment of migrants, for using criminalizing terminology and for engaging in a systematic process of 'othering'. When covering migration, journalists throughout the world likewise tend to ethnicize the social and economic issues which are at the root of migration trends.[31]

In the absence of any reporting guidelines dealing with sensitive issues eliciting high societal concern, the danger that journalists will use language which discriminates against minority groups further increases.[32] Particularly controversial is the application of the terms 'migrant', or even 'illegal migrant', to Russian Federation citizens of North Caucasian origin. Even the Kremlin-sponsored discourse lacks consistency on this issue. On some occasions, Putin argued that no citizen of the Russian Federation could be called a migrant;[33] on others he used the term 'migration' to describe the residency of North Caucasians in Moscow and other cities of central Russia.[34] More significantly, President Medvedev once directly compared North Caucasians 'who have moved to locations where ethnic Russians historically predominated' with migrants from North Africa and the Middle East in Europe, concluding that 'in contrast to Europe our migrants are internal, but the problems [they create] are similar'.[35] Such contradictory, even inflammatory, pronouncements are reported without reflection or interrogation.

Likewise, television news is prone to cover stories about Russia's *tsygane* (gypsies) as part of discussion of the impact of migration flows on Europe, even though Russia's Roma communities date back centuries and their members are citizens of the Russian Federation.[36] In the course of this book we critically reflect on different ways in which migration is understood and represented by Russian broadcasters and consider the potential social and political consequences of their terminological laxity.

During our recording period, news stories concerning 'migration' exhibited several striking features. From 2010 onwards, opinion polls have indicated rising resentment against non-Slavic nationalities.[37] The popular impression of sharply rising migrant flows into Russia does not, however, spring from any real increase in migration, at least not if measured on a Russia-wide scale. According to the World Bank, net migration to the Russian Federation actually decreased from 2.2 million in 2000 to 1.1 million in 2010.[38] Significantly, during our recording period, while the Russian print media were already featuring highly alarmist reports on migration's effects on Russia, state-aligned television was largely avoiding opportunistic exploitation of these widespread perceptions. The broadcasters seemed to have followed the Kremlin's general view of migration as essential to the Russian economy.

As Figure 2.4 demonstrates, in terms of frequency, on *Vremia* 'migration' was the second least covered topic after 'other religions', and on *Vesti* it generated less coverage not only than 'the Russian Orthodox Church' and 'separatist violence in the North Caucasus', but also than 'ethnic cohesion' issues.[39] During our first recordings between September and November 2010, migration-coded stories were simply absent on *Vremia*, and this was the period when the controversial deportations of Bulgarian and Romanian Roma from France were being criticized by the European Parliament and the European Commission. Whereas *Vesti* used the opportunity to claim much better conditions for the Roma, *Vremia* did not refer at all to the criticism provoked by the deportations.[40] Overall, *Vremia*'s coverage of migration remained minimal, with 0–2 relevant stories in 10 of the 12 months of recordings. The amount of coverage on *Vesti* was slightly greater and, in contrast to *Vremia*, it featured occasional reports on clashes between labour migrants and locals in Russia, particularly in Moscow.[41] Thus, migration was one of the categories in which differences between the two channels' approaches to reporting became particularly noticeable.

As we can see from Figures 2.11 and 2.12, both channels pay particular attention to migration-related issues outside Russia; on *Vesti* the amounts of domestic and international coverage of migration-related events were almost identical. During our recording period, the situation in Russia was positively contrasted with developments in Europe, where migration, it was argued, had brought about significant societal problems. Both channels linked the migration-related difficulties faced by the various countries to the theme of Europe's deeper-seated crisis of multiculturalism. *Vremia* and *Vesti* also claimed that the inevitable consequence of Europe's migration policies was a sharp rise in radical right-wing sentiments among the population and electoral success for the far right. The message here was that Russia ought to do better than copy western policies of managing diversity.[42]

The acute shortage of positive stories about migrants elsewhere in Europe has been noted by observers.[43] Nor did they feature prominently in Russian television news: following the pan-European pattern of reporting, positive examples were implicitly portrayed as exceptional and framed by the discourse of successful integration into the dominant culture. All such stories on Russian television news concerned EU or North American citizens who decided to move to Russia out of

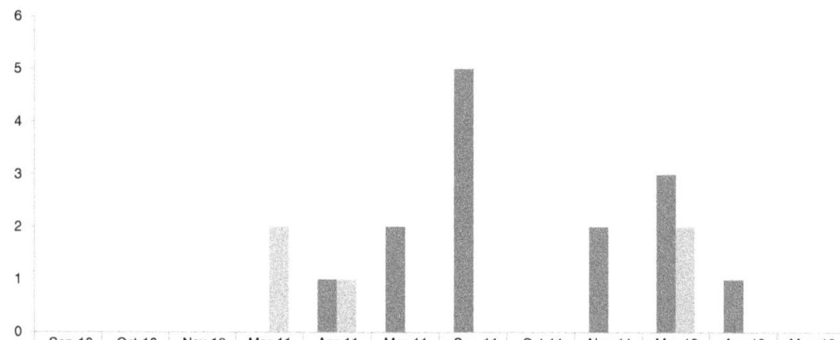

	Sep-10	Oct-10	Nov-10	Mar-11	Apr-11	Mar-11	Sep-11	Oct-11	Nov-11	Mar-12	Apr-12	May-12
Inside RF	0	0	0	0	1	2	5	0	2	3	1	0
Outside RF	0	0	0	2	1	0	0	0	0	2	0	0

Figure 2.11 Frequency of migration-coded stories inside and outside the Russian Federation over the total recording period: *Vremia.*

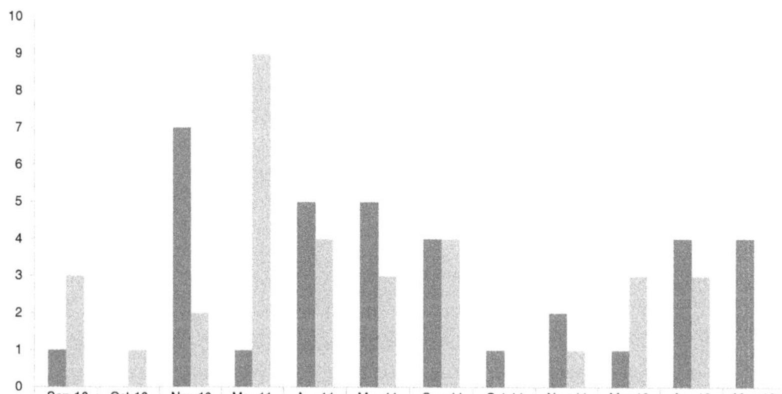

	Sep-10	Oct-10	Nov-10	Mar-11	Apr-11	Mar-11	Sep-11	Oct-11	Nov-11	Mar-12	Apr-12	May-12
Inside RF	1	0	7	1	5	5	4	1	2	1	4	4
Outside RF	3	1	2	9	4	3	4	0	1	3	3	0

Figure 2.12 Frequency of migration-coded stories inside and outside the Russian Federation over the total recording period: *Vesti.*

love for the country and its people. Particular publicity was given to stories of celebrity migration.[44]

As we will see in Chapter 9, the state-aligned broadcasters dramatically changed their coverage of migration-related issues during the first 18 months of Putin's third presidency. At the same time, certain differences between the channels have remained in terms of approaches to the topic. Rather than contrasting the ability to manage migration flows, since autumn 2012 *Vesti*, in particular, has begun to draw close parallels between developments in Russia and in western Europe.

Inter-ethnic conflict

The overrepresentation of negative examples related to western Europe was also noticeable in the coverage of 'inter-ethnic conflict'. Overall, more than half of the news items in this category concerned developments outside the Russian Federation. Most reports were of physical violence seemingly motivated by ethnic or religious hostility. As we will show, such motivations were not necessarily really the case, but were at times perceived as such by broadcasters. On *Vesti* and *Vremia*, conflicts in Europe were unfailingly linked to wider social and political issues, and especially to much-criticized policies of multiculturalism. For instance, *Vremia* reported at length on the shootings by a serial killer in Toulouse and Montauban which targeted French North African soldiers and Jewish civilians in early 2012, describing the event as a 'jihad at the heart of Europe' that once again demonstrated 'the complete ineffectiveness of the modern European state'.[45] The coverage of far-right Norwegian terrorist Anders Breivik's trial similarly linked the failure of European immigration policies and the resulting spread of far-right extremism in Europe.[46] Those states of the former Soviet Union with which Russia had troubled relationships, such as Ukraine, were also negatively represented. The Ukrainian police were particularly criticized for their allegedly lenient approach to 'Ukrainian Nazis', among whose victims were veterans of World Wars.[47] (This theme was exploited particularly actively during the Russia–Ukraine stand-off in 2014.) During the recording period, failed European multiculturalism was again contrasted with Russia's success in managing its multi-ethnic society. It was only during Putin's third presidential term that the state-aligned broadcasters began to argue that, in fact, there were more similarities than differences between Russia and western Europe in terms of problems arising from ethnic diversity.[48]

In contrast with their treatment of ethnic conflict abroad, the two channels systematically downplayed ethnic and/or racist aspects of violence in Russia and devoted little attention to them. Coverage of the trial of those accused of the 2009 murder of well-known Moscow lawyer Stanislav Markelov serves as an example. Markelov investigated criminal cases involving neo-Nazi groups, as well as human rights abuses perpetrated by Russian Army personnel in Chechnia. His murder was a high-profile case, which suggests that the broadcasters' reluctance to offer full details of the case was a deliberate policy rather than a mere oversight. *Vremia*'s single report on the trial indicated neither the nature of the cases

Markelov pursued as a lawyer nor the identity of his assassins.[49] *Vesti*'s more detailed coverage noted that Markelov was killed by neo-Nazis, but failed to mention his investigation of human rights violations in Chechniya.[50] Overall, when compared to data available from other sources, violence perpetrated by neo-Nazi and other extreme nationalist groups, including that against representatives of Russia's ethnic minorities and migrants from Central Asia, was systematically underreported by both channels.[51]

Extreme Russian nationalism is a highly sensitive issue for the Kremlin and, following the Manezhnaia riots, it began to take more stringent measures against such activities.[52] Previously, liberal critics of the regime had repeatedly accused the Kremlin of collaborating with Russian nationalists and of using radical nationalist groups to do the government's bidding.[53] Nationalists themselves regularly criticize the Kremlin for being too harsh toward ethnic Russian activists while allegedly displaying excessive lenience toward manifestations of extreme nationalism among the minorities.[54] Thus, the issue represents a major challenge for broadcasters.

It should be acknowledged that state-aligned television coverage of 'ethnic conflict' does include examples of responsible reporting. Our interviewees demonstrated a clear understanding that media reporting can inflame an already problematic situation.[55] So, in addition to paying little attention to the activities of extreme Russian nationalists, during the recording period the broadcasters also followed the Kremlin's position that certain conflicts, particularly those involving Russians and Caucasians, had social origins and had nothing to do with ethnicity, even if the public tended to see them as ethnically motivated.[56] Yet today broadcasters inevitably have to take popular perceptions into account and engage with ethnicized interpretations of cases which attract heated debate on the internet and other media.

The conundrum emerged in coverage of the incident involving a well-known Sambo master, Rasul Mirzaev, who got into a fight with another young man in Moscow in August 2011, as a result of which the young man died. The incident attracted much public attention not only because of Mirzaev's celebrity status, but also because he happened to be Dagestani and his opponent Russian. In the public discussion which followed, the case became ethnicized. The relatively light sentence Mirzaev received and his conviction for manslaughter rather than murder provoked outrage among Russian nationalists, who argued that this was yet another example of the state failing to defend the *russkie* from systematic abuse. While an inter-ethnic dimension was superimposed on the incident in certain talk-show discussions,[57] throughout most of the story's coverage news bulletins represented the confrontation between the two men as a private dispute which had nothing to do with their ethnic backgrounds.[58] Yet when Mirzaev was released from detention at the end of the trial, *Vesti* became somewhat less cautious. Russian nationalist activists' strong objections to the verdict were aired and the reporter demonstrated open sympathy for the victim's angry father who questioned the court's impartiality.[59] *Vesti*'s treatment of the case seems to have reflected a critical view of the trial's outcome taken by the news production team. In an interview with us,

Kiselev criticized the court's lenient treatment of Mirzaev and compared it with what he alleged to be preferential treatment of black people by law enforcement organs in the United States.[60]

Here we see how norms and perceptions prevailing in society at large can influence the frames through which events are interpreted in the media. Biases toward easily accessible and seemingly plausible ethnic frames which politicize social and political processes are common in the media of most countries, and nowhere more so than in Russia.[61] Mirzaev's case is just one such example. It finds its mirror image in the silencing, or de-ethnicizing, of racially motivated crimes committed by neo-Nazi groups. Reporting on separatist violence in the North Caucasus, to which we now turn, is likewise marked paradoxically by both the occlusion and the over-interpretation of ethnicity as a factor in the conflict.

Separatist violence in the North Caucasus

On the face of it, there is considerable continuity between the treatment of inter-ethnic violence in the Russian heartlands and coverage of the separatist insurgency in the North Caucasus, although, as we shall see, the situation began to change somewhat in the second half of our two-year recording period and beyond. For the most part, while the 'international terrorism' theme continued to surface sporadically, many of the violent incidents in the North Caucasus were reported as acts of crime, sabotage and banditry, summarily dealt with by law enforcement agencies, rather than as examples of terrorism. Direct references to ethnicity and religion were rare, and accounts of the anti-imperial rhetoric and separatist ambitions of the perpetrators rarer still; the term 'separatist', in all of its contexts – Russian and international – occurred only 28 times in total throughout the entire corpus. When causality and motives were broached at all, economic and social factors were at the forefront rather than the Islamist or political dimensions. To the extent that the link between Islam and separatist violence was acknowledged, the term 'Wahhabist' (*vakhkhabit*), with its distinctly foreign (Saudi) origins (11 occurrences), was preferred to 'Islamist' (0 occurrences). References linking insurgent leaders to al Qaeda and the broader War on Terror were occasional and perfunctory. This is in keeping with trends observed in the period between 2006 and 2008 as part of an earlier project.[62] There, too, systematic suppression of the goals of the secessionist movement during a period in which the radical Islamist influence upon that movement had been gradually increasing only underlined the Kremlin's paranoid fear that, by acknowledging the insurgents' true motives, its loyal broadcasters might legitimize those motives and foster the impression that it had lost control of the situation.

The lack of background detail and explanatory analysis in reports on terrorist violence in the North Caucasus extended beyond the taboo regarding exploring the stated goals of the culprits. Heavy with the technical lexicon of military operations, munitions and impersonal casualty numbers, these reports tended to be conveniently context-free. 'Militants' (*boeviki* was the favoured term), 'criminals' and 'terrorists' were routinely 'eliminated', 'destroyed', 'liquidated', 'neutralized' and arrested by members of the Special Forces National Anti-Terrorism Committee. The perpetual threat of indeterminate origin that the *boeviki* represented was invariably cancelled out by the equally constant authority, decisiveness and superior force of the regime as it dealt with each situation in turn. Introducing a characteristically contradictory report of May 2011, the *Vremia* newsreader declared that '[i]n our country, especially in the Caucasus, there are *boeviki* attacks nearly every week, in spite of the fact that overall the authorities have the situation under control'.[63] The events described occurred within a curiously disjointed temporality of discrete, self-contained incidents with minimal connection either to one another or to any external time line.

The approach adopted by *Vremia* and *Vesti* is not unique to Russian broadcasters. In news reporting around the world (with the BBC a notable, though by no means consistent, exception), war reporting in general tends to provide scant analysis of the circumstances under which conflicts erupt or of the motives of the participants.[64] In discourse on international terrorism in particular, the threat posed is serious, indeterminate and without motive, yet never so great that it cannot be entirely contained by government force.[65] But the legitimizing potential that the demonstration of such force offers is of particular value to the Russian state. This was reflected in our running order data. According to Russian newsmakers, among them a Channel 1 journalist whom we interviewed, the first three news items are considered the most important when establishing a news agenda.[66] In the Russian context, these items will tend to include major domestic and sometimes international incidents, and events, meetings and official occasions involving state leaders. The fact that stories coded as 'separatist violence in the North Caucasus' made it into *Vremia*'s top three far more frequently than those from other categories on either *Vremia* or *Vesti* is a clear indicator of their salience for Russia's primary state-aligned broadcaster (see Figure 2.6). Indeed, for *Vremia*, 'separatist violence in the North Caucasus' was the most dominant category across all three criteria of analysis (frequency, intensity and salience). Thus, for example, in terms of intensity it accounted for 24 per cent of the total coverage for our topic (see Figure 2.7). For *Vesti*, the corresponding figure is only 9 per cent (see Figure 2.8).

The most significant event in the 'separatist violence' category was an explosion in the North Ossetian city of Vladikavkaz on 9 September 2010, which claimed nearly 20 lives and injured over a hundred people. Both channels avoided making reference to the ethnicity of the suicide bomber or speculating about his motives. Instead, the respective reports provided detailed factual accounts of what had occurred, the number of victims and, most prominently, the efficient work of the authorities. Even when they eventually referred to the identities of the terrorists (all of whom carry Muslim-sounding names), the reporters left ethnicities and the purposes of their group unexplained.[67]

In the Vladikavkaz report, as in many reports belonging to this category, the emphasis in the newsreader's account was on the routinized reporting of technical details of the Special Forces operation. Often, however, the accompanying footage clearly (if inadvertently) revealed ethnic and religious content. A long story on *Vesti* recounting a large-scale special operation in Ingushetia in March 2011 claimed that Russian forces had captured terrorists involved in organizing the Domodedovo bombing in January 2011, in which nearly 40 people were killed.[68] The main focus was, as usual, on the rigour and decisiveness of the authorities, with allusions to ethnicity notable by their absence. However, the reporter's narrative was complimented by imagery of the Qur'an and footage of men whose long beards and strict Islamic attire connoted the fanaticism of al Qaeda, rendering the broadcaster's refusal to acknowledge the terrorists' aims and demands all the more curious and contradictory. The tensions were compounded when, not long after this report on the large-scale counterterrorism action, *Vremia* ran a Sunday feature on Ingushetia which presented the image of a republic whose calm stability was 'the result of constant and successful special operations'.[69]

Both the ethnicization of certain conflicts and the occlusion of inter-ethnic factors at work in others are part of what Carolyn Nordstrom has termed the 'politics of (in)visibility' in her research on post-conflict countries.[70] Nordstrom suggests that there is a meticulous, yet contradictory, editing of news relating to violence in these countries such that, depending on political interests, key aspects of the violence are alternatively foregrounded and suppressed. In the North Caucasus, the two strategies can be observed side by side. Thus, on the one hand, according to the generalized propaganda of normalization dominating state news coverage of the region since the mid and late 2000s,[71] Russia's intervention in the North Caucasus eventually brought peace and prosperity; on the other, particular reports on Ingushetia implicitly deliver the message that radical Islam remains a real threat, albeit one monitored and contained by Special Forces, whose perpetual state of alertness will ensure that Russia's future remains secure.[72]

Around the time of the shocking assault on Domodedovo airport in early 2011, we begin to witness a slow and gradual shift in emphasis. Following the comprehensive international coverage generated by the event, it became progressively more difficult to entirely suppress the threat posed by North Caucasian-based radical Islamism. Indeed, reporting on the Domodedovo assault itself was littered with references to and ominous images of the Chechen 'Black Widow' (*Chiornaia vdova* or *Shakhidka*) fanatic who was implicated in the attack. (Two Black Widows had carried out the earlier Moscow metro suicide bombing of March 2010; between October and December 2013, *shakhidki* were heavily implicated in several bombings on Volgograd's transport system). At this point, although the state media re-invoked the strategy of inscribing Russia into the global 'war on terror' that has been deployed at intervals since the 9/11 attacks of 2001, it co-existed in tension with the reverse strategy of occluding the role of jihadist ideology and portraying a region undergoing a protracted but successful process of normalization.[73]

Toward the end of our recording period, the balance of references to Islamist extremism in the North Caucasus slowly increased. This preceded a veritable deluge of scaremongering stories broadcast on Rossiia and NTV in 2012 and 2013 and linking the problem of illegal migrants in Moscow and St. Petersburg to jihadist groupings from the North Caucasus and Azerbaidjan planning terrorist acts in Russian cities.[74] The ever-cautious Channel 1 succumbed to the appeal of the anti-Islamist line less unambiguously than its two rivals, Rossiia and NTV. Thus, it accompanied its coverage of a series of arrests in St Petersburg in March 2013 with the carefully worded reminders of an Islamic scholar that the danger represented by those arrested 'lies in the fact that they are destroying the foundations of traditional Islam, dissolving it, and enticing our Muslims into their incomprehensible extremist organizations'.[75]

In addition to the renewed critical focus on North Caucasian (and other) 'migration' that occurred during the first 18 months of Putin's new presidency, the greater prominence of largely negative treatments of Islam in the North Caucasian context was also reflected in a spate of stories relating to the building of a large mosque in Moscow and to the wearing of the Islamic *hidjab* in the North Caucasian region of Stavropol, which is predominantly populated by ethnic Russians. The Stavropol story attracted particular attention because it centred on the pressure brought to bear on beleaguered Russian head teachers bravely fighting against the tide of local Islamism in insisting that girls at their schools remove their *hidjabs*, since they breached school dress code.[76] Between December 2012 and March 2013 alone, we counted a total of ten substantive reports on the topic from *Vremia* and *Vesti*. Our NVivo analysis of the texts of those reports generated 40 results for combinations involving the word '*hidjab*'.

The *hidjab* controversy acquired significant media resonance; the main evening bulletin of the populist NTV devoted as many reports to it as *Vremia* and *Vesti* put together during the same period (10), generating almost the same number of references to *hidjabs* as the two other channels combined (38). This is partly attributable to the controversy's links with a related issue which we touched on in the context of our earlier project on European television coverage of Islam, where we drew attention to a spate of Channel 1 reports in 2007 telling of a concerted attempt by elements within the local Muslim population in Ingushetiia to intimidate ethnic Russian residents, particularly school teachers, into leaving the region.[77] The *hidjab* issue added new potency to the fears generated by this development and seemed to chime with a growing sense that the North Caucasus was on the point of being lost to Russia – a sentiment, however, which the vast media event of the 2014 Sochi Winter Olympics, held close to the region, was intended to help dispel.

***

The aim of this chapter was to identify broad patterns of coverage in ethnicity-related news reporting and to provide some empirically based context for further discussion throughout the book. Above all, our analysis revealed that the Russian state's nation-building policy was replete with contradictions which manifested themselves in the approach taken to ethnicity-related questions. On the one hand, these reports presented ethnic and cultural diversity as one of Russia's uniquely positive qualities; on the other, with multi-ethnicity and migration proving to be a powder keg within the population at large and with xenophobic hatred growing, state broadcasters were caught between two options: (i) attempting to preserve ethnic cohesion by underreporting sensitive or inflammatory topics, and (ii) acceding to popular sentiments by echoing, and even fomenting, the prejudicial fears to which those topics gave rise.

Against the background of the Kremlin's concerns, Channel 1 and Rossiia are well aware of their responsibility to support state policy on diversity management. This was particularly visible in relation to migration issues, where state broadcasters initially differentiated themselves from other media outlets by exercised restraint rather than joining those outlets' anti-migrant bandwagons in order to achieve higher ratings. This approach changed in 2012, as we will discuss later.

Throughout the chapter, we noted certain discrepancies between the two channels. Rossiia, though state-owned, tends to be more provocative and swifter in responding to the public mood. Taking account of its more international audience, the part-privately owned Channel 1 tacks closer to the Kremlin's line and is more cautious about ethnicizing news stories. These differences manifested themselves in each channel's responses to the 2012 Pussy Riot scandal, whose indirect links to ethnic relations will become apparent. While even Rossiia was slow in reacting to the incident, Channel 1's *Vremia* did not take up the topic for over six weeks.

Notwithstanding the constitutional commitment to multi-confessionality, both channels consistently promoted Russian Orthodoxy as an unchallenged pillar of Russianness which transcends national and religious identities. Judaism, Buddhism and non-Orthodox Christian denominations and 'sects', by contrast, were all marginalized. Benefiting in part from the Eurasianist thinking underpinning elements within official rhetoric, Islam received more attention, though nothing to rival that accorded to Orthodoxy. The hysteria about 'radical Islam' prominent since our recording period finished had yet to develop in earnest, but it was foreshadowed in reactions to the terrorist attacks on the Moscow metro and at Domodedovo airport in 2010 and 2011 respectively.[78] Major incidents such as the Vladikavkaz bombing and Special Forces operations in Ingushetia were rarely reported in terms of ethnic or religious conflict, despite the importance attributed to such factors by actors in the region and by the Russian public.

One of several paradoxes that we noted was the dual function played by the emphasis placed on western Europe's reportedly catastrophic failure to handle migration flows and ethnic tensions and the perceived crisis within European multiculturalism. For while, against this backdrop, Russia's approach to diversity management could be presented in a much more positive light, the deadlock in Europe also provided a convenient alibi for the strong measures that Russia itself has been occasionally forced to take with respect to its own problems in the area of inter-ethnic relations.

The contradictions we have identified and the often unpredictable terrain we have mapped in this chapter are indeed cast into sharp relief when seen in the context of television news coverage of inter-ethnic relations in present-day western Europe as well as that of the preceding Soviet period, albeit not as Russia's state broadcasters would wish. In each case we can speak of strong continuities and significant differences. Thus, while the baton of Soviet state television's obligations as an instrument of Kremlin policy has been passed to its post-Soviet successor, the relationship between policy and broadcast output is now considerably more complex and less 'transitive' than in Soviet times; there is greater heterogeneity, more editorial autonomy and journalistic room for manoeuvre, more inconsistency and shifting in response to changing circumstances and a stronger sense of the need to account for popular opinion.

Equally, west European PSBs are undoubtedly grappling with similar issues to their Russian counterparts (increased migration flows, growing inter-community tensions, the breakdown of diversity management policies, the inexorable rise of right-wing populism). They too fulfil a powerful nation-building function within their respective establishments. But the post-Enlightenment principles and language of tolerance are far more deeply entrenched within their collective psyches than in their Eastern neighbours'. Moreover, their public service ethos, sheltered by mature democratic systems within which they represent the outer limit of a powerful 'fourth estate', is distinctly lacking in Russia. For that very reason, they tend to exhibit more consistency in their approach to diversity management issues. Their adherence to a relatively narrow band of consensual opinion on the subject is, ironically, stronger than that of either Channel 1 or Rossiia, though it is significant that the BBC has itself recently been pressured into acknowledging and redressing a certain perceived 'liberal bias' in its own coverage of immigration.[79]

The full significance of the broad-brush spatiotemporal juxtapositions will emerge in the analysis we undertake in later chapters. We begin that analysis, however, by focusing on Russian television coverage of the ill-fated Day of National Unity: an example of nation-building whose brief and curious history is very much of post-Soviet time and space.

Notes

1 In our data graphs, unless otherwise stated (i.e. when we are dealing in percentages), 'frequency' refers to total number of items within our corpus, 'intensity' to total hours of coverage and 'salience' to total number of items within the top 3 of the running order.

2 We draw here on Christopher Flood, Stephen Hutchings, Galina Miazhevich and Henri Nickels, *Islam, Security and Television News: A Comparative Study*, Basingstoke: Palgrave Macmillan, 2012.

3 We originally monitored *Vremia* exclusively, but after the first two weeks of recording we added *Vesti*. Thus, the data for *Vesti* collected in September 2010 is incomplete: only ten out of the month's 30 days were recorded.

4 For example, Glasgow Media Group's book on media coverage of asylum seekers, refugees and migrants in the UK draws on a television news content analysis of a total of only 15 reports in 2006 and 2011. See Greg Philo, Emma Bryant and Pauline Donald, *Bad News For Refugees*, London: Pluto Press, 2013. The BBC Trust's 2013 report on balance in its news coverage, which included a section on immigration, relied on a television news content analysis based on two months of recordings in 2007 and 2012. See BBC Trust, *Review of the Breadth of Opinion Reflected in the BBC's Output*, BBC Trust: London, 2013.

5 The recording period dates were: first, 1 September to 30 November 2010; second, 1 March to 30 May 2011; third, 1 September 2011 to 30 November 2011; and finally, 1 March to 30 May 2012.

6 In Chapter 9, we analyse the reporting during the first 18 months of Putin's third presidency, which followed the end of our recording period.

7 SOVA Center for Information and Analysis, 'Rasizm i ksenofobiia', 30 December 2010, online, available: http://www.sova-center.ru/racism-xenophobia/publications/2010/12/d20663/; 30 December 2011, online, available: http://www.sova-center.ru/racism-xenophobia/publications/2011/12/d23370/; 29 December 2012, online, available: http://www.sova-center.ru/racism-xenophobia/publications/2012/12/d26147/ (all accessed 29 November 2013).

8 *Vremia*, 19 July 2011, www.1tv.ru/news/social/180914 (accessed 11 November 2013).

9 Interviews with a Rossiia journalist, 29 March 2013; with a Channel 1 journalist, 29 January 2013; and with an NTV journalist, 31 January 2013.

10 The word Russian (as an ethnic denominator – *russkii*) appeared 1,483 times in the transcripts we collected during our recording period. The use of the term was very consistent. It appeared mostly in a context of culture and literature. Associations with the church and religion in general were also numerous. In contrast to this, the civic classification of Russian (*rossiiskii*) appeared slightly less often (1,035 times) and predominantly in more official contexts. The most frequent use was in combination with the Russian Federation, politics, diplomacy and the multi-national Russian state.

11 Interview with Shevchenko, 3 April 2013.

12 *Vesti*, 17 October 2011. We are not providing web-links to the *Vesti* web archive, because, in contrast to the situation with the *Vremia* archive, *Vesti* links tend to change frequently.

13 *Vesti*, 18 October 2010.

14 Interview with Kiselev, 27 March 2013.

15 Elena Chebankova, 'Contemporary Russian Multiculturalism', *Post-Soviet Affairs*, 28/3, 2012, 337.

16 Heather A. Conley and Theodore P. Gerber, 'Russian Soft Power in the 21st Century: An Examination of Russian Compatriot Policy in Estonia', *A Report of the CSIS Europe Programme*, Washington, August 2011.

17 See Chapter 8 for more details.

18 Ibid.

19 *Vremia*, 19 April 2012. Available online at www.1tv.ru/news/social/204898 (accessed 11 November 2013).

20 *Vremia*, 22 April 2012. Available online at www.1tv.ru/news/social/205068 (accessed 1 November 2012) and *Vesti*, 22 April 2012.

21 *Vesti*, 29 September 2011.

22 *Vesti nedeli*, 4 September 2011.

23 *Vesti nedeli*, 25 September 2011.

24 *Vesti*, 20 November 2011.

25 *Vesti*, 2 August 2011 and 21 April 2011.

26 'Russia: Unregistered Baptists Feel Under Pressure from State', 22 September 2008. Available online http://www.christiantoday.com/article/russia.unregistered.baptists.feel.under.pressure.from.state/21475.htm (accessed 29 October 2013).

27 *Vesti*, 1 March 2011.

28 Zoe Knox, 'Preaching the Kingdom Message: The Jehovah's Witnesses and Soviet Secularization', in Catherine Wanner (ed.), *State Secularism and Lived Religion in Russia and Ukraine*, Oxford: Oxford University Press, 2012, pp. 244–71 and Knox, 'The Watch Tower Society and the End of the Cold War: Interpretations of the End-Times, Superpower Conflict, and the Changing Geo-Political Order', *Journal of the American Academy of Religion,* 79/4, 2011, 1018–49.

29 For a detailed history of religion in Russia see Zoe Knox, *Russian Society and the Orthodox Church: Religion in Russia after Communism*, New York: Routledge 2005.

30 Bridget Anderson and Scott Blinder, 'Who Counts as a Migrant? Definitions and their Consequences', *Migration Observatory Briefing*, COMPAS, University of Oxford, UK, September 2013.

31 For a detailed discussion see Chapter 9.
32 Interviews with a Channel 1 journalist, 29 March 2012; and with a RIA Novosti journalist, 15 February 2013.
33 *Vesti nedeli*, 29 January 2012.
34 Vladimir Putin, 'Rossiia: Natsional'nyi vopros', *Nezavisimaia gazeta*, 23 January 2012.
35 *Vremia*, 11 September 2011. Available online at www.1tv.ru/news/polit/185022 (accessed 19 October 2012).
36 *Vesti nedeli*, 26 September 2010.
37 See, for instance, Levada Centre, 'Natsional'naia politika i otnoshenie k migrantam' 28 November 2012. Available online at www.levada.ru/28-11-2012/natsionalnaya-politika-i-otnoshenie-k-migrantam (accessed 20 May 2013).
38 The World Bank Data, 'Net Migration'. Available online at http://data.worldbank.org/indicator/SM.POP.NETM (accessed 20 October 2013). Net migration is the net total of migrants; that is, the total number of immigrants less the number of emigrants (including both citizens and noncitizens).
39 This stands in sharp contrast to the anti-migration campaigning the channels started in 2012 (see Chapter 9).
40 See the Introduction.
41 The number of stories on *Vesti* was 68 (cf. *Vremia*'s total of 19), filling 4 hrs 22'32" of airtime (cf. *Vremia*'s 1 hr 14'36").
42 *Vremia*, 31 March 2011. Available online at www.1tv.ru/news/world/173811 (accessed 19 October 2012); *Vesti*, 30 March 2011.
43 Anna Triandafyllidou, 'Migrants and the Media in the Twenty-First Century. Obstacles and Opportunities for the Media to Reflect Diversity and Promote Integration', *Journalism Practice*, 7/3, 2013, 241.
44 *Vremia*, 8 March 2011. Available online at www.1tv.ru/news/social/172230 and 6 September 2011. Available online at www.1tv.ru/news/social/184591 (both accessed 19 October 2012).
45 *Vremia*, 25 March 2012. Available online at www.1tv.ru/news/world/202466 (accessed 19 October 2012).
46 *Vesti*, 21 April 2012; *Vesti nedeli*, 22 April 2012.
47 Among them was *Vesti*, 9 May 2011.
48 See Chapter 9.
49 See *Vremia*, 6 May 2011. Available online at www.1tv.ru/news/crime/176081 (accessed 19 October 2012).
50 *Vesti*, 6 May 2011.
51 SOVA Center for Information and Analysis, 'Rasizm i ksenofobiia', 30 December 2010. Available online at www.sova-center.ru/racism-xenophobia/publications/2010/12/d20663/; 30 December 2011, available online at www.sova-center.ru/racism-xenophobia/publications/2011/12/d23370/; 29 December 2012, available online at www.sova-center.ru/racism-xenophobia/publications/2012/12/d26147/ (all accessed 29 November 2013).
52 See Chapter 4.
53 Vera Kichanova and Aidar Buribaev, 'Neupravliaemyi natsionalizm Surkova', *Slon.ru*, 16 May 2013. Available online at http://slon.ru/russia/goryachev_ivanov_surkov_povyazannye_obrazom-941946.xhtml (accessed 18 November 2013).
54 See, for instance, 'Tsigane, edinorogi i chechenskie bloggeri: pravitel'stvo utverdilo FTsP "Tolerantnost"', 26 August 2013. Available online at http://sputnikipogrom.com/politics/4126/pogroms_are_coming/; 'Vseobshchii sinopsis ili sistema mnenii' 15 October 2011. Available online at http://krylov.livejournal.com/2319664.html (accessed 29 October 2013); 'Natsionalisty nameknuli na novuiu "manezhku" iz-za osvobozhdeniia Mirzaeva', 27 November 2012. Available online at www.km.ru/v-rossii/2012/11/27/delo-rasula-mirzaeva/698206-natsionalisty-nameknuli-na-novuyu-manezhku-iz-za-osv (accessed 18 November 2013).
55 See, for instance, an interview with a Channel 1 journalist of 29 March 2013.

56 For a detailed discussion see Chapter 4.

57 For example, in the edition of Channel 1's talk show, *Pust' govoriat*, devoted to the tragedy, the supposedly barbaric cultural norms associated with Mirzaev's community became the focus of a heated debate: 22 August 2011, television programme. Available online at www.1tv.ru/prj/pustgovor/vypusk/10261 (accessed 22 September 2013).

58 As early as its first report, *Vremia* stressed that the conflict was of an 'everyday' character [*bytovoi*]: 19 August 2011, online television programme. Available online at www.1tv.ru/news/social/183116 (accessed 25 October 2013). *Vesti* re-affirmed this, for example, on 31 August 2011.

59 See, for example, *Vesti*, 27 November 2012.

60 Interview with Kiselev, 27 March 2013.

61 Rogers Brubaker, *Ethnicity Without Groups*, Cambridge: Harvard University Press, 2004, p. 174.

62 Flood *et al.*, *Islam, Security and Television News*, pp. 32–40, 118–19.

63 *Voskrosenoe Vremia*, 29 May 2011. Available online at www.1tv.ru/news/crime/177525 (accessed 30 August 2013).

64 See, for example, Richard Jackson, *Writing the War on Terrorism: Language, Politics and Counter-terrorism*, Manchester: Manchester University Press, 2005.

65 Ibid.

66 Interview, 29 January 2013.

67 *Vremia*, 13 September 2010. Available online at www.1tv.ru/news/other/161046 (accessed 19 October 2012).

68 *Vesti*, 30 March 2011.

69 *Voskrosenoe Vremia*, 10 April 2011. Available online at www.1tv.ru/news/crime/174419 (accessed 19 October 2012).

70 Carolyn Nordstrom, *Shadows of War: Violence, Power and International Profiteering in the Twenty-First Century*, Los Angeles: University of California Press, 2004.

71 Flood *et al.*, *Islam, Security and Television News*, p.125.

72 On the threat of radical Islamism in Ingushetia, see Galina Yemelianova (ed.), *Radical Islam in the Former Soviet Union*, London: Routledge, 2010, pp. 62–81.

73 For more on the tension between these two strategies, see Flood *et al.*, *Islam, Security and Television News*, pp. 120–1, 185–9.

74 Chapters 4 and 9.

75 *Vremia*, 4 March 2013.

76 *Vesti nedeli*, 21 October 2012; *Vremia*, 21 October 2012. Available online at www.1tv.ru/news/social/218107. See also another *Vremia* report on the same topic, 23 October 2012. Available online at www.1tv.ru/news/leontiev/218271 (all accessed 1 November 2012).

77 Flood *et al.*, *Islam, Security and Television News*, p.120.

78 As a mark of this hysteria, in January 2014, the RIA Novosti Press Agency released a surreal publicity poster entitled 'How to Recognise Terrorists and Suicide Bombers', consisting of a list of behavioural gestures indicating that someone is about to commit an act of terrorism. 'Kak vedut sebia terroristi-smertniki', 23 October 2013. Available online at http://ria.ru/infografika/20131023/972165703.html (accessed 15 April 2014).

79 Josh Halliday, '"BBC had Deep Liberal Bias over Immigration", Says Former News Chief', *The Guardian*, 3 July 2013.

3 Re-inventing Russia in television news commemorations of the 'Day of National Unity'

Mediation as fracture

A new day is born

Like religious belief, the sharing of foundation myths contributes to the coherence of national communities, binding them through their shared stake in the symbolic birth of the nation.[1] When the nation was becoming an international norm in the nineteenth century, states began setting aside a specific day for public commemoration of the birth of the nation as a single 'imagined community'. The urgency with which post-1991 Russia required a convincing myth of origin appropriate for the new realities it faced relates to the fact that, until the late twentieth century, the states with which today's Russian state claims direct historical lineage were supra-national, multi-ethnic communities, wherein the Russians occupied an ambiguous position. When the Russian Federation emerged as an independent state, its political leadership proclaimed it to be a post-imperial national state, though it remained a vast, ethnically diverse entity, the boundaries of which have proven difficult to agree.[2]

However, when, under Putin, Russia began to implement a co-ordinated nation-building strategy to counter the centrifugal tendencies, it was able to draw on two distinct advantages: first, the preceding centuries of Russian (and, latterly, Soviet) historical achievements; second, a compliant media system inured to serving as the propaganda tool of a powerful, centralizing state. One of the Kremlin's key motivations for imposing its grip on television broadcasting was to promote its own vision of a powerful Russian state with a unity of patriotic purpose and an enviable cultural history spanning the imperial and the Soviet periods. As early as 1999, Pushkin's bi-centenary year, the main state television channel mounted a campaign to re-install the poet at the heart of the nation's cultural history. In 2003, it used the St Petersburg tercentenary to celebrate the old imperial capital's re-established status as Russia's 'showcase' to the west.[3] Putin has overseen a steady ratcheting up of patriotic rhetoric surrounding the annual 9 May 'Day of Victory' celebrations, culminating in a year of saturation coverage of Soviet wartime achievements in 2005, the 60th anniversary of the end of the Second World War.[4]

2005 also saw the introduction of a 'Day of National Unity', celebrated with hesitant media fanfare on 4 November and repeated annually. The new holiday had an implicit ideological purpose pitting it against three competing dates: first, 12 December, which El'tsin had introduced as the 'Day of the Constitution' in 1993 to mark Russia's newfound commitment to democracy; second, 7 November, the Day of the October Revolution, re-named by El'tsin in 1996 as the 'Day of Accord and Reconciliation' to counter the divisive associations attached to the old Soviet ideology but still celebrated by the Communist Party; third, Russia Day, marked on 12 June, introduced in 1992 following the Russian Federation's declaration of sovereignty within the Soviet Union on 12 June 1990 but rejected by many for its implicit endorsement of the collapse of the Soviet Union. In inaugurating the Day of National Unity (henceforth DNU), Putin simultaneously called for the elimination of Constitution Day, a rallying point for liberals, as 'compensation' for Communists by removing what they saw as an illegitimate rival to the Day of the October Revolution.

Apart from the need to assert primacy in the nation-building calendar, DNU reflected concern over the deterioration of the North Caucasus situation. This, in turn, was mirrored by the growth of nationalist forces (encouraged by the 'mainstreaming' of figures like Zhirinovskii and the legitimization of concerns about the fate of ethnic Russians in light of demographic changes within the Russian Federation, as well as the legacy of Soviet policies toward nationalities which were perceived by many as anti-Russian). The latter development coincided with a steady increase in hate crimes against minority communities.

Unsurprisingly, therefore, despite the mass mobilization of youth movements in support of DNU, it soon threatened to become a site not of togetherness but of discord. Indeed, even before the establishment of DNU, 4 November had been marked by the notorious 'Russian March', a demonstration initiated by Russian nationalists involving the participation of the extremist Movement Against Illegal Immigration and even the Ku Klux Klan. DNU has witnessed a three-way split into 'those who celebrated (ultra-nationalist youth and the "United Russia" Young Guards political movement) and those who did not celebrate it (an older generation for whom 7 November remained the date and the holiday to celebrate)'.[5] Contrary to Durkheim's view of national rituals as promoting social integration, DNU provided 'crucial elements in the mobilization of bias'.[6]

The two linked historical events around which DNU's official symbolism is based are themselves problematic in terms of their capacity to instil unity. The ending of the Time of Troubles (*Smuta*) in 1612 when, led by the legendary Kozma Minin and Dmitrii Pozharskii, a popular army of volunteers (or *opolchentsy*) marched from Nizhnii Novgorod to Moscow to expel the Polish–Lithuanian occupiers and revitalize Russia is a rousing story (it is no coincidence that during the Ukraine crisis of 2014, Russian media sources referred consistently to East Ukrainian rebels as *opolchentsy*). However, its capacity for acceptance as marking the birth of modern Russian nation is limited. Moreover, the fact that the Russian Orthodox Church – which, before 1917, had celebrated the 'Day of the Kazan Mother of God Icon' on this date for 270 years – soon intertwined its own celebrations with those of the state by emphasizing that Minin and Pozharskii gained inspiration from the icon, strengthening DNU's claim to continuity with the pre-revolutionary era. But the Orthodox narrative clashed with the theme of Russia's multi-faith, multicultural make-up and its adopted mantra of 'Unity in Diversity'. This tension is reflected each year in the conflicting images of the Russian nation which characterize television coverage of DNU.

The significance of the holiday is evidenced by the fact that all three main national television channels (Channel 1, Rossiia and NTV) produced historical films for the occasion.[7] Since 2005, it has gathered pace in terms of the level of capital, actual and symbolic, that the government has invested in its success; there are now co-ordinated demonstrations across the Russian Federation on 4 November, tailored to the regions in which they take place. Since 2009, a concerted effort has been mounted to reclaim the Russian March from the nationalists; nonetheless, polling indicates that ordinary people have not adopted DNU. By 2010, the percentage of those polled who considered the holiday 'necessary' had dropped from 46 per cent to 38 per cent. The same poll revealed that those with no plans to celebrate the occasion numbered 71 per cent; in 2005 the figure was 73 per cent.[8]

National broadcasters must mediate between the values of the establishment and the perspectives of the viewing public. DNU presents a particular set of challenges to Russian television (the top-down, artificial 'imposition' of the celebration; the lack of viewer 'buy-in'; tensions surrounding the event's meaning; the potentially divisive symbolism it deploys). The present chapter explores how Channel 1 and Rossiia have addressed these challenges, focusing on their approach toward DNU's modulating emphasis on ethnic cohesion, its interpretation of the 'Unity in Diversity' mantra and the broader meaning of Russian nationhood. We trace shifts that have occurred over a seven-year period, dwelling at some length on 2010–12 since this period marked a key turning point in the coverage, and making reference to Russia's two other main national channels and to foreign broadcasters.

Our analysis is informed by Hobsbawm's notion of 'invented tradition' and we examine how television news contributes to the construction of DNU as an organic tradition with popular support rooted in spontaneous local rituals. Hobsbawm's recognition of television's importance to this process aligns with the first category of 'media event' outlined in our Introduction (the pre-planned collaboration of state television and its political masters in celebrating rituals of national communion which recur in the official calendar). It corresponds, too, to Dayan and Katz's concept of the 'media event', defined as those 'high holidays of mass communication' when 'broadcaster and state "resonate together" in common cause to celebrate historical events of national significance'.[9] We argue, however that DNU both conforms and fails to conform to this definition, attributing the failure to, first, the broadcasters' lack of participatory commitment to the 'spirit' of the occasion; second, the absence of a 'media metalanguage' enabling them to ground the DNU rituals in everyday practices 'beneath' official state discourse and simultaneously transpose those rituals to a transcendent level of universal values 'above' the level of the state; third, narrative discontinuities affecting the inter-ethnic cohesion aspect's integration with other components. In this sense, we suggest, DNU benefits from interpretation within the framework of later critiques by Dayan and Katz, which take into account the inherent capacity of all media events to expose discontinuity and conflict.

We link the discontinuities to perceptions of Russian identity fiercely contested by rival elites, and to the contradictions affecting the treatment of ethnicity within the Kremlin's official nation-building programme. This, we argue, constitutes a barrier to the performative role required of media event broadcasters. That role dissipates altogether from 2012, when the subtle and at times productive dynamic of the contradictions gives way to open polemic on one hand and a 'congealed', manufactured harmony on the other. By way of examples from the UK, we caution against overstating the contrasts with parallel media events in mature, multicultural democracies – concluding, however, that the potential for complete fracture is particularly great in Russia.

A media event like any other?

In many senses, post-communist Russia offers a textbook model of the nation-building prerogatives and media strategies of other states. It is difficult to pinpoint precisely how ritualized national days unify societies according to Durkheimian theory. Much clearer is the way that, whether successful or not, they all expose social cleavages and the unequal structuring of political and symbolic power.[10] Indeed, every media event reveals 'truths' about contemporary mediated societies which are otherwise invisible.[11]

Nor is Russia the only country whose turbulent history makes it difficult to agree which events should constitute its foundation myth. Germany has a similar problem.[12] The urgency with which the *opolchenie* narrative was imposed upon the official version of Russia's historical roots to cement the unification of its post-Soviet political community confirms Hobsbawm's assertion that invented traditions arise 'when a rapid transformation of society weakens or destroys the social patterns for which "old" traditions had been designed … or when such old traditions are eliminated'.[13]

The oxymoron at the heart of 'invented traditions' points to the requirement that, in order for such phenomena to endure, 'tradition' must ultimately suppress 'invention'. Russia's DNU is new, and it may be premature to dismiss its capacity for ever generating popular engagement – for 'ritual can come first in formulating experience', as the USSR's cross-societal celebration of 7 November demonstrates.[14]

The onus on the mass media to subordinate public experience to an elite-sponsored ritual finds expression in Dayan and Katz's argument that national celebrations of historical traditions are inseparable from their media representations. It is also implicit in their identification of the three criteria that media events must satisfy: *semantic* (they deal reverentially with sacred matters); *syntactic* (they interrupt the flow of daily life); *pragmatic* (they invite the response of a committed audience).[15] All three criteria aim to 'naturalize' the traditions as inherently worthy of respect, and as embedded in the national consciousness.

DNU meets the 'semantic criterion' in both its literal and its figurative senses. Russian television has long accorded 'reverential' treatment to 'sacred matters', ever since the Soviet period when deference to official rituals was central to its raison d'être. Unsurprisingly, the words of Medvedev and Putin feature prominently in coverage of the occasion from its inception. In 2010, one of two *Vremia* reports on DNU focused on a ceremony marking the occasion at which Medvedev made special awards to foreign guests to thank them for their contribution to strengthening Russia's international ties. The newsreader's opening reference to 'international friendship' was part of the sacred lexicon of Soviet official discourse and provides DNU with a historical lineage, strengthened by the re-sanctification of the beloved Russian language.[16] The re-appropriation of Orthodoxy within official patriotic discourse provides an opportunity for the sanctification of DNU in a more direct sense. The intentional foregrounding of the intersection of religious and secular commemorations is common for national days in other countries, particularly where Orthodox Christianity remains important.[17]

The media account of DNU's historical grounding grafted the *opolchenie* narrative onto the 270-year old celebration of the Day of the Kazan Icon of the Mother of God, a gesture which intensifies both the historification and the sanctification of the event. The intensity of the sacralization procedures acquires legitimization in the hallowed historical events to which they are directed, for, as in many 'media events', it is the foundational status of the myths at their heart which underpins the reverence paid to them: these are no run-of-the-mill narratives, but the stories which define nationhood. In 2010, reviving the nineteenth-century interpretation of the end of the *Smuta* as marking the birth of modern Russia,[18] a *Vremia* newsreader concluded: 'This event marked the beginning of the emergence of the Russian (*rossiiskogo*) state'.[19]

For countries that do not ground themselves in a royal lineage whose mystique rests in its indeterminate origins, foundation mythologies are essential. From the story of the founding of Rome by Romulus and Remus, to the forging of modern France in the fire of revolution, to America's Declaration of Independence, such myths have abounded in the history of nations and empires. Characteristic of a certain category of foundation myth is the story of a nation's coming to selfhood through the expulsion of an alien other. The *opolchentsy* myth belongs to this familiar category. A comparison can be made with the USA's Independence Day; just as the act of expulsion marked on the Day of National Unity has been (mis) appropriated by extreme Russian nationalists, so America's Republican right has reinterpreted Boston Tea Party mythology as a conservative 'call to arms' against the 'liberal establishment'. To consolidate their 'sacred status' and render them eligible for entry into the media event canon, successful foundation myths must reflect positive value systems that transcend the original act of expulsion. Latter-day Tea Partiers target not British imperialists but 'tax and spend liberals' who threaten America's true, Christian, way of life. Here the blending of the Minin and Pozharskii story with the Kazan Mother of God icon gains significance for, as Patriarch Aleksei's pronouncements on the eve of the very first DNU indicate, it links transitory history with the eternal spirit of the nation: 'It is no secret that the flags had icons sewn into them. And this was the standard which, with faith and hope, led the Russian people to the salvation of the motherland.'[20]

The 'eternal' dimension to the *opolchenie* establishes it as one link in a chain of sacred feats traversing Russian history. Thus, the most sacred victory of all – the defeat of fascism – has been invoked annually by both *Vesti* and *Vremia*.

If such historical linkages achieve unity through time, then the spiritual dimension 'draws out' and universalizes the 'unity of the peoples' principle posited as inherent in Minin and Pozharskii's feat. Every year, attention is drawn to the multi-faith, multi-ethnic make-up of the original *opolchentsy* in what is an anachronistic projection onto the original historical event of contemporary political and ideological concerns (cf. annual efforts in BBC coverage of war commemorations to associate the sacrifices of veterans of the two World War campaigns with more recent British military heroism). It demonstrates how media events celebrating specific historical moments ground themselves in an organically linked 'canon' of celebrations, projecting the transitory present into the 'Great Time' of the History of the Nation.

By associating the 'Unity in Diversity' principle with historical events, the organizers play an active role in consolidating the principle. For it is a defining feature of the mass media that they unite in one 'space' diverse individuals dispersed across a vast land mass. This is a generic practice shared by all media events, but particularly apposite in the case of DNU, where unity in diversity is both the 'means' and the 'object' of the celebration. *Vremia*'s report on the 2010 celebration centres on a collage of images from across the Federation, bound together by the newsreader's authoritative voice.

The inscription of the Orthodox Church into the DNU myth accounts for another dimension to the event: the attempt to connect it with the inherent 'generosity' of the Russian people. The original proposal that DNU should be marked by 'good deeds' is attributed by state television to the Inter-Religious Council of Russia, which represents Russia's four officially recognized 'traditional religions': Christianity, Islam, Judaism and Buddhism.[21] Yet from the start the main commentator on the proposal is Kirill, in 2005 still the Metropolitan of Smolensk and Kaliningrad, who on the eve of the first DNU celebration was interviewed at length on Rossiia about the meaning of the new holiday.[22] Following the Metropolitan's endorsement, *Vesti* claimed that 'Day of Good Deeds' (*Den' dobrykh del*) has emerged as an 'unofficial name' for the occasion and tells stories of a blood donor who was to save the life of a boy with leukaemia, a sports benefactor who was to finance the treatment of sick veterans and a woman who was to give money to support the Siberian tiger. The report included a comment from the benefactor that he would mark the occasion by 'drinking a little vodka, telling a few jokes, and having a little singsong', demonstrating the extent to which the new holiday has taken root in the popular rituals by which Soviet holidays were celebrated. It concluded by projecting this wave of spontaneous generosity across the entire nation.

In subsequent years, blood donation programmes were installed as one of the 'traditions' of the holiday until, in 2007, blood donation became the centrepiece of Channel 1's coverage, as youth leaders declared to the camera their commitment to its dual political and symbolic significance. '"We are all of one blood" was the name of the joint initiative taken by the youth movements *Nashi* and *Rossiia*

Molodaia. "Today will be International Donor Day, because blood has no nationality".'[23] In the 'one blood' motif the spiritualization of the *opolchenie* unifies the original act of 'sacrifice' in driving the Poles from Moscow, the 'unity in diversity' represented by the *opolchentsy* and Orthodox exhortations to give to charity, thus embodying Russians' inherent self-sacrificial generosity. The Great Time of the Russian Nation converges with the Great Spirit of the Russian People.

But the attention paid to the blood donation theme aligns DNU with media events elsewhere: for semi-spontaneous blood donation is an index of a coming together of official and non-official, people and state, ceremonial ritual and heartfelt, spontaneous tradition. The integration of state and people provides the 'vertical' subplot in its narrative structure to which the 'horizontal' equivalent is the unification of peoples, races and ethnic groups. Again, with its ability to speak in the 'popular idiom' and its close ties to the state's official agendas, the public broadcaster is uniquely placed to play the mediating role.

In later years, the two state broadcasters have cultivated the impression that ordinary people are committed to the official and not the extremist version of the celebration, conjoining the unifying ceremonials carried out by elites with enthusiastic popular endorsements. This function requires a second mediator located within the events themselves. In many western nations, that role is assigned to public figures shown participating alongside the ordinary masses, or to a presenter who thus succeeds in merging the two mediations. In DNU, the ranks of the semi-official youth organization are allotted the role, in a deliberately managed throwback to the Komsomol marches which many Russians recall with nostalgia. *Nashi* and *Molodaia gvardiia* dominate *Vremia*'s and *Vesti*'s coverage. In 2009, *Vremia* enacted vertical integration in the very structure of its report on a DNU initiative, as the account proceeded from the official realm, to that of the 'spontaneous' – but suitably compliant – *Molodaia gvardiia*, to the demotic voices (and informal text messages) of the people who, in closing the circle, 'write back to the centre': 'Text messages were posted on the giant screen … answers to the question "What future do you want for the country?" Of several thousand messages, the most popular was "Happy".'[24]

Such gestures embrace the same principle of circularity as that of the crowds of British mourners queuing to write in books of condolence on the occasion of dignitaries' deaths, whose surviving relatives thank them in formal messages of gratitude addressed back to the nation.

The vertical and horizontal subplots of nationhood are required to work in unison. In 2009, *Vesti* described the gatherings of *Nashi* supporters in Moscow, focusing simultaneously on the role of Russia's 'youth' in securing the country's future and the unity of the diverse ethnic groups making up *Nashi*: 'Twenty thousand people in multi-coloured costumes symbolising tolerant and diverse Russia.'[25]

As elsewhere, the gesture inserts the transitory present, with its fissures, into an 'eternal time' oriented both backward to a glorious past and forward to a bright future, when Russia will realize the unity in diversity that is integral to its national spirit. This is not only a Russian phenomenon: in 1998, when France hosted and won the World Cup, French television was awash with images uniting France's revolutionary history and larger-than-life photos of its multiracial football heroes gazing to a future in which France will erase the inter-ethnic tensions of the present and the colonial misdeeds of the past, to realize the spirit of unity embodied in its founders.

Day of National Unity: A media event apart

Although certain aspects of DNU set it apart from media events elsewhere, few national days can claim mass popular participation and popular consensus. Those of France, Norway and the United States are rare exceptions. What distinguishes these successful national days from their peers? Above all, Bastille Day in France, Constitution Day in Norway and US Independence Day all began in the nineteenth century, during the high days of nation-building. Even though the construction of nineteenth-century national narratives was not conflict-free, the number of groups actively competing to offer their interpretations of the commemorated events was smaller than today. This, alongside a greater concentration of power in fewer centres, made it easier for a narrow circle of national elites to formulate the nation's foundational myths.[26]

Successful national days are what Pierre Nora has called 'sites of memory', which 'by dint of human will or the work of time [have] become a symbolic element of the memorial heritage of any community'.[27] A particular entity can only embody national memory if 'a sense of historical continuity persists'.[28] DNU's main problem is that the nineteenth-century attempt to date modern Russia's birth to the end of the *Smuta* had been largely forgotten during the twentieth century.[29]

A second important element in successful national days is either the decentralized nature of the nation-state – where regions with strong identities enjoy long-established traditions of local celebrations, as in France and the United States – or relative homogeneity, as in Norway.[30] While adopting a top-down approach, the Russian authorities appreciated the importance of feeding local celebrations into the overall vision projected by its foundation myth. Yet, far from emerging spontaneously from the grassroots, local DNU celebrations were mostly organized by United Russia, pro-Kremlin youth groups and local governmental bodies. The narratives of Russian nationhood disseminated by the state media organizers of DNU leave little room for Russia's ethnic autonomies to establish genuinely local celebrations. Tellingly, *Vesti*'s and *Vremia*'s reports on DNU celebrations in Kazan, the capital of Tatarstan, focus on the commemoration of the Kazan icon of the Mother of God, although the Tatars, who make up the plurality of the population, are Muslim.

Let us now focus more closely on other distinct features of DNU, grouping them into three categories.

Apartness as non-participation

DNU scores poorly against the 'syntactic' or 'pragmatic' criteria for a media event. Arguably, its inability to address those criteria disqualifies it altogether from the status of 'media event'. DNU even falls down on the semantic criterion if we move beyond Channel 1 and Rossiia. In 2010, REN TV's round-up of the news at the end of the first week of November featured DNU as its third item, following a story of the shocking beating of a *Kommersant* journalist known for criticizing the *Nashi* movement. Its report began with a sarcastic description of Russia's 'newest and least comprehensible holiday'. Its focus was on the continued presence of extreme nationalists in the Russian March, whose slogans were difficult to tell apart from those of the *Nashi* supporters.[31]

REN TV's 'desacralizing' approach leads to DNU's failure to meet the 'syntactic' criterion, for even the 'supportive' channels fall short of interrupting their daily schedules, or the standard structure of their news bulletins, to embrace the festive spirit. DNU lacks the 'live coverage' dimension associated with 'media events'. However, DNU challenges the very distinction that media event theory makes, between the unpredictability, negativity and tension associated with 'reported news' and the spontaneous pleasure that comes from shared participation in a universally familiar 'celebratory ritual'.

For *Vremia* and *Vesti*, DNU must be accommodated within its regular bulletin format. While on 4 November they usually lead with the celebration, the reports form part of a typical sequence of domestic and international news items. The inclusion of multiple reports – the saturation coverage characteristic of media events – applies only sporadically, and is absent altogether from the popularizing NTV, which tends to feature just one perfunctory DNU report each year.

Failure against the syntactic criterion is the corollary of a failure to meet the 'pragmatic criterion'. The fact that schedules are not interrupted is a function of the absence of a committed audience likely to 'accept' an 'invitation' to suspend their regular viewing habits. In an apparent regression from the first year of celebration, when we witnessed ordinary people 'helping imperilled cats' in the spirit of the 'Day of Good Deeds',[32] by 2010 the 'spontaneity of joyous participation' was reduced to senior bureaucrats announcing charitable initiatives.

The lack of demotic support means that there are no vernacular rituals and practices grounded in an organic relationship linking popular culture and official discourse. Nor is there evidence of the temporary 'suspension of rules' permitting an erasure of distance between journalists and public and the participatory enactment of a holiday spirit: the newsreader wearing a Remembrance Day poppy (in 2010 Britain's Jon Snow generated controversy by refusing to perform this gesture when presenting *Channel 4 News*); the deviation from normal editorial style as the sounds of a military band marching along the Champs-Élysées on Bastille Day form the soundtrack for a montage of images from across France.[33] These points, at which the nation becomes both 'subject' and 'object' of discourse, at which the 'we' of shared nationhood is co-articulated by media and public, help solidify the ties that bind citizens together; they explain the central role of 'media events' in modern nation-building strategies. There are copious examples of such procedures in Channel 1 coverage of the Victory Day celebrations, reminding us of the dangers of 'essentializing' the Kremlin's difference from other nation-builders.

Also rare are instances of media traditions echoing the traditions they celebrate – those 'strategies that repeat themselves from media event to media event', compensating for the destruction of pre-modern 'aura' by re-injecting it into the intimacy of the living room:[34] the BBC's allocation of commentary on state occasions to David Dimbleby, whose father fulfilled the same duty in the 1950s and 1960s; its comical wheeling out of Peter Snow's old-fashioned 'swingometer' on every election night. Seven years of DNU coverage has delivered no such deviations from the generic norm of news reporting. Nor – unlike Victory Day, which in 2006 featured cameras attached to the cars of the political leaders as they saluted the troops – has it promoted any technological innovations, demonstrating that 'the power of media events lies in the rare realization of the full potential of electronic media technology'.[35]

Apartness as the fractured narrative of nationhood

A significant indication of the difficulties faced by DNU is the discontinuities in the narrative of Russian nationhood fostered by the event's media cheerleaders. The inclusion of multiple interpretations within an overall narrative is not necessarily a sign of failure; they can be co-opted into, be reconciled with, and thus reinvigorate the dominant narrative. At the heart of the current official vision of the *rossiiskaia* nation is a major tension relating to the ambiguous position of ethnic Russians within Russia's multi-ethnic, state-framed community. The ambiguity was already a focus of debates among Russian and non-Russian nationalist movements in the late imperial period and was exacerbated by the Soviet approach to managing multi-ethnicity.[36] In post-1991 Russia, it is revealed in the conflicting perceptions of Russia as either a nation-state of and for the ethnic Russian majority or a multi-ethnic homeland of multiple nationalities. Both perceptions are complicated by the difficulties that many members of the Russian elites still experience in distancing themselves from their state's imperial legacy and the perception that the larger the space a nation occupies, the greater is its internal strength.[37] This complex legacy profoundly shapes the interpretation of Russian nationhood in television coverage of DNU.

The tensions are traceable to the two power centres dominating the articulation of Russian nationhood.[38] The first is represented by the president, United Russia and *Nashi*; the second is the Orthodox Church, for which Patriarch Kirill is the main spokesman. DNU coverage demonstrates that both centres engage with a range of visions, yet until recently each had its own preferred vision which dominated the narratives of its key actors. Contradictions, now diminished, were previously only too apparent. On the one hand, DNU celebrates a nation which derives its strength from Russia's multi-ethnicity and multi-confessionalism, yet this is periodically contradicted by a mono-ethnic, populist vision promoting the Orthodox Church's 'unifying' role. In another conflicting set of representations, alongside a nation bound together by its members' affiliation with the Russian state, there also exists a vision which unites all Russian speakers across the world, irrespective of their citizenship. Finally, Russia is imagined as both a European nation and as a 'special world' (*'osobyi rossiiskii mir'*), bridging Europe and Asia.

The stress on multi-ethnicity was closely associated with the first locus of power and is central to presidential speeches and local celebrations organized by United Russia and *Nashi*. In 2005, Putin argued in a speech which dominated *Vesti*'s and *Vremia*'s headlines that it was 'particularly symbolic' that the *opolchenie* troops first gathered in Povolzh'e, the centre of multi-ethnicity in seventeenth-century Rus. Although the Orthodox dimension was prominent from the start, the importance to the DNU narrative of Russia's diversity is accepted by all parties. Thus, the proposal to introduce a Red Square parade of representatives of the different peoples of Russia as a key feature of DNU was articulated at a United Russia meeting in November 2010 by the Head of the Patriarchate's Department of the Relationship between the Church and Society, Father Vsevolod Chaplin.[39]

Russia's multi-ethnicity has complex origins, which lie in the country's imperial and Soviet past, but this controversial historical legacy is represented as unproblematic by DNU's key actors. For Patriarch Kirill, Russian imperialism is a source of pride and the means by which the Russian people carried out an important cultural mission. In an interview with *Vremia* on 6 November 2005, Kirill notes that as a result of the Time of Troubles 'Russia ceased to be a mono-national country'. While Putin comments on the contribution of non-Russian and non-Orthodox people to the *opolchenie* efforts, Kirill supports his argument by saying that the end of the *Smuta* engendered Russia's 'territorial expansion eastwards', bringing Russian culture to the shores of the Pacific Ocean. The expression used (*'postupatel'noe dvizhenie na Vostok'*) is a standard euphemistic description of Russian imperial expansion dating to the nineteenth century, and references to a civilizing mission in the East were a common justification of European colonialism in that bygone era.

Television coverage of celebrations organized by United Russia regularly stresses the presence within the *rossiiskaia* nation of representatives of the nationalities of the former Soviet Union. The 2010 DNU celebrations opened with an event called 'Peoples of Russia are Moscow's Treasure' which, as depicted by *Vesti* and *Vremia*, celebrated the activities of Moscow's 'largest diasporas' from the now independent states of Armenia, Azerbaijan, Georgia and Ukraine.[40]

DNU coverage helps clarify which aspects of multi-ethnicity Russia's current ruling elites perceive as making the most positive contributions to society. In depicting modern Russian diversity, *Vremia* and *Vesti* emphasize the importance of what are now officially termed 'the four traditional religions' of Russia in binding local communities through a set of shared moral values, such as the commitment to 'charitable deeds'. At annual Kremlin ceremonies, Russia's diversity is symbolized by the presence of the official leaders of these four religious communities.

Other ethnic distinctions are portrayed through the lens of dress, song and dance. Since 2006, one region has been selected in which local people are filmed performing folk songs and dances, dressed in the many national costumes of the Russian Federation and the former Soviet Union. This reduction of identity to ethnographic performance characterized the Soviet interpretation of nationality.[41] Yet many members of the Russian political elites, including Putin, have warned that the policy of ethnicizing identities hastened the collapse of the Soviet Union.[42] In attempting to counter the danger of such fragmentation, DNU organizers have modified the Soviet approach. In the regions, a carnivalesque changing of roles now takes place. In 2008 on *Vesti*, in the Astrakhan region a Tatar dance was performed by an ethnic Russian, while a Russian folk song was sung by a Kazakh woman wearing traditional Russian headgear. In 2010, the capital of the Khanty-Mansi autonomous region was the focus of celebration. However, the main commentator on both *Vremia* and *Vesti* is not a representative of the Khanty or Mansi communities, but an Azeri. In 2009, in the selected region of Ekaterinburg, ethnic Russians performed the *lezginka*, a traditional Caucasian dance.[43] Ironically, following the Manezhnaia riots, unauthorized public performances of the *lezginka* emerged as a symbol of the minority communities' refusal to subscribe to Russian norms.

DNU coverage reveals cracks in the narrative of Russia's multi-ethnic nationhood. Since 2009, *Nashi* has been entrusted with enacting societal support for the Kremlin vision of multi-ethnicity and with representing the people's rejection of nationalist xenophobes. In 2009, *Nashi* staged an alternative 'Russian March', subtitled 'Everyone Is Ours' (*Vse svoi*) and aiming to promote inter-ethnic tolerance. Yet, at the point at which the *Vremia* voiceover on the piece refers to 'young people of different nationalities', the camera alights incongruously on the faces of black African students carrying *Nashi* banners – hardly the 'unity in diversity' intended by the official slogan.

As early as 2005, another vision in which Orthodoxy constitutes the basis for Russian national identity was articulated. The commemoration of the Kazan icon of the Mother of God has become directly linked to DNU, as both *Vremia* and *Vesti* consistently remind their views that the leaders of the *opolchenie* prayed before the icon for victory. Orthodoxy's centrality to Russian identity is further elaborated through the narrative of the Kazan icon as a protector of Russia over the centuries.

In 2005 Kirill's status as a key DNU actor emerged in a televised interview in which he highlighted the Orthodox Church's role in the events of 1612.[44] Alongside the secular rituals in Red Square, every year an Orthodox Christian site provides the locus from which the DNU celebrations commence – most notably the Spaso-Evfimiev monastery in Suzdal', near whose walls Pozharskii is buried, and the Kazanskii Cathedral on Red Square, where the Patriarch conducts a Church service. Local celebrations are also dominated by Orthodox rituals.

In addition to contradictory interpretations of the *rossiiskaia* nation's distinctive features, there is ambivalence about its membership. The contradiction is introduced by the key actor representing the first power centre, the president. The DNU narrative which Putin and Medvedev project in their annual speeches, excerpts from which dominate the first evening bulletins on both channels on 4 November, stresses state patriotism as the force which engendered the modern Russian nation and brought together the different ethnic groups of Rus. Yet from 2006 onward, the nation is also represented as extending beyond the Russian Federation. That year witnessed the introduction of a now standard Kremlin ritual – a ceremony in the Kremlin at which the president bestows awards to 'compatriots' (*sootechestvenniki*) active in promoting Russian language and culture outside the state borders. They are part of what is now called, in the dominant political discourse, 'the Russian world' (*rossiiskii/russkii mir*) – helping to unite members of the nation not only across space but also through time, as among those 'compatriots' who receive the greatest publicity are figures connected with the imperial past.[45] The intentionally vague categories of 'compatriots' and 'the Russian world', and their origins in Russia's imperial past, have played a major role in shaping Russian domestic and international policy. Dramatic consequences ensued in the 2014 Ukraine crisis, when Russia's widely condemned intervention was justified by reference to the interests of its 'compatriots' in Crimea and elsewhere in eastern and southern Ukraine.

Kirill's emphasis on the uniqueness of '*rossiiskii mir*' co-exists with representations of Russia as a European nation. In an interview on the eve of the first DNU celebrations reported on both *Vesti* and *Vremia* on 2 November 2005, Kirill concluded his description of Russia's imperial expansion after the *Smuta* by saying that 'from a small European country... we created our own, great Russian (*rossiiskii*) world'. In contrast, the Sunday edition of *Vremia* on 6 November 2005 depicted Russia as a European nation, noting that with DNU Russia now meets the European average for the number of state holidays; he also draws parallels with other nation-binding events in Europe, such as England's Guy Fawkes Day.

DNU actors all posit the harmonious co-existence of the two conflicting 'horizontal' narratives of national unity, and of these 'horizontal' narratives with the 'vertical one' (the people united in loyalty to the state). On 4 November, the president attends Orthodox commemorations. Conversely, Kirill explicitly utilizes Kremlin discourse in his pronouncements when explaining the meaning of the new holiday. In 2005, he referred to the restoration of 'the vertical of power' and the emergence of a diverse 'civil society'.[46]

The illusion of harmony is undermined by a further narrative discontinuity along not only ethnonational but also political lines. A prime task for broadcasters is managing the relationship between the ritual and cyclical (those aspects of the occasion that are continuous with the past) and the novel and unexpected (those aspects which deviate from the norm, but only inasmuch as they refresh the viewing experience). For Victory Day in 2005, a focus on previously unknown facts about the original 1945 celebration was both entertainingly new and reassuringly familiar in terms of its war mythology. In the 2010 DNU, the 'news' about *Nashi* members leading a 'city clean up' of social undesirables fulfilled a similar function, albeit more artificially.

However, there are key points at which the hierarchy subordinating norm to deviation is disrupted. Prime among these flashpoints is the fraught relationship between the *Russkii marsh* and the official DNU events. The problem is exacerbated by the fact that *Russkii marsh* had appropriated 4 November long before DNU was mooted. In 2006 the tensions were particularly apparent. That year,

Vesti deployed a dual strategy. On one hand it assimilated marginal elements into an inclusive DNU spirit in which parties of all shades unite. Here, the 'deviation' represented by the right-wing and liberal-progressive counter-demonstration was subordinated to the norm represented by the mainstream within a single narrative.

On the other hand, it detached from that mainstream any deviant events, discarding them as representing a minor rupture ('hooligan elements') which it made no effort to re-assimilate: 'The initiators of the unsanctioned marches were detained, along with straightforward "hooligan elements"'.[47]

2009 proved to be a milestone in the DNU meta-narrative, as it witnessed a re-appropriation of the Russian March for official 'Unity in Diversity'. The very act of supplementing the title of the march to lay the stress on ethnic cohesion ('*vse svoi*') is an index of discord to be overcome. Little wonder, then, that the remainder of the *Vremia* report focuses on the articulation of a more manageable dialectic of division and unity in which deviant occurrences (a demonstration called by the LDPR) can be readily swept up into the normative account.

By 2010, a full cycle was completed. *Nashi* and the ROC are firmly established as lead actors in the narrative, the norm is reinstalled and 'new', deviant events are re-subordinated as manufactured 'demonstrations' by other groupings, diverse in their orientations and ethnic origins but 'united' in their commitment to building a cohesive Russia. But the conflictual process by which the re-subordination has been achieved points to the 'short circuiting' of deviation and norm in 2010; there is little that is 'deviant' about the annual mushrooming of ever more, conveniently compliant youth groupings. It also indicates the battle for control over the meaning of DNU, and the centrality of power to the whole media event.

The structure of DNU coverage had, by 2010, developed a consistent template. Rather than an organic narrative in which norms are breached then renewed, and where 'sacred' ritual and 'secular' news are integrated, each year the ceremonial and the vernacular, the ritual and the 'new', are separated in a tri-partite progression: from first, a report on a carefully managed series of demonstrations led by *Nashi* and United Russia, to second, a ceremony involving the Orthodox Church, to third, a ritual event or announcement filmed from the Kremlin.

Apartness and the absent metalanguage

There is a parallel, and related, problem: that of the absence of an effective broadcasting metalanguage capable of transcending official rhetoric in order to authenticate 'truths'. Such transcendence requires two separations: that of media from official discourse (hence the importance of media traditions distinct from those being celebrated) and that of official discourse from a set of banal practices in which that discourse can be grounded.[48] This dual separation allows the media to posit themselves as the site at which 'Russian-ness' (British-ness, French-ness, etc.) emerges as both 'less than' and 'more than' the state's version of what it is, at which official national identity is both grounded in the vernacular and subnational and realized in the transcendent and supra-national. The power of American-ness lies in its claim to incarnate the everyday truths of individual liberty and the aspiration to succeed, and to root them in the multiple biographies of extraordinary Americans such as the tennis-dominating Williams sisters who accord television the power to 'mediate' between the everyday values they represent and the 'eternal truths' of the American spirit which transcend their embodiment in transient governmental form.

The BBC's measured coverage of London's reaction to 7/7 was attributable to its grasp of the 'stoicism and common sense' underlying the 'British spirit'. This enabled it to embed that reaction in long-standing institutional practices like the precedence-based legal system and pragmatic suspicions of patriotic fervour and to distinguish it implicitly from both post-9/11 US hysteria, and the Blair government's protestations that the attack was unconnected with Britain's involvement in Iraq. Shots of Londoners, unbowed, continuing with their daily business, indicate a subtle mastery of the nation-building 'metalanguage'.

Because Russia's nationhood is so intensely contested and its nationalism is anything but banal, DNU lacks an effective metalanguage. Examples of vernacular endorsements of the 'essentially Russian' value of 'kindness' range from the comical (the woman who saved a cat) to the portentous (the *Nashi* campaign against drug dealers). Russia's authoritarian regime and supine broadcasting media, and the novelty of both the Russian state and DNU, help account for the weakened metalanguage. But what is also significant is the Russian state's meagre stock of 'banal' practices on which to draw.

The consequence is a laxity of control over, and thus instability within, that metalanguage, permitting tension to infiltrate the 'Unity in Diversity' formula. We referred earlier to the way in which Channel 1 foregrounds the blood donation metaphor (the basis for the 'Day of the Donor' appellation belatedly accorded to DNU in 2008) as a way of symbolizing unity in diversity and simultaneously focusing it on the Orthodox value of charity. The power of the image of young people 'giving their blood' has to do with its capacity for literally *em-bodying* and 'ethnicizing' the civic, multicultural ideal. But blood connotes racial purity and ethnic exclusivism, as recognized by Zhirinovskii, who seized upon the blood imagery for his own, 'othering', purposes: 'Europe is living and thriving. Because for 300 years the Russian army spilled its blood so that they could live in peace.'[49] Any contradiction with the previous item is lost on the newsreader who introduces Zhirinovskii's tirade in approving mode (though he substitutes the 'civic' term '*rossiiskii*' for Zhirinovskii's 'ethnic' term '*russkii*').

Also prominent is the tendency to expand the metaphorical value of the *opolchentsy* narrative itself – a gesture which de-historicizes rather than sanctifies the original events. To equate the assault on the Poles with *Nashi*'s twenty-first century campaign against social undesirables is both dubious and dangerous and the need to avoid DNU's exploitation for xenophobic purposes was recognized in 2005, when a *Vesti* report on a State History Museum exhibition commemorating the *opolchenie* highlighted reassurances given by the museum to a Polish television team.[50]

Despite the fact that current Russian–Polish relations are far from smooth, those who – according to the state-media DNU narrative – should be expelled from the nation are not the Poles but figures who symbolize Russia's most recent Time of Trouble, the 1990s, with which Putin's period has been contrasted. In a historical film shown by Channel 1 to mark the first celebration of DNU in 2005, the commentator, Mikhail Leont'ev, compares the treacherous Vasilii Shuiskii of the seventeenth-century Smuta with contemporary Russia's 'traitors', the oligarchs Berezovskii and Khodarkovskii.[51]

National symbols and narratives are always open to multiple appropriations, often by diametrically opposed political movements. St George's Day, like DNU, fell into disuse in recent centuries, only to be subjected to a campaign for revival in the present decade as a means of strengthening English national identity. One reason why the campaign failed is that the celebration has also been adopted by the extreme British National Party as a symbol of English racial purity. Nonetheless, as the co-ordinated unmasking suffered by the BNP leader, Nick Griffin, on the BBC's *Question Time* in 2009 demonstrated, misappropriations of national symbols by extreme politicians in the UK are subject to a far greater degree of media 'gatekeeping' than similar actions by the likes of Zhirinovskii in Russia (ironically, in such cases control is more rigid in the British media environment).

The BBC exuded confidence in this case due to its knowledge that the apparent deviation from its own codes of impartiality was, in the BNP case, endorsed by the overwhelming majority of its viewers. This indicated its capacity to occupy the transcendent, meta-lingual position from which the essence of 'British moderation and decency' is spoken. It contrasts with the anxiously dialogical stances adopted by *Vremia* and *Vesti* in relation to the extremists with whom the state battled in 2009 and 2010 for proprietary rights over the meaning of the Russian March. Unlike the BBC, which has little need to spell out the 'transcendent meanings' of the media events it leads, both *Vesti* and *Vremia* speak a meta-lingual idiom whose position of externality to DNU is pedagogical and polemical rather than transcendent and inclusive: 'In the mass celebrations everyone displayed remarkable unity … Twenty thousand people in multi-coloured costumed symbolized a tolerant, multicultural Russia.'[52]

Rather than embodying national cohesion, Channel 1 and Rossiia direct their attention toward combating competing perceptions of disharmony which find expression at the peripheries of the Russian media: that of a nation riven by ethnic tension and corruption (REN TV); that of a weak state in thrall to migrant cultures (various nationalist websites).

DNU in a comparative context

Given the relative novelty of DNU, it is not surprising that, to return to Hobsbawm's classic formulation, 'invention' prevails over 'tradition'. However, novelty cannot fully explain the difficulties which DNU encountered in establishing itself as a 'media event'. Part of the problem lies with the broadcasters' inability to 'mediate' between the predispositions and practices of ordinary Russian people and the preoccupations and policies of the Russian state. But the problem also reflects the fact that the perceptions of Russian nationhood are so overtly contested at the levels of societal and political elites. DNU coverage constantly exposes tensions at the heart of the nation-building project, between the *rossiiskaia* nation as a civic union of cultural and religious communities and an ethnocentric interpretation in which the Orthodox Church, the Russian language and culture and ethnic Russians are binding forces. It strives simultaneously to appease those groups which feel that ethnic Russians are not properly recognized and those who see in the growing power of ethnic Russian nationalism a major threat to the country. Another unresolved tension is that between the 'horizontal' visions of the nation as united in its ethnic diversity or through Orthodoxy and the Russian language on one hand, and the 'vertical' vision in which people and state are one on the other. Overall, the distribution of symbolic power favours institutions and groups which promote an ethnic Russian interpretation of nationhood, while ethnic minorities merely enact the script written for them by the Moscow elites.

Russia's difficulties in managing ethnic diversity in the face of globally induced mass population movements are shared across Europe. Controversies over the expulsion of gypsies and the wearing of religious attire in public places point to a common crisis in European values of tolerance from which Russia is not immune. David Cameron's recent pronouncement of the death of what he tendentiously calls 'state multiculturalism' could have been targeted at the Kremlin's increasingly hollow Unity in Diversity formula. But Cameron's loaded expression correctly identifies the passing of a brief (and mythical) moment coinciding with the attainment of the Kantian categorical imperative (the fact that Britain *is* a multicultural state becomes inseparable from the universal moral 'truth' of multicultural*ism* as a state policy). The denigratory term 'state multiculturalism' signals the irrevocable separation of 'is' from 'ought', as multiculturalism reverts to the status of an official mantra imposed against the grain of reality and of popular consensus (the desire for 'the ethnic other' to assimilate or disappear). This is the context upon which DNU was artificially grafted.

The struggle faced by Russia's state broadcasters in disentangling the ritualistic and predictable elements of DNU from the new and 'surprising' components (particularly in relation to the proprietary battle over the Russian March) resembles annual controversies embroiling the BBC over the extent to which it focuses on the spates of 'ethnic crime' that afflict the annual Notting Hill Carnival, rather than on its ritualized celebrations of diversity. Even the impression of a BBC positioning itself comfortably at the heart of a stable societal consensus is misleading. One should not forget the liberal outcry against the BBC's original decision to invite Nick Griffin onto *Question Time* in the interests of 'free speech', a decision seen by many as 'providing a platform for extremism at taxpayers' expense'. Discursive struggle is not confined to the Russian context.

DNU differs from similar occasions in mature democracies, but only inasmuch as it expresses in acute form the discursive conflict lurking beneath them all, thereby adding empirical weight to critiques of Dayan and Katz's theory. Referring to public disputes over the significance of Princess Diana's death, Nick Couldry attacks Dayan and Katz for 'automatically excluding from the definition of media events highly mediated situations that are contested'.[53] Questioning their division between 'true' media events based around pre-planned rituals and second-order phenomena centring on saturation news reporting of unanticipated disasters, he argues that 'the fact that societies are stable … does not necessarily mean that they have a set of shared values or… that it is these values … that holds them together'.[54] Rather than *disqualifying* it from media event status, the Russian state's aversion to live coverage renders DNU the *ideal tool* with which to expose the contradictions underlying media events per se.

In John Fiske's and John Hartley's estimation: 'A media event is a point of maximum … turbulence … It … invites intervention and motivates people to struggle to redirect some of the currents flowing through it to serve their interests.'[55] In Russia's case, the currents in question threaten the nation's cohesion as a multi-ethnic, multi-faith state.

At the day's end (the 2012 celebration and beyond)

On the face of it, 4 November 2012 offered little to differentiate it from previous years. Many of the same media rituals recurred for the seventh time: the obligatory reminders to viewers of the reason for the celebration; the ceremonial laying of flowers on the monument to Minin and Pozharskii; the reaffirmations of Russia's commitment to multi-ethnic harmony and of its greatness as a nation; the Kremlin awards to those who had advanced the international cause of Russian language and culture; the city parades and patriotic, slogan-carrying young enthusiasts; the complementary role of the Orthodox Church and the sonorous pronouncements of Patriarch Kirill.

The process of accreting historical facts, myths and events capable of incorporation into DNU mythology continued unabated in 2012. A *Vesti* report on 4 November made reference to a city-wide Crucession, or Religious Procession (*Krestnyi khod*), taking place for the first time in St Petersburg in honour of DNU.[56] A new figure was welcomed into the DNU canon by Patriarch Kirill: his distant predecessor, Patriarch Germogen, who headed the Orthodox Church at the time of the *opolchenie* and who, according to Kirill, was a pillar of strength during the troubled times – despite his imprisonment by the hostile forces occupying Moscow, Germogen refused to order the *opolchentsy* to retreat.[57]

Yet something unmistakable had changed, making 2012 a turning point in the brief history of the initiative. In fact, 2012 marks the end of any claim that the occasion might have had to constitute a 'media event' in the performative, actively nation-binding sense. For example, in 2012 *Vesti*'s coverage was noticeably Moscow-centric, with Nizhnii Novgorod (whose historical importance as the starting point for the *opolchenie* assures its place at the heart of the canon) the only city featured other than Moscow. Of 32 of Rossiia's news reports dealing with aspects of the occasion in total, two focused on Nizhnii Novgorod; there was also one on Ekaterinburg, but this was in the context of trouble at a nationalist march. *Vremia* featured nine DNU-related reports, of which two were Moscow-specific; the other seven treated Russia as a whole, with most of the emphasis on events in Moscow. The apparent switch of the nation-binding function from Rossiia to

Channel 1 is part of a wider blurring of missions observable over the course of the last decade. More noteworthy, perhaps, was the dearth on either *Vesti* or *Vremia* of vox pop interviews with 'ordinary people' in the streets. Absent, too, were the 'heart-warming' special reports on young patriots giving blood. And gone was the pretence that DNU is anything other than a carefully managed exercise in reaffirming the national narrative.

One of the most important shifts was the effort to separate DNU coverage from reporting on the Russian March. *Vesti* treated them largely as unrelated events and spread them apart within the bulletin running orders. This was a tacit admission of defeat over the struggle to absorb the March within DNU, or to exploit selected elements of the nationalist cause for official patriotism. Reports on the Russian March were consistently negative, focusing on breaches of the law, on the provocative slogans and on extremist calls to have Article 282 of the Criminal Code annulled (the article is directed against actions which arouse hatred or diminish human dignity).[58] All of them reminded viewers that the authorities had for the first time allowed the march to take place in the centre of Moscow, with the implicit suggestion that the marchers had not repaid that trust. One of the few reports in which the march is specifically linked to DNU quotes an 'expert' who accuses the participants of 'casting a shadow' over the March organizers and on a great national celebration, and of reviving memories of the Manezhnaia Square violence.[59] Again, the fact that tensions and contradictions in the approach to the Russian March and its relationship to DNU have receded and the lines of distinction are now more clearly drawn points to a loss of dynamic performativity; the contradictions were the corollary of the active hegemonic struggle to give voice to, assimilate and moderate currents of extreme opinion at the peripheries of the nationalist movement.

The second important shift was that Channel 1 and Rossiia appeared to have resolved any residual rivalry between competing state and ecclesiastical centres of power. Reflecting the broader convergence which had been occurring since 2012, DNU reports were crafted to emphasize the complementarity of church and state. Even the visual choreography of the reports detailing the now 'traditional' ritual of the laying of flowers at the monument to Minin and Pozharskii emphasized the two leaders' unprecedented affinity in their solemn approach to the monument.[60] Indeed, in one of the few examples of a residual performativity to the celebration, albeit one safely concealed within a historical narrative, an NTV *Novosti* report on Patriarch Kirill's prayer for the well-being of the Russian nation offered in the Kremlin's Uspenskii Cathedral quotes only one passage from Kirill's address – the one in which he refers to the Church's role in the Time of Troubles.[61]

The implication is that it was lack of faith which plunged Russia into the Troubles, just as faith (and by extension the Orthodox Church) saved it from ruin at the hands of its occupants. The symbolism of the setting – Uspenskii Cathedral: the very locus of church–state unity – is unmistakable; it was the Church which rescued the state from disaster during the Time of Troubles, and religious faith which continues to maintain the stability of the church–state concordat responsible for Russia's continuing status. However, the act by which the Church and the state acquired the power of unity remains safely encased in the historical narrative of the *opolchenie*; the context for Kirill's address merely symbolizes the unification, rather than enacts it.

The 2012 version of DNU strives precisely to 'stabilize' rather than to 'enact' national values and meanings. Acts of generosity epitomizing the spirit of national unity are perfunctorily described rather than displayed. One of the few reports which did feature extensive vox pop contributions in 2012 was broadcast on the *Vesti* evening edition of 4 November and was entitled 'Where are the nationalists marching to?' The contributors were no 'ordinary' people; they were leaders of various nationalist factions. When the National Democratic Party leader, Vladimir Tor, was allowed to promulgate to camera his call for a stricter migrant visa regime to ensure that Russian workers are not undercut, the reporter interceded, clarifying that the crowds were motivated not by these policies (which resemble the position of the Moscow mayor Sobianin), but by the extremist actions and criminality with which the Russian March is now associated.[62]

In 2012, the strenuous media efforts to incorporate the Russian March into the mainstream of DNU which characterized previous years were replaced by an equally concerted attempt to expel it to the peripheries by engaging with, and actively undermining, its credibility.

True to its populist leanings, NTV paid significantly more attention to the Russian March than the two official state channels. Its longest report began by underscoring the large number of towns and cities across Russia in which the March took place. It proceeded to list the wide variety of slogans under which the demonstrators marched, noting that they included 'We are not extremists – we are Russians', 'Russian power for Russia' and 'No to drugs and migration', and that marchers included campaigners against paedophilia as well as nationalists.[63] While not free of a critical tone, the reporters are far less condemnatory than their state channel counterparts. Mindful of the onus placed upon it to engage the interests of ordinary viewers alienated by the official ceremonials by privileging real, unpredictable news over pre-prepared ritual, NTV devoted several reports to an incident barely mentioned by Rossiia and Channel 1: the violent skirmish which was supposed to have taken place between nationalist and anti-fascist demonstrators at Dostoevskaia Metro Station in Moscow.

Even the strictly ritualized stabilization strategy pursued by Rossiia and Channel 1 retains a residual dynamism, indicating that the erasure of distinction between the two power centres is incomplete. This is evident in the unusual precedence given by *Vesti* to Kirill's pronouncements. Here it takes the form of an implied polemic with the opponents of 'traditional Orthodox values' which had been severely challenged in the aftermath of the Presidential election. Kirill's familiar address in the parallel celebration of the 'Day of the Kazan Icon of the Mother of God' recapitulates the sentiments he expressed when anathematizing Pussy Riot for their act of sacrilege in the Cathedral of Christ the Saviour. They are now re-inserted into the heart of the DNU celebration: 'We must ensure that never again and in no circumstances should the confusion (*smuta*) in people's minds, which today, too, is conditioned by a loss of religious faith, lead to civil unrest, revolution and the loss of national independence.'[64] By equating the *Smuta* with the atmosphere of protest succeeding Putin's re-election in 2012, Kirill almost supplants Putin as the state's key figurehead, simultaneously targeting: (i) the atheistic sentiments which supposedly motivated the 'punk prayer' performance; (ii) related calls for a change of regime; (iii) the decoupling of church and state; and (iv) the foreign forces purportedly behind the seditious activities.

The importance attributed to Kirill's exploitation of the ceremonial occasion to renew his anti-atheistic diatribe is signalled in the carving out from Kirill's address of a discrete report in which he extrapolates from his earlier defence of the religious 'right' to speak out against the blasphemers of Pussy Riot from the Orthodox view of sin: 'The Patriarch is convinced that the Church is called upon … to affirm traditional-moral values … "The Church will never be silent when it identifies sin and vice, especially if it is being disseminated … on a mass scale".'[65] But Kirill's barely disguised slight on Russia's present opposition occurs in the context of his paean of praise for Patriarch Germogen, which enables him to ground his diatribe in Russia's heroic past. The values acquire both the renewed dynamism and authenticity which comes through them having to be re-forged on the anvil of the latter-day struggle against sedition and the enhanced authority and stability accorded to them by being reconfirmed within the historical canon. The past is also deployed as a premonition of the appalling future that awaits Russia if Kirill's dire warnings about its present are not heeded.[66] The rhetorical similarity between Kirill's address and the conclusion to Putin's speech at the laying of flowers ceremony quoted in the preceding report points to a single script jointly co-authored by Patriarch and president.[67]

DNU 2012 can thus be delineated from earlier years through the increased linkage it posits between, first, the negative consequences of atheism; second, the seductive danger represented by 'revolutionaries' and 'false pretenders' in thrall to alien forces; and third, Russia's potential loss of independence. A Channel 1 documentary film made for the 2012 celebration by Aleksei Denisov and narrated by Fedor Bondarchuk, son of the famous Soviet director Sergei Bondarchuk, acquires new significance against this background. Its title, 'The Russian Time of Troubles: History of a Disease', gained further political significance against the backdrop of events immediately preceding 4 November 2012. The politicization and polarization of the contemporary point of reference is evident even when such politicization is being explicitly denied: 'Today on Bolotnaiia Square there were no politics at all; just a student festival … Young people came to Bolotnaia not just for entertainment; here one could also give blood.'[68]

The implied contrast is with the anti-Putin protests on Bolotnaiia Square of a year earlier. This example is emblematic of DNU 2012 coverage, especially that concerning the Orthodox Church's pronouncements: for its significance is to be found not in its highly predictable surface text – by 2012, DNU and its official media representations had reached an unprecedented level of redundancy – but within the hyper-polemical subtext that the new political situation accords to it, and at which it is consistently directed. Polemic (oriented to the denigration of detractors and oppositionists) had effectively replaced performance (designed to reassure supporters and attract new ones).

By 2013 the process was complete and any pretence that DNU was capable of fostering either popular support or uniformity of sentiment had evaporated. In abandoning the performative, nation-binding function of DNU, Russia's state broadcasters are merely reflecting the stubborn refusal of ordinary citizens to incorporate DNU into the rhythms of their own annual calendar. Given the symbolic capital invested in DNU, it is unlikely that it will be discarded. What is clear, however, is that, even as (productive) tensions between the elites who crafted the initiative dissipate, at the operational level the mediation of inter-ethnic harmony tends increasingly toward polarization and polemic. Ultimately, it portends the kind of deep fracture that we explore in our next chapter.

Notes

1 David McCrone and Gayle McPherson (eds.), *National Days. Constructing and Mobilising National Identity*, Basingstoke: Palgrave Macmillan, 2009.
2 Vera Tolz, 'A Future Russia: A Nation-state or a Multi-national Federation', in Wendy Slater and Andrew Wilson (eds.), *The Legacy of the Soviet Union*, Basingstoke: Palgrave Macmillan, 2004, pp. 17–38.
3 Stephen Hutchings and Natalia Rulyova, *Television and Culture in Putin's Russia: Remote Control*, London: Routledge, 2009, pp. 57–74.
4 Stephen Hutchings and Natalia Rulyova, 'Commemorating the Past/Performing the Present: Television Coverage of the Second World War Victory Celebrations and the (De)construction of Russian Nationhood', in Birgit Beumers, Stephen Hutchings and Natalia Rulyova (eds.), *The Post-Soviet Media: Conflicting Signals*, London: Routledge, 2009, pp. 137–57.
5 Dennis Zuev, 'The Russian March: Investigating the Symbolic Dimension of Political Performance in Modern Russia', *Europe-Asia Studies* 65/1, 2013, 102–26 (122).
6 Ibid.
7 Viktor Yasmann, 'Russia: New Russian Holiday Has More Behind It than National Unity', *Radio Free Europe*. Available online at www.rferl.org/content/article/1062670.html (accessed 15 May 2013).
8 Fond obshchestvennogo mneniia, '4 noiabria – Den' natsional'nogo edinstva', 30–31 October 2010. Available online at http://bd.fom.ru/pdf/d43dne10.pdf (accessed 15 May 2013).
9 Daniel Dayan and Elihu Katz, *Media Events: The Live Broadcasting of History*, Cambridge: Harvard University Press, 1992, p. 16.
10 McCrone and McPherson (eds.), *National Days*.
11 Nick Couldry, *Media Rituals. A Critical Approach*, London: Routledge, 2003, p. 63.
12 Vera Simon, 'Reunification Day – Day of German Unity?', in David McCrone and Gayle McPherson (eds.), *National Days*, pp. 151–65.
13 Eric Hobsbawm and Terence Ranger (eds.), *The Invention of Tradition*, Cambridge: Cambridge University Press, 1993, p. 4.
14 Mary Douglas, *Purity and Danger*, London: Routledge, 1984, p. 64.
15 Dayan and Katz, *Media Events*, p. 14.
16 *Vesti*, 4 November 2010.
17 Victor Roudometof, 'Toward an Archaeology of National Commemoration in the Balkans', in M. E. Geisler (ed.), *National Symbols, Fractured Identities: Contesting the National Narrative*, Hanover: University Press of New England, 2005, pp. 35–59.

18 Paul Bushkovitch, 'The Formation of a National Consciousness in Early Modern Russia', *Harvard Ukrainian Studies* 10, 3/4, 1986, 355–76.
19 *Vremia*, 4 November 2010. Available online at www.1tv.ru/videoarchive/4625 (accessed 15 November 2013).
20 *Vesti*, 3 November 2005.
21 *Vesti*, 2 November 2005.
22 *Vesti*, 3 November 2005.
23 Ibid.
24 *Vesti*, 8 November 2009.
25 *Vesti*, 4 November 2009.
26 McCrone and McPherson (eds.), *National Days*, pp. 6–17; Pierre Nora, 'Between Memory and History: Les Lieux de Mémoire', *Representations*, 26, 1989, 7–24.
27 Pierre Nora and Lawrence D. Kritzman (eds.), *Realms of Memory: Rethinking the French Past,* vol. 1, New York: Columbia University Press, 1996.
28 Ibid., p. XVII.
29 The official Soviet narrative gave greater prominence to Rus's victory over the Mongols in 1380, Russia's war with Napoleonic France and Soviet victory in the Second World War than to the *Smuta*.
30 Richard White, 'National Days and the National Past in Australia', *Australian Cultural History* 23, 2003, 59.
31 *Nedelia* (REN-TV), 6 November 2011.
32 *Vesti*, 4 November 2005.
33 *Journal de Vingt Heures* (France 2), 14 July 2008.
34 Dayan and Katz, *Media Events,* p. 26.
35 Ibid., p. 15.
36 Terry Martin, *Affirmative Action Empire. Nations and Nationalism in the Soviet Union, 1923–1939*, Ithaca: Cornell University Press, 2001.
37 Tolz, 'A Future Russia', in Slater and Wilson (eds.), *The Legacy of the Soviet Union.*
38 Media events are organized by particular power centres and these shape the ways in which the events shape reality. See Nick Couldry, Andreas Hepp and Friedrich Krotz (eds.), *Media Events in a Global Age*, London: Routledge, 2010, p. 4.
39 *Vesti*, 3 November 2010.
40 *Vesti* and *Vremia*, 3 November 2010, *Narody Rossii – bogatstvo Moskvy.*
41 Martin, *Affirmative Action Empire*, pp. 443–7.
42 Tolz, 'Russia: Exiled, Submerged, Restored', in Simon Dixon (ed.), *Oxford Handbook of Modern Russian History*, Oxford: Oxford University Press (forthcoming).
43 *Vesti*, 4 November 2009.
44 *Vesti* and *Vremia*, 2 November 2005. Available online at www.1tv.ru/news/social/55309 (accessed 15 November 2013).
45 *Vesti* and *Vremia*, 4 November 2006. Available online at www.1tv.ru/news/social/61431 (accessed 14 April 2014).
46 *Vesti*, 2 November 2005.
47 *Vesti*, 4 November 2006.
48 Michael Billig, *Banal Nationalism*, London: Sage, 1995.
49 *Vremia*, 4 November 2008. Available online at www.1tv.ru/news/world/69095 (accessed 14 April 2014).
50 *Vesti*, 2 November 2005.
51 Arina Borodina, 'Smutnyi gosudarstvennyi prazdnik', *Kommersant*, 10 November 2010. Available online at www.kommersant.ru/pda/kommersant.html?id=1534900 (accessed 20 November 2014).
52 *Vesti*, 4 November 2009.
53 Couldry, *Media Rituals*, p. 65.
54 Ibid.
55 John Fiske and John Hartley, *Reading Television*, London: Routledge, 1996, p. 8.

56 *Vesti*, 4 November 2012.
57 Ibid.
58 Ibid.
59 Ibid.
60 Ibid.
61 *Novosti* (Rossiia), 4 November 2012.
62 Ibid.
63 Ibid.
64 *Vesti*, 4 November 2012.
65 Ibid.
66 Ibid.
67 Ibid.
68 Ibid.

4 Ethnic conflict and television news coverage of the December 2010 Moscow riots

Managing the unexpected

The previous chapter demonstrated that state-endorsed media projects to promote inter-ethnic harmony via elaborately planned celebrations like the Day of National Unity are highly fallible tools. Broadcasters need above all to respond appropriately to sudden outbreaks of conflict and to handle media events of the unexpected kind, making sense of them within accepted policy frameworks. The Day of National Unity's failure to attain the full status of media event in the first sense is, in part, attributable to the fractured narrative of nation that the project has been attempting to articulate; the contradictions in official media accounts of the shocking events considered in this chapter owe more to the state's fading capacity to assert its dominance over the strictly managed pluralism it had permitted within the national broadcasting sphere. Just as performance is eventually replaced by polemic in coverage of the Day of National Unity, so here we will see managed pluralism ceding place to a managed populism, betraying the limits of the Kremlin's grip on events and foreshadowing the altogether more ominous atmosphere that was to prevail in 2012 – our focus in Part III.

Even if, as Chapter 2 illustrated, the two main state-controlled channels tended to underreport conflicts perceived by the public as inter-ethnic and to avoid discussions of societal xenophobia, major incidents which provoked wide public resonance and required the authorities' intervention could not be ignored. Furthermore, in the current information-rich environment, no media organ can completely disregard popular expectations regarding coverage of much discussed events, particularly if they involve ethno-racial violence. Such events have the power to challenge the Kremlin's discourse about Russia as a state-framed civic nation which derives its strength from the historical experiences of multiculturalism. They tend to highlight particularly strongly a tension at the heart of the state-sponsored nation-building agenda, between a discursive promotion of the concept of the civic Russian multi-ethnic nation and a simultaneous systematic utilization of ethnic Russian nationalism as a tool of political self-legitimation and popular mobilization.[1] In short, such events test both the Kremlin's policies and the state media's ability to endorse them. A situation in which a fast reaction to unanticipated developments is required can lead journalists to fall back onto preconceived ideas and stereotypical perceptions, which might have been modified had more time for reflection been available. Therefore, from the coverage of dramatic, quickly unfolding events which raise sensitive issues we are likely to learn a lot about the fragile nature of state television's balancing act: its need to negotiate the contradictory rhetorical discourses of political leaders on the one hand, and wider public (and diverse media) perceptions and representations of inter-ethnic relations on the other. Television's ability to perform this balancing act was tested severely during the ethno-racially motivated riots in Moscow in December 2010, whose coverage will be the focus of this chapter. The riots were to become a milestone in the troubled history of inter-ethnic relations in post-Soviet Russia, and a select member of the canon of such major disturbances now invoked by the media whenever ethnic tensions spill over into violence.

The riots were sparked by the 6 December murder in the capital of Spartak football fan Egor Sviridov, during a late-night altercation between Sviridov and his friends and a group of men from the North Caucasus. Subsequent mass demonstrations, culminating in a violent riot which began next to the Kremlin walls on Manezhnaia Square on 11 December, seemed to have been provoked by the fact that the police quickly released the detained North Caucasians, including Aslan Cherkesov, who had allegedly killed Sviridov. The release was widely perceived as a manifestation of police corruption. The fact that Cherkesov was from the North Caucasian republic of Dagestan allowed some people to view what originally seemed to be an ordinary, if violent, street fight through a lens of ethnic conflict. The rioters therefore targeted their anger not only at the police, but also at people whom Russian media typically refer to as being of 'non-Slavic appearance'. Numerous shocking beatings occurred.

It took the re-arrest of the original suspects, interventions by Medvedev and Putin on 12 and 13 December respectively and Putin's appearance at a Sviridov memorial meeting, all broadcast on prime-time television, before calm was fully restored in the capital and other cities where protests in solidarity with Moscow rioters took place.[2] The riots forced the two main state-controlled channels to confront the growing xenophobia among the country's Slavic majority, to which, other than in their hesitant and minimal coverage of the 2006 Kondopoga disturbances, they had hitherto turned a blind eye, but which the Manezhnaia events underscored. Since the late 1990s this xenophobia has been reflected in opinion polls.[3] In its ugliest and most destabilizing form, it has – as the SOVA statistics quoted in Chapter 2 indicate – been manifested in an alarming rise of neo-Nazi hate crimes targeting ethnic minorities across the country.[4]

The growth in neo-Nazi extremism and inter-ethnic tension is not unique to Russia and its coverage presents challenges to broadcasters elsewhere. Many European countries have experienced such problems, often in the context of a reaction to the 'threat' posed by Islamic fundamentalism. Most North Caucasian republics are predominantly Muslim, but in 2010 the religious dimension of the tension in the central Russian cities was less pronounced than in other European locations. This is one of the reasons why the Manezhnaia case study gains significance in the comparative context, which will never be far from the discussion of this chapter.[5] Far-right forces in the UK, France, Scandinavia and elsewhere now target what they term the 'Islamicization' of European societies.[6] As we will see later in the book, federal broadcasters began systematically exploiting the theme of the 'Islamic threat' almost two years after Manezhnaia.[7] In contrast, their coverage of the 2010 riots made no reference to Islam.

Prior to Manezhnaia, state television had tended to contrast Russia's reportedly positive historical legacy of ethnic diversity with European multiculturalism, to which solely negative outcomes were attributed.[8] But in the immediate aftermath of the riots, talk shows on the Rossiia channel proposed that the Russian situation now be viewed within the international context, acknowledging that Russia shared with Europe and the rest of the world difficulties in managing ethnic diversity in the face of global mass population movements. In particular, on Rossiia's *Poedinok* talk show broadcast on 20 January 2011, which was dedicated to the Manezhnaia riots, Vladimir Zhirinovskii argued that Russia's multi-ethnic ideal supported by the Kremlin did not exist. Those who had destroyed it, Zhirinovskii claimed, were exclusively North Caucasians, particularly those from Dagestan, Chechnia and Ingushetia.[9] Zhirinovskii further argued that most other countries in which one ethnic group constituted 80 per cent of the population would be seen by their leaders as mono-ethnic. In Russia, however, where ethnic Russians enjoyed such a majority, leading politicians insisted on representing the state as multiethnic, thus benefitting minorities at the expense of the majority. The show's moderator did not challenge Zhirinovskii's radical claims and at times even agreed with them. While permitting the elaboration of such extreme views at such length was at the time unusual for Rossiia, other media had long been cultivating similar sentiments, both shaping and reflecting public opinion. (As we will see later in the book, such arguments resurfaced on Rossiia in 2012.)

There is a patent mismatch between the realities of the societal xenophobia represented by Zhirinovskii and the official rhetoric of multicultural harmony. The challenges that this disjunction poses to the media's ability to promote cohesion agendas both in Russia and in Europe[10] constitute the backdrop against which television news coverage of the Manezhnaia riots is analysed here.

Inevitably, actors interpret complex issues with the help of the conceptual apparatuses available to them. When journalists have to cover swiftly unfolding events, they tend to use these apparatuses particularly crudely, inadvertently exposing their contradictory and problematic aspects. In the Russian context, where factual and opinionated reporting are poorly differentiated, dramatic events encourage even relatively careful state-controlled broadcasters to advance radical arguments, which would otherwise be articulated in a more nuanced form.

We aim to identify and analyse the conceptual apparatus used by national television broadcasters in handling the media event that was Manezhnaia, suggesting that this apparatus to a large extent determined the ways in which they represented ethnicity and nationality-related issues. Many Russian observers highlight similarities between the discourses of Putin's era and those of their Soviet predecessors. In relation to discourses of nation and ethnicity specifically, we emphasize the impact of Soviet policies on nationality, demonstrating, however, that the conceptual apparatus utilized systematically by state broadcasters during the riots is not bound by its Soviet origins. It now reflects a new, post-Soviet take on the concepts of nation, ethnicity and race, in which Soviet interpretations are modified through the influences of late imperial Russian intellectual traditions and western approaches to societal diversity. This apparatus bears the impact of interpretative lenses (or prisms) which possess a mythic resonance. Its operation simultaneously at the official, (pseudo-)academic and unofficial, demotic levels of the public sphere accounts for its power and durability.

In addition to Channel 1 and Rossiia, we analyse relevant news reports by NTV and REN TV. The reasons for this comparison are three-fold. First, when covering dramatic events, the main state-aligned channels now have to take account of what they are each reporting, as they are competing for audience share and advertising revenue. The comparison will thus allow us to explore the dialogical nature of the two main channels' coverage of controversial issues. Second, in contrast to Channel 1 and Rossiia, NTV and REN TV had acknowledged rising inter-ethnic tensions as a serious problem for Russia prior to Manezhnaia, offering detailed coverage of the issue.[11] We need therefore to assess whether these differences in the pre-Manezhanaia approaches to covering ethnic tensions led to divergent outcomes during the riots. Finally, REN TV's *Nedelia* news programme was overtly critical of the Kremlin and strove to subject official positions and interpretations to critical scrutiny. Comparing its coverage with that of Channel 1 and Rossiia could therefore shed light on those aspects of the coverage which reflected the two main channels' subordinate position.

The hour-long weekend news bulletins of each channel provide the main sources for this chapter, but it also draws on the complete range of coverage across the channels viewed throughout the two weeks in which Manezhnaia dominated the headlines. We focus particularly on the weekend editions for two reasons. First, our comparative approach required careful selection of bulletins for in-depth scrutiny to ensure that the basis for comparison was consistent. As stipulated by the qualitative paradigm we follow,[12] we derived that selection from an inductive reading of the entire corpus, which revealed in the weekend editions the 'characteristic rhythms and patterns' that, for Martin Harrison, differentiate weekend bulletins from weekday news and which, as Espen Ytreburg argues, distil into more 'assertive' and more 'dramaturgically' defined trajectories the disparate narrative fragments to emerge from weekday broadcasts.[13] Second, the fact that, in conforming to Ytreburg's thesis, the Russian weekly overviews indicated the 'settled' view adopted by each channel in relation to the breaking events to which they spontaneously reacted over the preceding days meant that they were pertinent to our central concern with narrative coherence (the degree to which the various accounts of the disturbances which the channels provide are consistent within and between one another).

Proceeding by channel, we show how the transformative process (more or less 'linear' in nature, according to each channel's relative need to cleave to shifts in the governmental and/or popular accounts) leads to partial convergence around a common line, while the contradictions within and between broadcasters remain unresolved and a coherent narrative fails to emerge. We adopt the principle that consensus, and the power relations that it reflects, is never more than the provisional fixing of dominant meanings within a complex system of antagonisms liable to disrupt and reconfigure those meanings at any point.[14] Post-Soviet Russian public discourse, we contend, offers a specific, and particularly vivid, demonstration of that principle.

Our core argument, which has three interlinked components, can be briefly summarized as follows: first, rather than adhering to the edicts of a univocal state machine, post-Manezhnaia broadcasting reveals multiple fault-lines, and their partial convergence around a single narrative which congeals the emergent media event into a settled form reflects less an imposed Kremlin version of events than the restricted logic of the available conceptual apparatus and a perceived need to reflect the public mood; second, the emphasis within that apparatus on the perceived clarity and fixity of ethnic boundaries – a legacy of multiple sources which influence the post-Soviet discourse of nation, ethnicity and race – leads both to over-interpretation of the inter-ethnic dimension to the crisis and to its occlusion; third, other European broadcasters are not immune to this paradox, which relates also to the contingent nature of ethnicity as a category.

The four interpretative lenses

Conflicting information about the Manezhnaia disturbances was refracted by the four channels through a set of consistently utilized interpretative lenses reflecting official myths, (semi-) academic theories and popular interpretations of events: first, 'the friendship of the peoples'; second, 'ethnic criminality'; third, 'culture conflict' or 'inter-ethnic strife'; and fourth, 'conspiracy of power' theories. Possessing distinct genealogies, these interpretative devices have, as we show, acquired new life in the contemporary discursive environment. The second and third lenses, in particular, reflect distinct racializing worldviews which, while avoiding the articulation of crude biologically determined hierarchies, naturalize ethno-cultural differences and transpose to culture some of the prejudices commonly associated with biologically defined race. Such perceptions are typical of what scholars call the 'new racism'.[15] In fact, in most situations, public discourses poorly differentiate the ethno-cultural and racial aspects of identity, and attempts to disentangle them founder. In Russia, demotic, media and semi-official pronouncements (such as those of Duma deputies and regional and local politicians) often describe the cultural specificities of minority groups as being 'in their blood' or 'in their genes'.[16]

Coined by Stalin in 1935 and used until the late Soviet period, the 'friendship of the peoples' notion emphasized pan-Soviet unity, highlighting the central role attributed to Russians in achieving it. It signalled a major shift in the USSR's nationalities policies, away from the earlier approach of fostering the national self-expression of non-Russian minorities while stigmatizing ethnic Russians as members of an 'oppressor-nation'. Yet, despite the fact that Stalin's new slogan reversed the Russians' role in sustaining the unity of the state-framed multi-ethnic community, it did not presuppose the transformation of the Soviet Union into a Russified nation-state, stressing instead the multi-ethnicity and multiculturalism of the Soviet community of peoples.[17] In view of the formula's original meaning, it sits awkwardly with the current Russian government's attempts to construct a more unified national identity among citizens of the Russian Federation than the Soviet approach had allowed. Simultaneously criticized as Russification in disguise by nationalist activists in the non-Russian Union republics and as a licence to exploit the RSFSR for the benefit of the non-Russian nationalities by Russian nationalists, Stalin's formula was discredited by the end of the Soviet period. Significantly, it was only within the context of the December 2010 riots that the 'friendship of the peoples' metaphor suddenly resurfaced in the coverage of the two main state-controlled channels.[18] The resurrection of this slogan, which lost its power under Gorbachev, seems to indicate a lack of a clear vision among broadcasters and political leaders regarding the causes of and solutions to the Manezhnaia violence.

The genealogy of the 'ethnic criminality' concept connects it with late nineteenth-century racial theories, particularly those of Italian scientist Lombroso's school of criminology, which linked particular anthropological types to criminality. Lombroso's ideas acquired their own life in Russia as, in the late imperial period, a debate took place among Russian anthropologists about whether those ideas could be applied to entire ethnic groups, particularly the peoples of the Caucasus.[19] Significantly, the works of the pre-revolutionary experts occupying the most extreme position on the issue are currently being republished in Russia. Particularly often repeated today is the hypothesis – first articulated at the turn of the twentieth century, and since the 1970s revived in the ideology of the European New Right – that social norms and social deviance were determined by the indigenous people (*korennoe naselenie*) of a given territory, whereas migrants (*prishloe naselenie*), deprived of links to their native soil, more readily displayed an inclination towards criminality.[20]

Despite the fact that the official Soviet approach stressed the social origins of crime, the shift in the 1930s to an essentialist view of the nation permitted the assigning of immutable characteristics to every member of a national group subjected to repressions.[21] Furthermore, in the Soviet Union of the 1970s, in the context of the diminishing power of Marxist–Leninist ideology, leading Soviet ethnographers and certain historians were already claiming that an 'ethnos' possessed 'psychological specificities' (*psikhologicheskii sklad*) which determined 'certain typical features of behaviour' of its members. In this context, some legal scholars also began to concur that individuals' psycho-physiological features could, at times, influence behaviour more than social circumstances.[22] The post-1991 period has witnessed a proliferation of publications claiming a deterministic link between ethnicity and/or migrant communities on the one hand and criminality on the other, often on the basis of tendentious statistical data. In the media, as in various official state documents and academic texts, even seemingly neutral references to 'ethnic criminality' in terms of the occasional stratification of criminal groups along ethnic lines are frequently marred by racial undertones.[23]

The 'culture conflict' notion was likewise first alluded to by European theorists of race in the late nineteenth century, and then reformulated in more politically neutral terms in North American sociology of the 1930s. This approach suggests that in complex, multicultural societies, the practices followed by some groups (particularly immigrants) are liable to antagonize the dominant culture. Under such circumstances, migrant behaviour may be viewed as offensive and even criminal by wider society.[24] In turn, the ideologists of the New Right in Europe have re-appropriated the 'culture conflict' notion in order to argue that the dominant European culture, which they represent as homogenous and static, is under threat from the incompatible cultural norms of migrant communities.[25] In today's Russia, the notion is in wide circulation, along with references to its multiple interpretations in the west.[26]

In current public discourses, 'culture conflict' theory also overlaps with the Soviet-era idea of 'inter-ethnic strife' (*mezhnatsional'naia vrazhda* or *rozn'*).[27] The latter defines a type of activity forbidden in Soviet and post-communist Russian legislation.[28] As we will see, the two main state channels reflected the Soviet tendency to, spatially and temporally, externalize 'inter-ethnic strife' (as occurring elsewhere and/or to be resolutely avoided in the future). As deployed in today's Russia, 'culture conflict' theory and the 'inter-ethnic strife' concept are based on an essentializing understanding of ethnic boundaries as clearly definable and fixed. Recent research into the origins of ethnic conflict has shown that narratives about 'ethnic criminality' have powerful potential to act as a catalyst for violence against 'the ethnic other', as they project onto 'the future victims of violence the very impulses entertained by those who will victimize them'.[29] The 'culture conflict' argument, with its dehumanizing tendency, has its own conflict-generating power, albeit in a less pronounced form.[30]

Finally, events which represent a sea-change tend to stimulate the appearance of conspiracy theories, suggesting that the events are not at all that they seem.[31] In Russia, conspiracy theories have historically performed important functions.[32] Thus, in Stalin's period the notion of conspiracy was systematically used by the government to delegitimize political opponents and justify repression. This legacy is still alive in contemporary Russia, and its impact has been increasing lately. The Manezhnaia disturbances were at times refracted in news coverage through a conspiratorial lens. Let us now examine how this, and the other lenses, shaped that coverage.

Channel 1 and Rossiia

Channel 1 and Rossiia were hesitant in responding to the Sviridov affair, as they attempted to closely follow the interpretations set out by representatives of law enforcement organs and political leaders. At the same time, owing to its more explicitly domestic remit, Rossiia's *Vesti nedeli* differed from Channel 1's Sunday *Vremia* in the complexity of the angles it adopted and in the wider range of voices heard. The first reports on Sunday editions of *Vremia* and *Vesti* of 12 December attempted to downplay the racist nature of the violence. The reporting was dominated by comments from representatives of the Moscow police and the Minister of the Interior, who used euphemisms such as 'non-football slogans' (*nefutbol'nye lozungi*) in describing the crowds' overtly racist chants and described 'left-radical youth' and unspecified 'extremists' in reference to members of extreme Russian nationalist groups participating in the disturbances. Any possibility that racist views could be shared by members of football clubs (that is, members of 'our' 'ingroup') was categorically denied. It was only towards the end of the 12 December coverage that, after showing footage of crowds of people outstretching their arms in Nazi salutes, the reporters on both channels began to use the expression 'radical nationalist groups' and mention attacks on 'passers-by who did not look Slavic'. To some extent, *Vesti* went beyond what *Vremia* acknowledged at the time, concluding its first Sunday report with an interview with Rossiia journalist Nikolai Svanidze, who admitted that on Manezhnaia, 'Nazi slogans were everywhere'.[33]

The most detailed coverage occurred in the *Vremia* broadcast of Sunday 19 December which followed Medvedev's and Putin's speeches, both of which suggested that what had happened in Moscow on 11 December should be treated with the utmost seriousness. The leaders condemned manifestations of extreme ethnonationalism, yet balanced this criticism – which was implicitly aimed at the Slavic majority – with reference to problems arising from violations of societal norms on the part of unintegrated minorities and migrants. At the same time they framed their general analysis of the riots with criticism of local authorities' frequent failure to adequately address conflict situations.[34]

For the media, a week after the riots, it was difficult to deny the racist frenzy that had gripped the Manezhnaia crowds and to avoid acknowledgement of the extent of the violence directed against non-Russians. Even though the speeches by the country's two main leaders seemed to have indicated the emergence of a more 'settled' official narrative about the events, the *Vremia* bulletin of 19 December continued to frame its coverage in a highly contradictory way. Apparently following the official line, it started with the insistence that 'the catalyst for the mass outrage... was a story not about nationalism, but corruption' of law enforcement organs. The lack of detail about what exactly constituted corruption in this particular case lent the discussion a populist-conspiratorial colouring.

However, the report then changed tack and began showing YouTube clips featuring shaven-headed youths attacking, without any visible provocation, people with darker skin; the attackers were referred to by *Vremia* as 'fascists' or 'skinheads'. It next juxtaposed this footage with an equal number of clips depicting young Caucasians behaving aggressively and in a highly asocial manner on Moscow streets. The shifting of the blame to Caucasians was reinforced by the Orthodox Patriarch Kirill, who, unlike the political leadership, openly invoked 'ethnic criminality'. For Kirill, rather than Caucasians reacting to racist attacks, Russians had been 'provoked' by the criminal 'radicalism' of the minorities. He willingly acknowledged that there was indeed a 'majority' crowd acting upon 'understandable', even if prejudicial, sentiments. Such differences between Kremlin politicians, church leaders and state journalists indicate tensions within a post-Soviet establishment lacking firm ideological underpinnings.[35] In conclusion, the reporter argued that 'the number of crimes caused by extreme nationalism' (*prestupleniia na national'noi pochve*) 'was growing and growing'. So, in contradiction to its opening statement, halfway through the *Vremia* report the disturbances began to be interpreted in the context of ethnic relations in Russia. In the depiction of these relations, the label of 'extreme nationalism' was applied to activities of a very different nature and the main victims of the violence were assigned a large part of the blame for what happened in Moscow on 11 December.

Within this contradictory framework, *Vremia* attempted to explain the origins of the racially motivated violence. Despite strong criticism of the manifestations of such violence, the terminology used indicated how pervasive the power of a racializing worldview had become in Russia, to the extent that certain words and concepts which, in their original meaning, had nothing to do with ethnicity and race were now ethnicized and racialized. At the beginning of the 19 December bulletin, in the initial context of Manezhnaia as a spontaneous, popular response to an incident of corruption in the law enforcement organs, the moderator claimed that it was quite common for frustrated youths to start directing their anger at 'strangers' (*chuzhie*). According to the *Vremia* reporter, 'it is easy to provoke hatred of strangers', and so in Moscow 'visitors (*priezzhie*) were beaten up'. The original meanings of the words *chuzhie* and *priezzhie* have nothing to do with ethnicity or race: they refer respectively to those who are not family members or friends and those who have recently moved to a particular area. *Vremia* unreflectively adopted the currently popular trend of using such terms to label non-Slavs, particularly people from the Caucasus and Central Asia, without giving any consideration to the appropriateness of applying words with a strong distancing and 'othering' effect to fellow citizens of the Russian Federation, such as North Caucasians.

After eventually acknowledging that the Manezhnaia riots revealed serious problems with ethnic relations, the programme-makers evidently felt compelled to give a positive example of Russia's traditional 'friendship of the peoples' in a story about an Armenian boy, Gagik, who was set upon by adult males during the riots and was defended by two Russians, Lesha and Sasha. In a performative evocation of the 'friendship of the peoples' myth, the reporter commented that Lesha and Sasha did not care 'what physical appearance their friend had'. In its original meaning, the Soviet myth was supposed to mark the happy coexistence of different ethno-cultural traditions; in the *Vremia* coverage, however, this was reduced to a mere reference to ethno-racial distinctions.

In the context of *Vesti*'s more dialogical approach to reporting (its tendency to pitch its reporting as an implicit response to views which it feels the need to rebut), a reference to the North Caucasians' – albeit partial – responsibility for the Manezhnaia disturbances appeared earlier than was the case on *Vremia*, being alluded to as early as the 12 December broadcast. Evoking the 'conflict of cultures' theory, Svanidze, whose normally liberal inclinations appeared to have been set aside, spoke of 'people of alien religion and alien culture' who needed to be 'taught the local norms of behaviour' if exacerbation of ethnic tensions was to be avoided.[36]

Both channels included the Manezhnaia riots in their surveys of the main events of the year, shown on 26 December.[37] In these final annual broadcasts, both channels, in different ways, returned to their original tactic of downplaying the overall significance of the events and their implications for the state of ethnic relations in the country. *Vremia* distilled its Manezhnaia narrative down to the simple story of an 'everyday street fight', purging any references to ethno-racial unrest. *Vesti*, in turn, claimed that – the riots notwithstanding – the 'friendship of the peoples' was still flourishing in Russia, whose citizens were proud of their legacy of fighting against fascism.

The approach taken in the final annual broadcast of *Vesti nedeli* could be seen as a polemical rebuttal of the line taken by alternative media (for example REN TV), that ethno-cultural pluralism in Russia had acquired destructive forms, and of the nationalist opposition's argument that Soviet 'friendship of the peoples' was based on the exploitation of Russians by non-Russians.[38] Appropriately, it fell to Putin to perform the final act in the 'friendship of the peoples' drama. Providing no context, the *Vesti nedeli* report depicted Putin's meeting with the Russian national martial arts team – whose members, viewers were reminded, came from throughout the former USSR.[39] In this officially sanctioned perception of sport as a mirror of Russia's multi-ethnicity there was no place for ethnically motivated violence. But the broadcast suddenly featured a challenge to the rhetoric of harmony. Following a report on Putin's post-Manezhnaia actions, details were given of the arrest of a retired GRU (Main Intelligence Directorate) colonel, Anatolii Kvachkov, whose son participated in the riots. The report revealed that Kvachkov senior was accused of 'wide-ranging nationalist subversive' activities. His arrest was widely reported in the print media and on the Internet,[40] leaving *Vesti nedeli* little choice but to cover it. Significantly, Rossiia acknowledged that Kvachkov aligned himself with the seventeenth-century *opolchentsy* movement, whose march on Moscow to drive out the Polish occupiers dominated the symbolism of the officially instigated Day of National Unity celebration marked a month earlier.[41] There could be no more vivid illustration of the difference between the two kinds of media event, nor of the path leading from official 'unity in difference' to the 'unity via the expulsion of difference' principle which has come to dominate the coverage since 2012, and which we will analyse at the end of the book.

Overall, the two main government-backed channels failed to articulate an 'authoritative discourse' about the Manezhnaia riots. Their coverage was unable to resolve the contradictions between the representation of Russia as a place where a historically rooted 'friendship of the peoples' was still flourishing and narratives of 'ethnic criminality' and 'conflict of cultures' reflecting popular prejudices that were clearly shared by the reporters and representatives of Russia's elites whom we interviewed.

NTV

Of all the four channels, NTV most consistently reproduced ethnic and racial prejudices, as it relentlessly stereotyped minority communities according to the 'ethnic criminality' lens. In NTV's Sunday news programme, *Segodnia*, 'ethnic criminality' emerged as the main cause of the disturbances in particular, and ethnic tensions in Russia in general; the coherence of the channel's narrative was bolstered by selective quoting of authoritative figures and biased visual representational techniques. NTV's preferred 'investigative' mode (a legacy of its earlier enthrallment with western news formats) accorded its reporters the leeway to exceed the bounds of the approved sources to which Channel 1 and Rossiia were tied, and to indulge in free-ranging, populist interpretations.

The first Manezhnaia report on *Segodnia*, of 12 December, unashamedly adopted a viewpoint close to that of the Spartak crowds, righteously incensed by the release of Sviridov's assailants, whose North Caucasian origin was stressed by the reporter.[42] Similar to the first reports on Channel 1 and Rossiia, NTV's only mention of racist slogans referred to an 'unfair accusation' from unknown sources; the violence in the centre of Moscow was attributed to unspecified 'left radical and nationalist organisations', that is, marginal actors on the fringes of the dominant 'ingroup'. This downplaying and externalization of ethno-racially motivated violence continued during the subsequent weeks, with the programme of 26 December describing the behaviour of Manezhnaia rioters as a mere 'emotional outburst'.[43]

The reports of 19 and 26 December were unambiguously framed by a narrative about 'problems with unintegrated diaspora communities', whose deviant behaviour and criminal actions, described as rooted in the specificity of their cultures, were presented as the main cause of social tensions across Russia. In the course of these reports, the question of 'who is guilty' for the riots was raised on several occasions, with the answer always starting with discussion of 'visitors' (*priezzhie*) – 'including those from the Caucasus', who tended 'to behave outrageously and criminally' (*naglo i kriminal'no*).[44] Moreover, the North Caucasian republics of Russia, from where these 'guests' came to Moscow, were depicted as areas where 'criminal gangs have merged with law enforcement organs' and where even members of the political elites behaved so offensively that neighbouring Russian regions had to set up special police units to deal with North Caucasian political leaders when they visited.[45] With the apparent intention of bolstering the credibility of its coverage, NTV interviewed members of law enforcement organs in Moscow so that they could provide statistical 'evidence' of 'ethnic criminality'. It was furthermore the criminality of ethnic minorities, the coverage suggested, that was largely responsible for incidents of corruption in local administrations and law enforcement organs across Russia. As the reporter on the 19 December programme put it, adopting the kind of conspiratorial tone set to saturate post-Manezhnaia public discourse: 'National diasporas in large cities have become criminalized; they are in cahoots with local authorities, the police…'

The selection of quotes from Medvedev's and Putin's speeches was also tendentious, as the greatest gloss was given to their remarks about problems with unintegrated 'migrants'. The coverage of instances of ethnic tensions involving Russians and North Caucasians other than Manezhnaia always adopted the viewpoint of the Russian majority, as confirmed in the use of the parenthetic phrase, 'As the local people say…' On the rare occasions when members of minorities were given a voice, they were quoted as reaffirming the majority viewpoint, arguing that their fellow North Caucasians 'bring shame on the country in which we live'.[46]

Visual techniques further reinforced the representation of North Caucasians as the guilty party. The 19 December report on the aftermath of Manezhnaia, sensationally and conspiratorially entitled 'Battle with the Forces of Darkness', opened with images of people in Moscow being checked for the possession of weapons. The hitherto unspecified voiceover was accompanied by footage depicting young Caucasians denying that they held weapons. But the film cut to a weapons haul. The reporter then 'rebutted' the young man's verbal denial (complimenting the 'rebuttal' of the visual edit): 'And yet they (weapons) do exist. These knives, hatchets, pistols – this entire arsenal – have been confiscated from Russian citizens marching towards each other on Moscow streets.'[47]

The reporter's initial even-handed phrasing of a 'general' problem was undermined by visual examples weighted against Caucasians. Similarly, on 26 December, in the coverage of Putin's meeting with football fans at which the Prime Minister gave his warning that 'inter-ethnic strife' posed a real threat to Russia, the camera dwelt on the darker-skinned participants, as if hinting that Putin's admonitions were directed at them.[48] This was certainly the reporter's own view, as he concluded that, despite Russia's history of peaceful coexistence of different nationalities, in the North Caucasus as well as in the capital 'a new generation has grown up, who apparently needs to be taught anew what Russia is and what norms of behaviour its people have'. In a further indication that this 'new generation' consisted of people from the Caucasus and migrants from elsewhere, the report ended by criticizing the St Petersburg educational authorities for producing a Russian-language textbook for children of minorities that was based on folk stories from the Caucasus. The reporter suggested that instead it would have been better to teach them – as everyone else – Pushkin's fairy tales.

NTV's position at the boundaries of approved discourse renders the interpretative substratum from which state media outlets constructed their narrative particularly susceptible to re-inflection by populist voices that cannot, post-1991, be ignored. Its former status as a repository of 'progressive' infotainment-style television formats only aids that cause ('commercialism' and 'liberal democracy' are no longer synonymous in Russia). NTV's threshold position also allowed it to offer coverage with much greater coherence than Channel 1 and Rossiia, conflating – to potentially devastating effect – the 'ethnic criminality' narrative with criticism of the corruption of Russia's law enforcement organs – which are themselves also, it was implied, a victim of such criminality. As we shall see in later chapters, this explosive theme was to resurface with a vengeance.

REN TV

REN TV's weekly news programme *Nedelia* made the clearest attempt to interpret the riots as part of a broader inter-ethnic problem, and was the most explicit in its criticism of extreme Russian nationalism and manifestations of racism. The same interpretative prisms used by Channel 1 and Rossiia were often put to opposite uses. *Nedelia* projected the view that 'friendship of the peoples' was dead in Russia, as the country was gripped by an intense 'conflict of cultures'. *Nedelia* also made clear that the Manezhnaia riots were not just a one-off event, as *Vremia* and, to some extent, *Vesti* attempted to claim. Instead *Nedelia* spoke about 'the epidemics of inter-ethnic conflicts' afflicting Russia, for which multiple failures of the government and law enforcement organs were responsible. Paradoxically, the channel's attempt to offer an open discussion of the causes and consequences of the riots, using western-style semi-documentary and 'investigative' modes of reporting, seemed to exacerbate its reliance on concepts and theories replete with ethno-racial prejudice.

Nedelia's first broadcast on 11 December made no attempt to conceal the racist nature of the riots, treating them as a major challenge to the Kremlin.[49] The report on Manezhnaia was placed first in the running order, in contrast to *Vremia* and *Vesti* on that day, which covered the riots towards the end of their news bulletins.[50] *Nedelia*'s report was dramatically entitled 'The Spartacus Uprising', evoking the fan club to which Sviridov belonged and the largest Roman slave uprising of the first century bce.[51] The historical semantics of the Spartacus myth convey the idea of a mass protest against authority. Risks are posed by awakening these associations while populist anger at alleged collusion between the authorities and minorities and migrant communities rages. That it is the self-styled 'liberal' *Nedelia* programme which courts such danger, misappropriating the rhetoric of civic protest, is as indicative of the fluid uncertainties of Russia's media environment as is Rossiia's focus on the appropriation of the *opolchentsy* narrative by the anti-government opposition.

In this report, *Nedelia* seemingly endeavoured to avoid concessions to the right, curtailing references to the possibility that the behaviour of the so-called diaspora communities could also be regarded as problematic. The issue of uncontrolled 'migration from the periphery to large cities' was briefly noted as causing tension, but was accompanied by a reporter's observation that 'visitors (*priezzhie*) acquire weapons in order to defend themselves from aggressive aboriginals (*aborigeny*)'.[52] The ironic, and deeply dialogic, use of the word 'aboriginals' in relation to permanent residents of cities in European Russia at this stage seemed to indicate the channel's understanding of the problematic nature of the term *priezzhie*.

In the subsequent broadcast of 18 December, however, the *Nedelia* coverage began to reflect a shared perception of Caucasian responsibility for social tensions in Russia's big cities. Dwelling on the causes of the riots, *Nedelia* attributed a dominant explanatory power to the 'conflict of cultures' lens. Within this context, *Nedelia* suddenly began to 'balance' its opposition to ethnic Russian nationalism with a concern about the asocial behaviour of North Caucasians.[53]

As with *Vremia* and *Vesti*, *Nedelia*'s hybrid of sources and ideas failed to gel. The over-reliance on the 'conflict of cultures' lens, which lacked a firm ideological mooring, produced particularly striking contradictions in REN TV's coverage. While strongly condemning the racist tone of the demonstrations, the programme itself promoted a racializing worldview invisible to the moderator and reporters, who subscribed to a narrow definition of racism, limiting it to explicit instances of the employment of Nazi symbolism.[54] (Indeed, the problematic nature of 'new racism', which claims a deterministic link between ethno-cultural distinctions and social issues, is recognized only by a small number of liberal commentators in Russia.[55])

In the *Nedelia* broadcasts, Slavs and Caucasians were depicted as two neatly demarcated groups with immutable behavioural norms. Like *Vremia* and *Vesti*, *Nedelia* suddenly began to racialize the words *priezzhie* and *gosti* ('visitor' and 'guest'), using them as a collective definition of anyone non-Slavic, irrespective of their citizenship or length of residence in Moscow. Likewise, the expression *korennoi moskvich* (indigenous Moscovite) was applied by *Nedelia*'s moderator, Marianna Maksimovskaia, solely to ethnic Russians/Slavs, though strictly speaking the word *korennoi* (indigenous) applies merely to a long-term resident of a particular location.[56]

Contrary to the Kremlin-sponsored discourse of a pan-Russian civic nation, whose multi-ethnic population is bound together by common efforts to build a strong state, REN TV's coverage of the December riots represented Russia's different ethnic groups as separate nations (*natsii*), among whom only Russians were identified with the country as a whole. Post-Soviet discourses of Russian nationhood have failed to overcome conflations of ethnicity and nationality and the linking of ethnically defined nationality to discrete territorial space that was characteristic of the Soviet era. It was this that caused REN TV, like other broadcasters, implicitly to place Dagestanis, Chechens and Ingush permanently residing in Moscow under the umbrella term of *priezzhie*. North Caucasians invited to speak on the programme seemed to have internalized their own 'othering'. This was reflected in the arguments of a Chechen student activist living in Moscow: despite the fact that he had moved to the Russian capital as a child, the student accepted the external definition of himself as an outsider whose personal behaviour was responsible for shaping a collective image of his nation (Chechnia) in the eyes of the host (ethnic Russian) society: 'Those who take out knives put their nation into shame ... In other cities... we should demonstrate [the best sides] of our culture, of our nation.'[57]

REN TV eschewed the other channels' tendency to quote liberally from Medvedev's and Putin's speeches. But extreme nationalists and communists on one hand, and spokesmen for North Caucasian communities on the other, were interviewed extensively. Also notable by their absence were members of the liberal opposition.[58] *Nedelia*'s west-leaning credentials were, by 18 December, outweighed by the emerging cross-channel consensus, with its grounding in domestic strains of anti-liberal, popular and pseudo-academic discourses.

Nonetheless, the liberal critique never faded entirely. On 18 December, an attempt by Vladislav Surkov, First Deputy Head of the Presidential Administration, to blame the liberal opposition for teaching nationalists to stage unauthorized demonstrations was dismissed. In an interview with Maksimovskaia in the same bulletin, Ingush President Iunus-Bek Evkurov blamed the media for the state of inter-ethnic relations, arguing that they regularly highlighted the ethnicity of Caucasian criminals while ignoring that of neo-Nazi skinheads.[59]

On 18 December, Maksimovskaia claimed that one of the most active participants in shaping Putin's youth policy, Vasilii Iakimenko, carried direct responsibility for the rise of xenophobia because, as leader of the *Nashi* movement, he helped divide the country into 'ours and theirs', further implying that such outcomes were not necessarily an oversight;[60] meanwhile Ruslan Aushev, the former president of Ingushetia, described the authorities' toleration of neo-Nazism in central Moscow as suspicious.[61] No attempt was made to reconcile hints at the possibility that the authorities stood behind the riots with the suggestion that they were genuinely terrified by them. Indeed, conspiracy theories, including suggestions that Putin or the exiled oligarch Berezovskii were involved in organizing the riots, ran rampant within the oppositional media, penetrating the peripheral space within official discourse which REN TV occupies.[62] The absence of robust political meta-narratives permitted such fantasies to coexist side by side.

***

In now recapitulating the key elements in our argument and linking them to one another, and to the European comparative dimension, we must consider what the weak or non-existent meta-narratives and the multiple divergences and convergences between national broadcasters meant for the state-aligned media's management of inter-ethnic flare-ups such as Manezhnaia. We have shown that the transformations of the four interpretative lenses through which they all viewed the riots can be traced through contradictions expressed diachronically, via shifts in the Manezhnaia narrative over two weeks, and synchronically, via ideological and terminological conflicts present throughout. For Rossiia and Channel 1 the diachronic axis is foregrounded, as these official outlets struggle to recalibrate their narratives in line with the shifting perspectives of their political masters and with popular opinion. On NTV's *Segodnia* and REN TV's *Nedelia*, the respective populist and liberal orientations are pre-set and, despite concessions to the official line or the perceived popular consensus, the synchronic axis prevails. This difference points to a further, 'meta-discursive', dimension to the contradictions and one of this book's recurring themes: that of a 'mediasphere' structured as a spectrum running from centre (Channel 1), through Rossiia, which leavens its official line with a strictly managed pluralism, to a periphery serving as a two-way filter to extra-official realms: demotic-nationalist (NTV) and liberal-progressive (REN TV's *Nedelia*).

The transformative process results in an apparent convergence around a common line related to a widely perceived problem with unintegrated minorities and migrants. Rather than the imposition of an unambiguous Kremlin view, it reflects the point at which state pronouncements intersect with 'popular consensus'. The convergence is partial and tensions between and within the different narratives, including that of the state itself, remain unresolved. All of the accounts are fractured from within and retain a strongly dialogic aspect. Thus, particularly when tested by events such as the Manezhnaia riots, the official discourse is incapable of re-integrating the different layers of its mythological substratum and struggles to 'mainstream' the multiple voices (nationalist, liberal, populist–conspiratorial, etc.)

that it confronts. Its power to assert its dominance over a Russian public sphere characterized by relative discursive plurality is further undermined by the range of contradictory ideological frameworks that it utilizes in confronting major political and social issues. Indeed, in their final broadcasts of 2010, the two main state-controlled channels ultimately resorted to denying the full scale of the conflict which they had earlier acknowledged. NTV's consistently populist approach enabled it to maintain a more coherent line – which, however, came dangerously close to inciting ethno-racial prejudices, even if paying lip service to the 'friendship of the peoples' narrative its subordinated position compelled it to adopt. Notwithstanding its explicit condemnation of ethnic Russian nationalism and of the authorities for encouraging it, REN TV's *Nedelia* paradoxically reinforced perceptions of impenetrable ethnic boundaries and irreconcilable inter-ethnic differences to a still greater extent than the two main state channels, thus highlighting particularly clearly the power of the pre-existing interpretative apparatus, which is used unreflectively when journalists are faced with quickly unfolding dramatic events.

The main point of consensus shared across all the channels remained the unreflective equation of ethnic Russians alone with the Russian Federation as a whole – a view that contradicted the Kremlin's ideal of a multi-ethnic civic Russian nation (*rossiiskaia grazhdanskaia natsiia*) and the Russian constitution. The perception that ethnic Russians have (or should have) a special status within the Russian Federation is, however, widely shared, and so Zhirinovskii's reiteration in the *Poedinok* talk show of 20 January 2011 of the demand that the constitution be revised to acknowledge this special status was predictable. The ultranationalist Zhirinovskii used bold, even radical language to express his demand. In contrast, by using a particular unreflective terminology, *Nedelia*'s liberal moderator merely implied that Russians were different from (and superior to) other national groups in the Russian Federation.

In their reproduction of ethno-racial stereotypes, Russian television practices are not dissimilar to those of west European media. The tendency to emphasize negative characteristics of ethnic 'outgroups' and downplay or deny those of the dominant 'ingroup', the selective application of ethnic labels in coverage of crime and inter-ethnic conflict construed from the perspective of the Slavic majority and the equation of cultural and biological differences find parallels in reporting in Europe as a whole.[63] There are significant differences between the media of different European countries and between individual media outlets within these countries regarding the degree of terminological laxity and the levels of crude stereotyping in coverage of issues related to ethnicity and race. Some outlets are much more effective than others in masking 'ingroup' bias.[64] Television coverage of the Manezhnaia riots demonstrated a very high degree of laxity and a strong inclination towards stereotyping with little attempt to conceal any biases. Thus, the label of 'visitor' was systematically applied to citizens of the Russian Federation whose difference from the ethnic Russian majority is solely ethno-cultural. North Caucasians are not first or second-generation 'migrants' from another country; the region of North Caucasus was incorporated into the Russian state in the course of the nineteenth century. In that sense, Russian citizens from the North Caucasus are different from North African migrants in west European countries. Yet in both the Russian and western European cases, the stereotyping and racializing media coverage betrays attitudes rooted in Europe's colonial past. North Caucasus was Russia's colonial domain, subjected to discrimination and 'othering'. The Soviet integrationist project did not overcome this legacy.

The terminological laxity seen here reflects deep uncertainties about how the Russian national community should be defined, which we also noted in the coverage of the Day of National Unity. In the broadcasters' treatment of the Manezhnaia riots, the uncertainties which came to the fore are connected with the legacy of Soviet policies on nationality. They are based on a highly ambiguous approach to the relationship between Soviet (civic) identity and peoples' ethnonational identities, as well as to squaring the dominance of the Soviet Union's most powerful contingent (the *russkie*) with recognition of the ethnonational separateness of non-Russian minorities of differing statuses. Thus, the prejudicial application of terms like 'visitors', 'guests' and 'diasporas' to people who are bona fide citizens of the Russian (*Rossiiskaiia*) Federation is the direct consequence of the division of first Soviet, then Russian, federal space into multiple territories inhabited by 'titular nationalities'. This terminology also reflects Russia's imperial legacy, which the Soviet experience uniquely reshaped, but did not overcome. So the identification of the Slavic population alone as the 'indigenous population' of Russia as a whole and of its capital city, which was common in the late imperial period, has been revived post-1991.

Key to our interpretation of the discursive crisis to which the lexical confusion points is the notion that the interpretative prisms we identified operate at, yet cut across, different levels of the public sphere: vernacular ('conspiracy of power'); intermediate/academic ('ethnic criminality'); official (the 'friendship of the peoples'). Meanwhile, the notion of inter-ethnic disharmony straddles the official/intermediate divide. It can manifest itself in 'virtual' form: inter-ethnic strife portrayed as a potential, but avoidable, consequence of the disturbances, and realizable only 'elsewhere'; but it also exists in a realized version (the 'conflict of cultures' represented by REN TV's *Nedelia* as the actual cause of the disturbances).

The differential location of the four prisms generates multiple interpretations of the official discourse within which they interact. For instance, Channel 1 and Rossiia draw at times on the vitality of conspiratorial sentiment to *minimize* ethno-nationalist and racist undertones and portray the disturbances as a singular act of resentment at an aberration resulting from nothing more than endemic legal inertia. Yet the linkage elsewhere of conspiratorial thinking to 'ethnic criminality', which receives its most lurid expression in NTV's invocation of 'dark forces', inevitably fosters *over*-interpretations of those same undertones.

The dangers represented by ambiguities and contradictions in the discourses of national unity and ethnic conflict are symbolized in the ambivalent semantics of the Spartacus revolt evoked by REN TV – for in its contemporary setting, the revolt is at once a justifiable populist march on corrupt officials in conspiratorial collusion with 'ethnic criminal groupings'; a terrifying outburst of anti-migrant rage; and a profound threat to the integrity of the Russian state. Equally ambiguous is the *opolchentsy* narrative. Central to the artifice by which, since the introduction of an official Day of National Unity in 2005, Russia celebrates its multi-faith, multicultural society, the story of the victorious uprising against a 'foreign invader' galvanizes popular opposition to that very unifying project, as reflected in the story about the arrest of Colonel Kvachkov which Rossiia linked to its coverage of Manezhnaia.

The fact that, in the coverage we analysed, the 'foreign invader' remained a culturally hybrid mass gathered under the ethno-geographical umbrella of *Kavkaz* (the Caucasus) demonstrates the explanatory power of Rogers Brubaker's insistence on 'performative ethnicity', according to which 'by invoking groups, [ethno-political entrepreneurs] seek to evoke them', and on groupness as an 'event' rather than a phenomenon or a mere 'construction'.[65]

Brubaker argues that, because of the performativity of ethnic groupings, what is reported as ethnic conflict 'may have more to do with thuggery, opportunistic looting and black-market profiteering'.[66] An instructive contrast therefore emerges between Manezhnaia and the rioting that shook London in August 2011 before spreading throughout the UK. The latter commenced as a protest against the police's shooting of a young black man and refusal to deliver justice to his family but degenerated into what was presented in all public representations and in popular perceptions as a looting spree devoid of inter-ethnic content.[67] The former began as a football fans' protest against police ineffectiveness which mutated into ethnically saturated violence. The radical contingency of ethnicity as a category (its emergence as a determining force in one situation but not another, ostensibly similar, one) is but one explanation for the difference, but if it is a factor at all, then the Manezhnaia disturbances might tell us something important about how the 'crisis of multiculturalism' is playing out in post-communist space.

One legacy of the Manezhnaia events has been a noticeable increase in attention to inter-ethnic relations within official discourse. In January 2012, as we shall see in Chapter 7, Putin chose to dedicate one of his 'election manifesto' articles to the 'national question'. Here, he made extended reference to collusion between corrupt law enforcement organs and migrants, linking it to the 'radicalization of the host society'. But he also depicted in alarming words the 'crisis of multiculturalism' afflicting west European societies, which he presented as being unable to cope with large migration flows – comparing 'the west' unfavourably to Russia, where 'culture conflicts' are easier to manage given the fact that 'for centuries Russia has existed as a multi-national state.'[68] The narrative struggles that Manezhnaia revealed as a 'crisis management' type of media event contradicted Putin's optimistic assessment. As we will later demonstrate, in the first year of Putin's third presidency, Rossiia in particular began to reject the earlier interpretation of the riots as a one-off episode, instead aligning itself with REN TV's *Nedelia* in linking them to the nation's crisis in inter-ethnic relations and frequently resorting to the populist approach adopted by NTV in December 2010. Nonetheless, when, in October 2013, mass inter-ethnic violence flared at the Biriulevo market in Moscow following the murder of an ethnic Russian by an Azeri migrant, even *Vesti*, which was by now leading the broadcast news agenda (*Vremia* was noticeably reticent), refrained from earlier anti-migrant excesses, fearful of provoking still worse violence.[69] By this point, however, as we shall see in Part III, state-aligned television news had effectively abandoned its favoured diversity management tool – the complex, and often contradictory, hybrid of interpretative lenses at work in coverage of Manezhnaia – in favour of a co-ordinated, univocal campaign of 'othering' in which the fears and prejudices prevalent at the margins of approved discourse drown out the dying echoes of the 'friendship of the peoples' myth. Curiously, those very margins are also the site of a range of alternatives to the ominous drumbeat of xenophobic alarm. It is these alternatives that Part II will now explore.

Notes

1 Marlène Laruelle (ed.), *Russian Nationalism and the National Reassertion of Russia*, London: Routledge, 2009; Alfred Evans, 'Putin's Legacy and Russia's Identity', *Europe-Asia Studies*, 60/6, 2008, 899–912; Stephen Hutchings and Natalia Rulyova, 'Commemorating the Past/Performing the Present: Television Coverage of the Second World War Victory Celebrations and the (De)construction of Russian Nationhood', in Birgit Beumers, Stephen Hutchings and Natalia Rulyova (eds.), *The Post-Soviet Media: Conflicting Signals*, London: Routledge, 2009, pp. 137–57.
2 *Nedelia* (REN TV) from 19 December 2010 reported that there were demonstrations in support of the Moscow rioters in five cities across Russia.
3 Marlène Laruelle, *In the Name of the Nation. Nationalism and Politics in Contemporary Russia*, Basingstoke: Palgrave Macmillan, 2009, pp. 35–48.
4 See also Martin Laryš and Miroslav Mareš, 'Right-Wing Extremist Violence in the Russian Federation', *Europe-Asia Studies*, 63/1, 2011, 129–54.
5 For the rise of neo-Nazi extremism in Europe, see Martin Schain, Aristide Zolberg and Patrick Hossay (eds.), *Shadows over Europe: The Development and Impact of the Extreme Right in Western Europe*, New York: Palgrave, 2002; Bert Klandermans and Nonna Mayer, *Extreme Right Activists in Europe: Through the Magnifying Glass*, London: Routledge, 2006.
6 See the UK's far-right English Defence League's mission statement. Available online at http://englishdefenceleague.org/ (accessed 9 April 2012).
7 See Chapter 9.
8 Christopher Flood, Stephen Hutchings, Galina Miazhevich and Henri Nickels, *Islam, Security and Television News: A Comparative Study*, Basingstoke: Palgrave Macmillan, 2012, pp. 125, 157.
9 *Poedinok* (Rossiia), 20 January 2011.
10 Antonis A. Ellinas, *The Media and the Far Right in Western Europe: Playing the Nationalist Card*, Cambridge: Cambridge University Press, 2010; Flood *et al.*, *Islam, Security and Television News*; Simon Cottle, *Ethnic Minorities and the Media: Changing Cultural Boundaries*, Buckingham: Open University Press, 2000; Ella Shohat and Robert Stam, *Unthinking Eurocentrism: Multiculturalism and the Media*, London: Routledge, 1994.
11 See, for instance, the following relevant reports on REN TV: 'National'nyi otvet' in the programme *Reporterskaia istoriia*, 12 September 2010; 'Molodym zdes' ne mesto' and 'Tam, gde byla voina' in *Nedelia*, 26 September 2010; 'Russkii samosud' in *Nedelia*, 30 October 2010.
12 Gaynor Lloyd-Jones, 'Design and Control Issues in Qualitative Case Study Research', *International Journal of Qualitative Methods*, 2/2, 2003, 1–20.
13 Espen Ytreburg, 'Moving out of the Inverted Pyramid: Narratives and Descriptions in Television News', *Journalism Studies*, 2/3, 2001, 357–71 (362).

14 See Ernesto Laclau, *Hegemony and Socialist Strategy: Towards a Radical Democratic Politics*, London: Verso, 1985, p. 135.
15 George Fredrickson, *Racism. A Short History*, Princeton: Princeton University Press, 2002, pp. 151–70; Unni Wikan, *Generous Betrayal: Politics of Culture in the New Europe*, Chicago: Chicago University Press, 2002.
16 Viktor Shnirel'man, *'Porog tolerantnosti'. Ideologiia i praktika novogo rasizma*, vol. 2, Moscow: Novoe literaturnoe obozrenie, 2011, p. 278.
17 Terry Martin, *The Affirmative Action Empire. Nations and Nationalism in the Soviet Union, 1923–1939*, Ithaca: Cornell University Press, 2001, pp. 432–61.
18 According to our project data – all news reports relating to inter-ethnic cohesion issues broadcast on *Vesti* and *Vremia* from September to December 2010 – the formula was not used once between the beginning of September and the second week of December 2010.
19 Marina Mogilner, *Homo imperii. Istoriia fizicheskoi antropologii v Rossii (konets XIX-nachalo XX vv.)*, Moscow: Novoe literaturnoe obozrenie, 2008, pp. 358–96; Daniel Beers, *Renovating Russia. The Human Sciences and the Fate of Liberal Modernity, 1880–1930*, Ithaca: Cornell University Press, 2008, pp. 103–15.
20 E. V. Erikson, 'Ob ubiistvakh i razboiakh na Kavkaze', in V. B. Avdeev (ed.), *Russkaia rasovaia teoriia do 1917 goda*, vyp. 2, Moscow: FERI-V, 2004. On the European New Right see Douglas Johnson, 'The New Right in France', in Luciano Cheles, Ronnie Ferguson and Michalina Vaughan (eds.), *Neo-Fascism in Europe*, London: Longman, 1991, pp. 234–44. Works by Alain de Benoist, an ideologist of the New Right in France, have been translated into Russian on many occasions since 1991.
21 Victor Zaslavsky and Robert Brym, *Soviet-Jewish Emigration and Soviet National Policy*, New York: St. Martin's Press, 1983; Terry Martin, 'The Origins of Soviet Ethnic Cleansing', *The Journal of Modern History*, 70/4, 1998, 813–61; Valery Tishkov, *Ethnicity, Nationalism and Conflict in and after the Soviet Union*, London: Sage, 1997, pp. 3, 8, 71–9.
22 For a comprehensive discussion of this issue, see Shnirel'man', *Porog tolerantnosti*, vol. 1, pp. 251–90.
23 Shnirel'man, *Porog tolerantnosti*, vol. 2, pp. 109–42.
24 Tzvetan Todorov, *On Human Diversity: Nationalism, Racism and Exoticism in French Thought*, Cambridge: Harvard University Press, 1993, pp. 57–60.
25 Alain de Benoist and Charles Champetier, 'The French New Right in the Year 2000', *Telos*, 115, 1999, 117–44.
26 See for instance V. N. Burlakov, *Kriminologiia. Uchebnik dlia vuzov*, St Petersburg: Izdatel'skii dom Piter, 2004.
27 The phrase translates more accurately as 'inter-nationality strife', but covers primarily ethnic conflict. The elision reflects the imprecise use of the term *natsional'nost'* in Soviet and post-Soviet terminology and indicates the confused policy and lexicon surrounding race, nationality and ethnicity in the USSR.
28 Article 29 of the Constitution of the Russian Federation; part 1 of Article 282 of the Criminal Code of the Russian Federation. Available online at http://m3ra.ru/2010/12/01/inciting-ethnic-hatred/ (accessed 28 December 2011); USSR law of 2 April 1990 'Ob usilenii otvetstvennosi za posiagatel'svo na natsional'noe ravnopravie grazhdan'. Available online at: www.bestpravo.ru/ussr/data01/tex10976.htm (accessed 28 December 2011).
29 David Horowitz, *The Deadly Ethnic Riots*, Berkeley: University of California Press, 2001, p. 77.
30 Richard Arnold, 'Visions of Hate. Explaining Neo-Nazi Violence in the Russian Federation', and Alexander Osipov, 'Minority Questions: Ethnicity, Discrimination, and Extremism in Russia', both in *Problems of Post-Communism*, 57/2, 2010, 37–49 and 50–60 respectively.
31 Peter Knight, *Conspiracy Nation: The Politics of Paranoia in Postwar America*, New York: New York University Press, 2002.

32 V. E. Bagdasaryan, *Teoriia zagovora v otechestvennoii istoriografii vtoroi poloviny 19-20 vv.*, Moscow: MSU, 1999; L. Gudkov (ed.), *Obraz vraga*, Moscow: OGI, 2005.
33 *Vesti nedeli* and *Voskresnoe Vremia*, 12 December 2010. Available online at www.1tv.ru/news/social/166893 (accessed 20 January 2014).
34 *Voskresnoe Vremia*, 19 December 2010. Available online at: www.1tv.ru/news/social/167400 (accessed 20 January 2014).
35 See Stephen Hutchings, Galina Miazhevich, Chris Flood and Henri Nickels, 'The Impact of "Islamic Extremism" on TV News Representations of Multiculturalism, *Intégration* and *Mnogonarodnost*: A Comparative Analysis', *Russian Journal of Communication*, 1/1, 2008, 43–70.
36 *Vesti nedeli*, 12 December 2010.
37 *Vesti nedeli* and *Voskresnoe Vremia,* 26 December 2010. Available online at www.1tv.ru/news/social/167879 (accessed 20 January 2014).
38 See for instance an interview with Zhirinovskii on REN TV's *Nedelia* of 18 December 2010.
39 The meeting took place in St Petersburg on 22 December 2010.
40 'Polkovnik Kvachkova arestovali kak terrorista', 23 December 2010. Available online at www.lifenews.ru/news/47624 (accessed 27 May 2011); *Nezavisimaia gazeta*, 24 December 2010; and *Kommersant*, 25 December 2010.
41 As *Vesti nedeli* pointed out, Kvachkov's movement styled itself 'the Minin and Pozharskii People's Volunteer Mission', invoking the legendary duo whose names were intoned repeatedly during the November rituals.
42 *Segodnia* (NTV), 12 December 2010.
43 *Segodnia* (NTV), 26 December 2010.
44 *Segodnia* (NTV), 19 December 2010.
45 Ibid.
46 Ibid.
47 Ibid.
48 *Segodnia* (NTV), 26 December 2010.
49 *Nedelia* (REN TV), 11 December 2010.
50 Arina Borodina, 'Poboishche proshlo vtorym ekranom', *Kommersant*, 14 December 2010.
51 On the role of Spartacus's uprising in Soviet/Russian historical mythology see Vera Tolz, *Russian Academicians and the Revolution: Combining Professionalism and Politics*, Basingstoke: Macmillan, 1997, pp. 80–1.
52 *Nedelia* (REN TV), 11 December 2010.
53 *Nedelia* (REN TV), 18 December 2010.
54 Laruelle, *In the Name of the Nation*, p. 38.
55 Shnirel'man, *'Porog tolerantnosti'*, vols. 1 and 2.
56 S. G. Barkhudarov *et al.* (eds.), *Slovar russkogo iazyka*, vol. II, Moscow: 1958, p.133. However, between the 1860s and 1917, the expression '*korennoi narod*' was applied in official and popular discourses specifically to the Russian population. This usage, rejected in the Soviet period, seems to be influencing today's utilization of the word '*korennoi*'.
57 *Nedelia* (REN TV), 25 December 2010.
58 This is despite the fact that in earlier REN TV broadcasts about ethnic conflict such figures as the leader of the 'Solidarity' movement, Boris Nemtsov, acted as commentators. See *Nedelia* (REN TV), 6 November 2010.
59 *Nedelia* (REN TV), 18 December 2010.
60 Ibid.
61 REN TV's programme *Reporterskaia istoriia*, broadcast on 19 December 2010, systematically explored conspiratorial explanations of the Manezhnaia riots.
62 Contributions to REN TV's audience forum ranged from claims that the riots were organized by Putin in order to strengthen his new bid for the presidency to

suggestions that they were the work of the oligarch Boris Berezovskii, as part of his plot to destroy the Russian state. See comments on 12, 17 and 28 December 2010. See also Kirill Rogov, 'Temnyi vsadnik na belom kone', *Novaia gazeta*, 20 December 2010.

63 John Downing and Charles Husband, *Representing 'Race': Racisms, Ethnicities and the Media*, London: Sage, 2005; Teun van Dijk, *Racism and the Press*, London: Routledge, 1991.

64 Russell King and Nancy Wood (eds.), *Media and Migration. Construction of Mobility and Difference*, London: Routledge, 2001.

65 Rogers Brubaker, *Ethnicity Without Groups*, Cambridge: Harvard University Press, 2004, p. 10.

66 Ibid., p. 19.

67 Post-riot headlines in the UK focused on issues of consumerist greed, youth amorality, parental indiscipline and police timidity. For an analysis of the British and foreign media coverage see Hugh Muir and Yemisi Adegoke, 'Reading the Riots: Investigating England's Summer of Discontent', 8 December 2011. Available online at www.guardian.co.uk/uk/2011/dec/08/were-the-riots-about-race (accessed 11 April 2012).

68 Vladimir Putin, 'Rossiia: national'nyi vopros', *Nezavisimaia gazeta*, 23 January 2012.

69 Its coverage of the incident was somewhat disjointed, on one hand remaining vague and circumspect about what had actually occurred at the market, but on the other turning – unusually – to arch-conservative Rossiia journalist Arkadii Mamontov, an outspoken critic of migration, for its on-site report.

Part II

Difference at the margins

5 Re-working Russian diversity

The 'marginal' role of television fiction

As our discussion of talk shows in the wake of Manezhnaia revealed, television's role in the (mis)management of diversity policy is far from restricted to news bulletins. The assumed separation between 'real news' and non-news programming is, anyway, something of a fallacy.[1] In fact, news and non-news broadcasting have always existed in a symbiotic relationship. Current events, first reported in news bulletins whose language and discourse place them close to the centre of the official sphere, are therapeutically 'worked through', or re-worked, in non-news genres at the margins of that sphere.[2] Meanwhile, news bulletins themselves import from the margins those styles, editorial practices, narrative structures and myths which will enliven their reports (the term 'news *story*' is revealing).[3]

The interconnectivity between news and non-news broadcasting has particular significance for the media's representation of inter-ethnic relations in the context of televised drama. News broadcasters strive to give vivid and dramatic expression to the problems that issues of inter-ethnic tension generate, but also to distance themselves from the inflammatory stereotypes that circulate in the discourses inflecting popular television fiction. Equally, television dramatists are drawn to the raw emotion and controversy of topical inter-ethnic conflicts, yet seek to avoid becoming news stories in their own right by overstepping the boundaries of permissibility in this sensitive area.

A compelling example of a case in which the boundaries were breached is provided by the controversy over the Russian reality television serial, *Shkola*, which we analyse below as a media event in which 'the media' themselves become the source of the 'event'. The serial featured nationalist extremism among teachers and skinhead violence against North Caucasian pupils – themes whose long-standing threat to burst from the peripheries to the centre of the news agenda was realized at the end of the year in which *Shkola* was released via the Manezhnaia riots.

By dint of its capacity to tap into popular mythologies, fictional television like *Shkola* serves as a gateway at the periphery of the spaces of official culture, through which subcultural prejudices and counter-discriminatory discourses rarely expressed even on talk shows might enter, and via which those prejudices and discourses are, in turn, processed, managed and responded to by mainstream media practitioners. It is the site of a permanent circulation and contestation of ethnicity-related meanings.

The above principles have particular relevance for the post-Soviet media environment, with its continuing legacy of government control on one hand, and with its manifest failure to articulate societal consensus and increased sensitivity to inter-ethnic tension on the other. As we have seen, official discourses offer their own contradictory perspectives on inter-ethnic tension and the need for tolerance. Our purpose here is to re-examine the circulation of official and demotic approaches to ethnicity on the territory of televised fiction. We seek to evaluate whether the seemingly marginal status of that territory exacerbates the capacity for ethnic prejudices to thrive, or whether it furnishes new tools with which to manage them (whether deliberately or as the subliminal effect of an impersonal process) or rebut them.

We rely on the overarching approach to media cultures developed by Jesus Martin-Barbero, whose account of cultural hegemony draws on the same neo-Gramscian principles as Laclau's, positing power as control over the definition of what passes for 'consensus' and emphasising 'cultural circulation'.[4] While it cannot be applied wholescale to post-communist societies lacking societal consensus, such a conception of hegemony helps us address the encounter of official Russian approaches to the management of inter-ethnic relations with popular and subcultural strands of opinion. It avoids the twin dangers of first, assessing the Russian political class's grip on power solely on the basis of the force (juridical, political, economic) at its disposal, and second, dismissing subcultural forms as peripheral to the exercise and questioning of that power.

We identify three strategies, each tied to a particular fictional format. First, we describe a *mediation* process in which current issues informing constructions of ethnic difference are explored within a variant on the drama series popular in Russia: the gypsy romance. This facilitates the filtering of 'otherness' through idealized imaginings of an inclusive selfhood in which that 'otherness' is appropriated and emptied of threat. Through the principle of narrative transgression, it also allows characters to renegotiate the boundaries of 'self' and 'other'. Via our reading of *Shkola* and the controversies it raised we then explore the longer, serial format as a locus within which xenophobic extremism can be *moderated* (or alleviated) by competing discourses of official multiculturalism and universal tolerance. We turn finally to the comedy series *Nasha Russia*, investigating whether humorous portrayals of ethnic minorities (the main cause of the scandal evoked by the series) re-confirm popular prejudices, or whether the format in which those prejudices are expressed *modalizes* (or 'de-literalizes') them. We locate the programmes in their broader contexts, addressing popular mythologies in the case of the gypsy serials and societal reactions and cultural references in that of *Shkola* and *Nasha Russia*.

The programmes treat issues drawn from the news bulletins analysed in other chapters, positioning themselves at varying distances from the official discourses represented in those bulletins. The gypsy serials were broadcast on Rossiia, with its government-imposed remit to give positive expression to Russia's ethnic diversity. *Shkola* aired on Channel 1, which adheres closely to the Kremlin line but, since the government is technically a minority shareholder, is more attentive to advertising income and viewing figures, and part of whose remit is to address

Russian speakers abroad who are inured to non-Russian, critical representations of their nation of origin. *Shkola* was, significantly, made by a director whose prior work dealt with problems affecting Russia's alienated youth. *Nasha Russia* is a product of TNT, a commercial channel aimed at the 16–30 market.

The spectrum of approaches (from Rossiia, through Channel 1, to TNT) indicates that, far from serving merely as a locus of vernacular racism, popular cultural forms also contribute to a tentative recasting of inter-ethnic relations. Channels and programme genres susceptible to the influences of peripheral cultural forms and discourses are likely to contribute most to this process; even a gypsy romance is further from the approved line than a Rossiia news broadcast. Thus, the strategies we identify serve not solely to attenuate extreme positions or to legitimize approved responses to them, but also as a means of interrogating these responses. It is no coincidence that two of the programmes we examine were the subject of scandal.

We supplement the overarching theoretical emphasis on popular culture's work in negotiating societal consensus with tools geared to each genre. In the case of the gypsy series, we apply narrative theory to explore how fictional characters can challenge established boundaries of national self and ethnic other. Our approach to *Shkola* relies on Martin-Barbero's own identification of the *telenovela*'s capacity to embrace a controlled dialectic of audience and producer and to provoke, but ultimately evade, official censure. For *Nasha Russia*, we adapt Todorov's and Ducrot's definition of modality as the distinction between the 'content' of language and the speaker's attitude to that content to describe the sources of the humour the series generates.[5]

We demonstrate that the peripheral space occupied by fictional and comedic genres proves to be fertile territory for the renegotiation of national and ethnic identities, official and subcultural discourses. Taken as a whole, our spectrum of genres and channels refutes any assumptions that the renegotiation process offers either univocal support for the Kremlin approach to 'the national question' or enthusiastic endorsement of the populist attitudes to ethnic minorities which that approach ambivalently targets.

The gypsy as vanishing mediator

From **tabor** *to screen (and back)*

When illustrating the symbiosis of news and non-news genres, one might turn to portrayals of North Caucasians who, as we have seen, have been a focal point in the recent rise of inter-ethnic tensions in Russia. Given the low official size of Russia's Roma population and their relative absence from the domestic news agenda, it is seemingly odd, then, that it is the image of the gypsy which provides the more compelling manifestation.[6] The paradox deepens when one considers discontinuities between that image's 'news' and 'non-news' manifestations. On one hand, gypsies cede little to North Caucasians in terms of the negativity that accompanies their presence in the news (both are tarred with the brush of 'ethnic criminality'). On the other, the romanticized image of the Russian gypsy as exotic outsider has a long pedigree stretching to Pushkin's *Tsygane*.[7] In this variant, we encounter a gypsy whose innate creativity, restless spirit and mysterious affinity with nature offer a counterpoint to the dry rationality of modern, urban civilization. There is, of course, an orientalist version of the North Caucasian whose lineage can also be traced to Pushkin,[8] but it was never as readily adaptable to mainstream customs as that of the gypsy. Moreover, the mystique of the Caucasus has declined precipitously since the Chechen wars.

Like many stereotypes, that of the gypsy is internally contradictory and includes notions of dishonesty and a sinister association with the supernatural (popular fears of the 'gypsy curse' and 'the evil eye' – the negative corollary to the celebrated affinity with the mystical).[9] So endemic is the myth in its complete form that news reports must erase those tensions if they are to benefit from the powerful resonance of the prejudicial attributes. Equally, if popular drama drawing on rural gypsy mystique is to renew itself with the vitality of the socially authentic, it must square the romantic idyll with the harsh realities of narcotics, poverty and sexual exploitation blighting the contemporary *tabor*.

Precisely because the figure of the gypsy confronts in stark form the challenges of the news/non-news dynamic, it also epitomizes the potentials which that dynamic holds. A key factor facilitating this role is that intra-television mediations (from news to non-news and back) are themselves the function of broader mediations linking official and vernacular discourses, state policy and popular sentiment, transnational form and local variant.[10] The gypsy community's status as a pan-European symbol of the excluded outsider is important here.[11] Few of the features of gypsy mythology discussed so far would not apply throughout Europe. Yet in each nation, the myth acquires local attributes. The Roma's historical status as an itinerant people with a transnational identity is a linear enactment of its semantic quality as mediator of the global and the national, a quality it shares with the Jews.

One peculiarity of the Russian variant of the myth is the extent of its assimilation into mainstream tradition.[12] For this reason, it is ideally placed to explore relationships with alterity, particularly in fictional forms which accommodate such experimentation: the *tsygan* is not of the self, yet nor does s/he belong unambiguously to the other. The *tabor*'s associations with a traditional lifestyle featuring a strong, authoritarian figurehead (the baron) and antiquated attitudes to women mean that it can serve, too, as the locus for an exploration of changing gender relations. Thus, the Russian myth of the gypsy lifestyle highlights the arranged marriage. In its extreme version, this evokes fascinated revulsion (girls of 11 undergo weddings to men twice their age); in its moderate form, it offers comfort to males yearning for traditional femininity.

Latter-day images of the gypsy as beggar and thief, combined with the squalor of camps beset with social problems, mean that ethnicity and class are conflated. Alternations between 'Roma' and 'gypsy' (*tsygan*) as appellations reflect this confusion of the cultural (tradition, way of life) and the ethno-racial (blood lineage, language, physical appearance). The alternation of cultural and ethno-racial characteristics is linked to a final mediatory function, for the *tsygan*'s exoticism and nostalgic appeal can be deployed to articulate an impossible, and thus disembodied, desire for the absolute difference of the ethnically 'other'. By eliding the image with characteristics relating to mundane lifestyle practices, that other becomes attainable, and so embodied. Many drama plots revolve around romances between Russians and gypsies in which the alien and the different are gradually familiarized. We relate the gypsy's multi-level mediatory function to the circulatory dynamic linking these plots, and thus the 'real' to the 'imaginary' in Martin-Barbero's understanding of mediation.[13] As fictional genres situated in the realm of desire, the dramas invariably bear the influence of the exotic imagery of 'wild' gypsy dancing, of the gypsy's fortune-telling powers and of the various cinematic realizations of gypsy mythology which punctuated the Soviet period.[14] The latter are important not least because they also broach the fraught issues of Russian national identity with which the gypsy theme is intertwined.

The new generation of gypsy dramas to flood Russian television can also be traced to the early 1990s arrival on post-Soviet screens of the Latin American 'soap opera' form, with its melodramatic narratives of provincial girls finding love and social advancement in new, urban environments. Post-Soviet social upheavals offered a fertile environment for the serial genre and Russian television generated its own variants, including those centred on gypsy themes. That the three Rossiia dramas we treat reflect similar thematics is evident from their titles: *The Enigmatic Tomorrow of Fate* (*Sud'by zagadochnoe zavtra*), *The Gypsy Girl with the Way Out* (*Tsyganochka s vykhodom*), *Love and Separation* (*Liubov' i razluka*). Of these, the most recently made was *Love and Separation* (2011, directed by Aleksei Kozlov). The previous year saw the broadcast of *The Enigmatic Tomorrow* (directed by Vasilii Moskalenko). *The Gypsy Girl* is the earliest of the serials (2008, directed by Aleksei Rudakov) but, like the others, it was showing in 2010–11 when the Roma were prominent in the news. We treat them in reverse order.

Love and miscegenation

The plot of *Love and Separation*, which ran over eight episodes, treads familiar ground. A young provincial teacher, Anna, marries a middle-aged police chief, Trekhin, but falls in love with a dashing gypsy, Lacho. Lacho's community expel him from the *tabor* for refusing to marry the gypsy girl chosen for him by the baron, but he persists in pursuing Anna, getting her pregnant. She refuses an abortion and gives birth to a son whom Trekhin removes, driving Lacho away and attempting to have him murdered. Meanwhile, Trekhin's own wayward son accumulates debts and kills his father for refusing to help him, but blames Lacho. Finally, Anna proves Lacho's innocence and the couple are reunited.

Lacho's orientalized exoticism (his swarthy features, passionate nature and love of horses) renders him the archetypal forbidden object of desire. The dangers which such desire holds are attenuated by the fact that the desiring subject is a provincial girl, herself objectified from the male viewpoint, and by Lacho's de-ethnicizing ejection from his own community. The narrative pattern of union, separation and reunion dramatizes in linear form an alternation between identification with and distanciation from 'otherness'. As Anna's relationship with Lacho deepens, she 'overcomes her feeling of alienation' (to quote the serial's website),[15] echoing the process by which viewers accommodate themselves to him. The identificatory viewpoint which 'sutures' the audience into the fictional reality is suspended between Lacho as the desiring male and Anna as the representative of Russian selfhood.[16]

But the narrative also permits an unfolding of the prejudicial fears of constitutive 'otherness'. Lacho is criminalized during the story. His initial pursuit of a married woman sets the tone. In his campaign to rescue Anna from Trekhin, he sets fire to a hotel owned by Trekhin's unscrupulous business associate, who also has designs on Anna. Finally he is wrongly accused of Trekhin's murder and endures intense suspicion from Anna's confidante, Anfis'ka, who serves as a cipher for anti-gypsy prejudice.

However, just as Anna's illicit desire for 'the ethnic other' is attenuated by the state of transition into which he has been thrust, and because, extra-diegetically, the romance is fictional, so Anfis'ka's fear of gypsy criminality is ameliorated by the fact that, intra-diegetically, Lacho's purported crime is a fabrication. It is further alleviated by the negative portrayal of Trekhin as a vindictive police chief immersed in shady business deals. In the intricate game of alternation between romantic ideal and social reality, identification and counter-identification, police corruption and collusion with the world of crime are balanced with the Roma community's arranged marriages and underage sex.

Central to the dénouement is the birth of Anna's illegitimate son. It is he who cements their love, precipitating Trekhin's descent into self-destructive vengeance. He is the embodiment of the miscegenation which is the semantic equivalent to the oscillation process. Moreover, he is a vehicle for exposing the social reality of ethnic prejudice. When he begins school, he is taunted with the racist nickname '*tsyganionok*' (little gypsy boy) as the serial directs its critical lens at anti-Roma sentiment. The serial exploits within its narrative the very structuring contrast between idealized image and grim reality of which it is itself a manifestation. When on the way to his trial, Lacho asks the guards if he should read their fortune, subjecting their racial stereotyping to ridicule. Yet the drama propagates those same stereotypes. Anna is herself subjected to the unnerving stare of the gypsy girl over whom she was preferred by Lacho, and is fearful that she has been given 'the evil eye'.

Ultimately, *Love and Separation* abandons the principles it adopts in its surface narrative (that of a love story targeting anti-Roma prejudice). Lacho's expulsion from his *tabor*, his love for Anna and his gradual de-ethnicization confirm his status as the 'exception that proves the rule' about gypsy criminality, for exceptionality is the driver of narrative (stories engage us to the extent that they transcend the predictable). Unsurprisingly in light of these tensions, the story fails to tie up some of its loose ends: the consequences of Lacho's expulsion; the subplot involving Trekhin's corrupt business associate.

The Enigma of Origin

The Enigmatic Tomorrow shares with *Love and Separation* a concern with identity, explored through the relationship between gypsy and Russian ethnicities. That concern relates to the contradictory way in which Russian television co-articulates culture, race and ethnicity. Here, though, gypsies play a secondary role. Even their *tabor* is located at the peripheries of the village in which the action occurs. Their very marginality means that the desires and fears they represent emerge with greater clarity. Set in a Cossack village, the 15 episodes revolve around a soldier, Ivan, who returns from the front to find that his fiancée, Nastia, is to marry Kostia, whom she does not love. The jealous Kostia has Ivan beaten: an act witnessed by a gypsy boy, Sanakai, and his betrothed, Rita. Their marriage is in turn imperilled by the gypsy baron, who has promised Rita to a man from another *tabor*. When they rescue Ivan, he reciprocates by reassuring the police that the gypsies are not to blame (suspicion has been aroused by the rival *tabor*'s criminal activities). Nastia's refusal to marry Kostia enrages him, yet he ignores a warning from the gypsy fortune-teller, Lachi, that he has 'chosen his fate'. Meanwhile his associate, Kurchavyi, tries to endear himself to Kostia by incriminating Ivan in the burning of a stable. When Kostia finds out, Kurchavyi kills him with a weapon stolen from Ivan, who is arrested – a deception witnessed by the gypsy couple, who expose Kurchavyi's crime, reuniting Ivan with Nastia. A joint wedding (Nastia and Ivan, Rita and Sanakai) is organized at the *tabor*, which is celebrating the discovery of the remains of a previous generation, executed by Nazi occupiers. However, Nastia had earlier given birth to a stillborn baby, surreptitiously replaced by another woman's child. The woman returns for her son, to Nastia's horror. This twist was intended as a link to a non-existent sequel.

Cossack tradition has its own place in the repository of archetypes on which Russian national identity draws.[17] The serial is replete with scenes of bareback riding and an *ataman* (who, in Cossack regalia, takes a delegation to the *tabor* to complain about its purported criminal activities). Cossacks are closer than gypsies to the mythological core of Russian-ness in the many redefinitions to which it has been subject since 1991, but not coincident with it, enabling them simultaneously to distance viewers from the gypsy 'other' (now a further stage removed from Russian-ness) and to bring them into alignment with it (Cossack and gypsy lifestyles are equated through their shared exoticism).

This process is reinforced by the bifurcation of the gypsy characters into positive and negative manifestations (Sanakai and Rita; the rival *tabor*). Within the Cossack subject position, the viewer is thus encouraged to incorporate gypsy 'otherness' within the boundaries of the self, then to expel it, in a process pointing towards a resolution of differences and a revitalization of Russian selfhood. The fact that the positive variant is itself subject to accusations of criminality (until Ivan intervenes, Sanakai is held responsible for the beating he receives) facilitates a partial viewer identification with the gypsies, for our position as voyeur matches that of Sanakai and Rita. Conversely, when Ivan is wrongfully detained, the prior arrest of the innocent gypsies casts him as a victim of discrimination, and viewer identification with 'one of our own' is retrospectively conferred on the gypsies.

The Enigmatic Tomorrow facilitates a complex interplay between self and other, halted by the denouement. The joint betrothal is enacted within the space on which the gypsies have located their ancestral origin. The presence of the Cossacks on that land points to a figurative resolution of the biological subtext underlying inter-ethnic tensions. In *Love and Separation* the gypsy male's passion for a Russian woman enables him to mediate between the two ethnic groups, offering a 'metonymic' solution to the mediation problematic in which the contiguity of gypsy and non-gypsy spheres leads to their mutual 'contamination'. In *The Enigmatic Tomorrow*, the two spheres remain separate. However, the echoes between them point to a 'metaphoric' resolution. In order to cement the two halves, a further plot mechanism is deployed. From the beginning, when Lachi foresees Ivan's danger-strewn future path, through her prediction of Kostia's demise, to her warning about Nastia's tragic fate as a mother, the fortune-teller shapes the plot. Since viewers are privy to the authenticity of her powers from the point at which Lachi's first premonition about Ivan's fate proves true, their position of epistemological authority (by identifying the pattern of prediction and enactment, they 'foresee' the narrative future) is 'focalized' through Lachi's viewpoint.[18]

Yet *The Enigmatic Tomorrow*, too, fails to resolve its tensions. The discordant child-swapping line collapses, meaning that while the collective 'paternity' of the gypsies is settled, that of Nastia's child is not. The integration of gypsy and non-gypsy identities proves illusory.

No Way Out?

The Gypsy Girl with the Way Out appears free of such anomalies. A gypsy girl, Ligita, who lives with her stepmother, father and half-sister, is courted by a Russian boy, Lesha. But her hand is promised to Misha, son of the baron, Latso. Lesha, already receiving unwanted attention from Ligita's jealous half-sister, Sandra, is beaten by Latso's henchmen, but Ligita cannot stomach the repulsive Misha. Her dancing talent offers her a 'way out' and, while participating in a dancing contest in Moscow, she falls for Ivan Golovin, a wealthy ballet master who perceives her as a potential star, if he can tame her wild gypsy talent. The remainder of the series recounts their love affair and Ligita's burgeoning career, hampered by the jealous Sandra who becomes involved in a drugs operation run by her baron in collaboration with Shandor, shady owner of a gypsy dance troupe. Shandor himself desires Ligita, and she can only bring her love for Ivan to a happy conclusion when Shandor dies in a car chase. The series concludes with a marriage and a reunion between Ligita and her blood mother, for whom she searches in a series of 'interludes'.

Laura Keosaian, who plays Ligita, features in numerous gypsy serials. Her Armenian nationality and appearance conform to the romantic stereotype of the gypsy beauty. They also reinforce that stereotype's capacity to negotiate Russia's relationship with its 'black' (in the vernacular Russian usage) Central Asian and

Caucasian others. Ligita's relationship with Ivan mirrors that between Anna and Lacho in *Love and Separation*. In the later series the gypsy male (Lacho) occupies the liminal position between the *tabor* and the non-gypsy realm; Ligita, by contrast, is the gypsy female, driven from her *tabor* for an illicit act of miscegenation. The trope of desire for the forbidden other is another recurrent feature, as is the splitting of the gypsy realm into positive and negative hypostases.

Gypsy Girl foregrounds a structural phenomenon recurring in all the dramas. If stories are constructed around the breaches of normality necessary to generate significance, there are profound ideological implications for a narrative which traces the fate of a gypsy assimilated into the non-gypsy realm.[19] This becomes the exception which proves the 'rule' of non-assimilability. Nor is it accidental that access is granted via desire. The universalism of romantic love, uniquely capable of conquering difference, operates against a background in which such love is 'out of the ordinary'. Accordingly, the depictions of gypsy 'difference' (the criminality, the alien practices) are represented as the 'everyday' backdrop to the main action. Within fictional narrative's paradoxical logic, it is by *not* highlighting such phenomena, and by supplanting them with the exceptionalism of universal love, that they acquire truth value.

Gypsy Girl furnishes two riders to the logic of narrative exceptionalism. The first relates to the drama's focus on performance. Since Ligita is an aspiring dancer, and since the plot centres on her rehearsals, her exoticism (her intuitive feel for sensuous bodily rhythm) is recast as mundane 'normality', while her exceptionality is naturalized as authentic. The second rider derives from the subplot depicting the mutual mistrust between the two communities. When Lesha's mother discovers his passion for Ligita, she tells him: 'These people have their own laws and their own traditions. We don't belong with them'. His brutal beating confirms the wisdom of this warning. Here, gypsy barbarism and anti-Russian prejudice is cast as normal, in opposition to Ligita's atypical success in escaping from the *tabor*'s oppressive regime. Thus, the narrative structure reconciles the two opposing components of gypsy mythology.

The narrative rhythms of Ligita's expulsion from and reconciliation with the gypsy world point to the familiar process of de-ethnicization and re-ethnicization. Her accession to the realm of non-discriminatory 'true love' represents the victory of universal sameness over ethnic difference. But in the discordant finale, Ligita's reunification with her blood origins reconfirms the difference and the lack of narrative integration between the two plot lines.

Loose ends

All three dramas indicate that the markers of difference which gypsies bear must, paradoxically, be highlighted, yet concealed. Equally, ethno-racial features (the irreducibility of ancestral ties; the sensuousness of the body) and culture are at once elided and separated. This is as true within the Russian self (of which the gypsy is one hypostasis) as it is between that self and the irreducibly other (of which the gypsy represents an ameliorated version). It renders gypsies particularly suitable for their mediatory role in negotiating the treacherous waters surrounding post-Soviet articulations of the race/ethnicity/national community relationship, according them privileged value in the nation-building struggle and demonstrating that fictional romance offers distinct advantages over grandiose state–media collaborations such as the Day of National Unity as part of this struggle.

Serializing national cohesion

The telenovela *dialectic*

By contrast, such was the controversy surrounding Valeriia Gai-Germanika's 69-part *Shkola* (*School*), shown on Russia's Channel 1 in 2010, that it was nearly banned. In a post-totalitarian society beset by intergenerational 'culture clash', along with the rise of ethnic and other societal tensions, promoting cohesion is paramount. *Shkola* appeared to do the cause no favours. Filmed in a real school using hand-held cameras, it laid bare the problems afflicting Russian education: racism, casual sex, depression, drug-taking, bribery and unremitting disorder exacerbated by dismal teaching.

However, unlike the gypsy dramas, *Shkola* adopted the *telenovela* format, whose dialectical logic was crucial to its ability to moderate the subcultural excesses it appropriated from the cultural periphery by bringing them into contact with both official discourses of the centre and the global rhetoric of universal rights. Its prolixity created space for extended discursive encounters, enabling it to 'get close' to its audience and to modify its course in light of reader responses.[20] The dialectical 'to-ing and fro-ing' mirrored what Martin-Barbero terms the serial's 'provocation–pacification' dynamics, enabling it to 'sti[r] people up and denounce[e] the atrocious contradictions of society, but in the same process … to resolve these problems without moving people to action'.[21] This process was linked with tensions in the serial's structure. Its open-ended immersion in ongoing 'eventness' was identified with its tendency to incorporate social conflict. Its need to conclude its elongated narratives provided the 'pacification' dimension, aiding it in 'resolving' the issues it raised and moderating the anxiety they generated.

A peculiarly Russian controversy

Gai-Germanika's own provocative reputation was sealed by her feature film *Everyone Dies but Me* (2008), an uncompromising exposé of the lives of Russian teenage girls which dismayed the Russian cultural establishment. However, Channel 1 felt she promised a boost to its flagging performance in the ratings war. Iurii Gevorkian noted that 'Channel 1 understood that it needed to change something … It took the first illogical step towards the new viewer by releasing *Shkola* … It wanted to attract a new, young audience.'[22]

Shkola began showing in 25-minute episodes on 11 January 2010. Initially broadcast in the prime-time 18.30 slot and at 23.30, it was discontinued during the Winter Olympics – causing many to assume that it would never reappear – but then restored at the later time. The serial attracted vociferous criticism from Duma deputies, who demanded that Channel 1's director explain himself.[23] The Kremlin-backed party United Russia split into two over the matter.

Channel 1 remained undeterred, issuing a robust online statement rebutting its detractors.[24] On 12 April, Putin intervened:

> I haven't seen this serial ... and if you haven't watched it then you don't want to say, as in Soviet times, 'I condemn Solzhenitsyn, although ... I haven't read him'. ... We know of the problems and we are trying to react to them.[25]

In an exemplary illustration of *Shkola*'s conformity to Couldry's emphasis on the 'thickened', multi-platform nature of media events, many bloggers saw *Shkola* as part of a deliberate plan to provoke indignant support for the teaching profession – view that was inadvertently endorsed in a Channel 1 discussion of the serial by Aleksandr Isaev, president of the Duma Committee on Social Policy, who described the programme as a 'present' to the Year of the Teacher.[26]

Ratings confirm the interest elicited by the show. Viewers in the 18–30 age range accounted for 22–23 per cent of the entire television audience during the early evening showing, but *Shkola* attracted significant viewer shares among people up to 55 years old.[27] Numerous online forums show viewers divided into three categories: those who welcomed the serial's honesty; those who found it insulting to Russia's young people; and those who felt that it understated realities.[28] A smaller number adopted Gai-Germanika's line: that the serial was an artistic film dealing with universal issues, not an attempt to represent Russian schools.[29] On 21 June 2010, Channel 1 announced that, contrary to demand, there would be no series 2.

A lesson in provocation

Viewers had never seen anything like *Shkola* before. The script was heavily improvised, adding to the air of authenticity. The dialogue was riddled with foul-mouthed youth jargon. Pupil misbehaviour was uncannily convincing, as were the extra-curricular cannabis-smoking sessions. The serial features an extended gay kiss, gratuitous sex and close-ups of drunken pupils vomiting in each other's faces.

The action portrays the parallel lives of the pupils of 9A, which only the physics teacher, Natalia Nikolaevna, can control (she has persuaded pupils to rehearse *Romeo and Juliet* at an after-school drama club). Its central character is Ania Nosova, the disturbed granddaughter of the history teacher. She is rejected by other pupils, one of whom disseminates erotic photographs of her. Ania seeks solace in a group of 'emos' (a darkly pessimistic youth subculture she encounters outside of school), and later discovers an incipient attraction to Orthodox mysticism. Her pursuer harries her with anonymous text messages, then begins voyeuristically filming her. She steals the camcorder, but sinks into depression, before being sent to a psychiatric hospital where she commits suicide, videoing her last, desperate words.

Also prominent is Vadim Isaev, who is beaten by his drunken father and is a racist bully at school. His xenophobic comments about immigrants torment Timur, a quiet Muslim boy from Dagestan. Isaev frequents local skinhead gatherings but falls foul of inter-gang rivalry and mends his ways. However, he is encouraged by the geography teacher, Arsenii Ivanovich, a nationalist who establishes an extra-curricular group to promote his anti-migrant beliefs and is eventually sacked by the acting head teacher. Timur, meanwhile – a paragon of virtue – dissuades his cousin from kidnapping Isaev's brother in revenge for his abuse and begins a relationship with Sonia, a modest pupil whom he invites to his brother's Muslim wedding celebrations, but who is spurned by the other guests.

The finale focuses on the aftermath of Ania's death, which shocks the community into placing its worldly concerns into perspective. The penultimate episode features a Soviet-style appeal by Kharitonova, the acting head teacher, calling for parents and teachers to 'listen to children, to try to understand them, and to be their "friends"' (episode 68).[30]

(De)construcing the nation

The much-quoted 1934 *Pravda* headline, 'The Whole Country is Watching *Chapaev*', might have applied to *Shkola*. Like that socialist realist classic, Gai-Germanika's social realist shocker gripped Russia. Despite its grim aesthetic, *Shkola* too belongs to a vast project aimed at unifying a fragmented nation by provoking shared outrage at misplaced patriotic optimism, but was also an, albeit controversial, contribution to Russia's Year of the Teacher. In its very reversal of the logic of Day of National Unity or Manezhnaia (here television itself catalyses broader societal fractures and convergences), *Shkola* was every bit as much a media event as those phenomena.

The serial addresses issues of social cohesion and national (dis)unity. The central trope is that of intergenerational alienation: the inability of adults to 'manage' their teenage charges. The theme operates not only on the level of plot (parent–child; teacher–pupil), but also 'performatively': the dominant reaction among older viewers was 'We don't recognize our children in these monsters'; a prevalent response from youngsters was 'At last, something which tells the truth!' Several plot lines revolve around boys treating girls as objects of sexual gratification. Class animosities are aroused through the image of a BMW-owning New Russian. Finally, the serial explores inter-ethnic tension, and the relationship between popular xenophobia and official patriotism.

Like the Shakespearean tragedy of inter-tribal conflict contained within it, *Shkola* aims to overcome a litany of 'othernesses'. Its trajectory enacts metatextually what is narrated at the textual level: just as the multiple conflicts are ameliorated, so the impression of alienation created by the initial negativity cedes place to a rhetoric of mutual empathy. The (partial) triumph of this 'adult' rhetoric of reconciliation signalled in Kharitonova's emotional address to parents is confirmed in the excruciating scene at Ania's graveside which further develops the girl's stirrings of Orthodox sentiment.

The closing episodes evoke a Soviet past still relevant to the serial's older viewers, but bolstered by the post-Soviet supplement of Orthodox virtuosity. Ironically, given its initial press, *Shkola* concludes by re-establishing the Soviet cults of the Child and of School which it had seemingly dismantled, but has eventually 'purged' of sentimental falsehood. Thus, *Shkola* both challenges official optimism about national harmony and builds support for the symbolic importance of the pedagogue, revitalizing the officially endorsed values of tolerance and social cohesion.

(Mis)managing inter-ethnic conflict

The struggle to overcome adolescent recalcitrance is central to the chain of conflicts which *Shkola* enacts, for these are the challenges that the state must meet to ensure intra-societal cohesion. Paradoxically, this enables the school to function metonymically as a displaced peripheral effect of the state's confrontation with alterity, and simultaneously as that confrontation's central, metaphorical model.

The attention devoted to Timur indicates the 'national question's' privileged position in the chain. Until the closing stages (when he is befriended by Sonia), Timur remains at the margins of the collective. He sits on his own, suffering the same ostracism endured by Ania. The two fates unfold in parallel: Timur proceeds from loneliness to friendship, a touching romance with Sonia (a blond girl of archetypally Slavic demeanour) and 'reconciliation'. Ania's journey ends, via the alterity of the emo subculture, in alienation and death. The intersection of ethnic exclusion (a 'peripheral' theme) and teenage angst born of social exclusion (the 'central' narrative line) guarantees the penetration of centre by periphery.

The modelling function is evident in the school's approach to managing inter-ethnic tension. History lessons follow a familiar sequence: from the revolution, through Stalin's purges, to Perestroika. The pupils' ignorance of historical landmarks highlights the challenges faced by the state in its efforts to instil national pride. During the lesson on the Second World War, Vadim mutters objections to criticism of the Nazis (episode 15). In the literature lesson pupils guffaw when one of them recalls that, as a 'part negro', Pushkin was famed for his gigantic penis (episode 7), revealing why adolescent subculture proves fertile territory for the xenophobic right. Geography is taught by Arsenii Ivanovich, whose account of regional structures characterizes Moscow as overrun with migrants, echoing the bending of official nation-building to extremist ends (as will be shown in Part III, since 2010 Arsenii's position has moved much closer to the mainstream).

When the conflict between Vadim and Timur flares up (episode 13), the permanent head announces the need to deal with 'national disharmony' (*natsional'naia rozn'*) – a term of Soviet origin still used in the state media. The intuitive Kharitonova inflects state harmonization policy with a less formalistic idiom; when she sacks Arsenii Ivanovich, her voice quivers with anger as she declares that she will not endure fascism (episode 62). Her spontaneous liberal tolerance provides a welcome fillip to institutional rhetoric.

Kharitonova's initial position outside the school hierarchy enables her to import the unsullied peripheral values of liberal tolerance to the heart of the authority structure; this, too, is why her peroration calling for mutual understanding transcends dry officialese. But that same extra-institutional position to which she must return undermines the school's modelling function, reasserting its role as a periphery at which normal rules do not apply.

In the classroom, centre–periphery contamination works to opposite effect. Vadim's clash with Timur simmers throughout. On one hand, it is depicted in detail drawn from the broader social context: frequent use of the insults *chernozhopye* (black arses) and *churki* (literally, wooden stubs); Vadim's chanting of the slogan 'Russia for the Russians' at a group of Tajik workers (episode 51). Less authentic is the manner in which the conflict is confined to two individuals, which sits uneasily with Gai-Germanika's efforts to convey the 'drip-drip effect' of racism in its recurring, everyday banality. Vadim's constant racist posturing is credible, but the fact that it is isolated threatens to turn the character into a mere cipher.

Universal empathy

Social contexts are, nonetheless, carefully observed. Vadim's obsession with Tajiks reflects popular perceptions, emphasizing the illogicality of his hostility to Timur who, as a Dagestani, is a Russian citizen. Timur's characterization, too, is grounded in his cultural background. His Muslim greeting to his relatives, and the use by his family of an incomprehensible Dagestani language, authenticate an 'otherness' that threatens to alienate viewers alongside Vadim.

But universal goodwill trumps social circumstance. Vadim and Timur begin the serial as friends and end with an implied reconciliation, as *Shkola*'s embrace of a universalist rhetoric of good and evil accounts for the submersion of the 'national question' within a network of surrounding issues. Vadim's everyday racism merges with the rest of Class 9A's insubordination; Arsenii Ivanovich's resentment of Kharitonova's challenges to his fascist sympathies reflect his jealousy over her appointment as acting head. The integration of socio-political antagonism with teenage alienation sees 'mutual understanding' alleviating both the conflict threatening to tear the Russian state apart and those private hells afflicting individuals' young lives. The prominence acquired by the universalist rhetoric is facilitated by the serial format. The visceral horror of the early episodes captured an audience, but the tone was subsequently recalibrated following the intensity of their reaction.

Whose tolerance, whose patriotism?

The ambivalence of the humanist triumph points to a mutual contamination of discourses. Vadim's skinhead prejudices enter the orbit of Arsenii Ivanovich's nationalism, which in turn appropriates the vigour of the former. Meanwhile, Ania's death prompts the replacement of the official agenda of opposition to 'national disharmony' with Kharitonova's spontaneous humanism, whose proximity to the wellsprings of adolescent rebellion authenticates her liberal empathy. Ania's incipient Christian mysticism, and the school's embrace of funeral ritual, reveal the 'reverse contamination' of liberal humanism with the Orthodox piety of formal nation-building. And in Kharitonova's empathy for '*nashi deti*', the Soviet idealized myth of the Child resurfaces.[31]

But does the fact that Kharitonova's Soviet-style sermon to parents after Ania's death has no subsequent impact on the pupils suggest that official discourse has failed to 'hegemonize' the humanist message? Does the continuing resistance of adolescent subculture to assimilation by the formal pieties of the centre leave space for Arsenii Ivanovich's nationalist extremism, now expunged from the institution, to realign itself with populist xenophobia?

The centre–periphery dialectic permits multiple voices to circulate, testing out their respective capacities for imposing unity (whether that of the adolescent subculture, the cohesive state institution, the multicultural federation or the ethnically cleansed nation). It also enables conflicting discourses (official patriotic/liberal oppositional) to compete for ownership of the director's universalism.

(Mis)placing race

We have emphasized *Shkola*'s focus on the transcendence of multiple 'othernesses', identifying inter-ethnic cohesion as the privileged link in a chain which includes gender, classroom and family issues. The emphasis on Vadim's 'everyday racism' as part of generalized class rebelliousness and the attribution of Arsenii Ivanovich's nationalism to professional rivalry reinforce the 'normalization' strategy.

But the fact that Vadim's racism lacks peer support opens up a gap between strategy and realization. The gap is a corollary of the metonymy/metaphor tension. Do events represent the displacement to the subcultural periphery of phenomena occurring at the centre (social tensions pitting migrant cultures against indigenous Russians)? Or is the Timur–Vadim conflict an individualized model of a larger Manichean duel? The invocation of Shakespeare's universalist discourse of tolerance cannot resolve this tension and is at odds with the visceral shock created by *Shkola*'s naturalistic detail.

Ultimately, the serial struggles to 'place' the national question. Its deliberately engineered web presence (it was released in full by Channel 1 online, free of charge) merely projects the problem into unpatrollable virtual space, where the intergenerational divide replicates itself in paratextual form – a phenomenon fully consistent with the 'thickening' process Couldry associates with media events. On one hand, adolescent bloggers debate issues relating to their age group, emulating the language and mindsets of their fictitious heroes in a symbiosis facilitated by a shared subcultural idiom and treating the characters as real acquaintances:

> Frigging hell! (*Blin*!) Why do we have to wait so long for the next episode? What a nightmaaare! (*Kaaashmaaar*!)[32]

> Friiiging hell! (*Bliiiin*!) She's really driven me crazy (*zabesila*), your Budilova. She gets pregnant and runs straight away to Pif [a male character in the serial] and straight away she needs him. What's all that about? … And that geography teacher – he's a rat.[33]

On the other hand, a much smaller number of adult commentators recognize the inter-ethnic problem as 'the most dangerous'.[34] Thus, an anonymous online interviewer asks Gai-Germanika whether the inclusion of words such as 'black-arsed' was really necessary and suggests that 'The nationalist theme in this film will shock society much more powerfully than what the youngsters are on about now: sex and so on'.[35]

Channel 1's answer to the question of where to 'place' the national question is to 'displace' it into a blogosphere of communities (liberal–oppositional; adolescent–subcultural; traditional–Soviet leaning) whose mutual disjunction mirrors the communicational malfunction targeted by the serial. The reason for the displacement is partly the political constraints placed upon the broadcaster, and partly the lack of a framework within which controversies might be aired. These issues came together in an exchange between Aleksei Venediktov, chief editor of the Moscow Echo radio station, and Irina Iarovaia, a United Russia deputy, which took place during an edition of Channel 1's *Sudite sami* discussion programme:

> Iarovaia: There should be no censorship. Today it is important to talk about the social contract (*obschestvennom dogovore*). If we speak of unified value orientations (*edinykh tsennostnykh orientirakh*), then they must be uniform for the mass media, too, and for schools. Whereas we have this strange situation in which the mass media can be immoral.
>
> Venediktov: So you are suggesting introducing uniform opinion (*edinomyslie*) in Russia?
>
> Iarovaia. No, a uniform culture.
>
> Venediktov: We have a very diverse culture in our country, a multi-faith society. There is no uniform culture.
>
> Iarovaia: Russian culture – we have that in common. If you've got pro-American values, then you'd better tell us.[36]

Despite this chasm and the sharp disagreements that it reflects, *Shkola* represents a gesture towards a centre–periphery dialectic through which a more stable consensus might be negotiated. The chastening dénouement and the attenuation of socio-political conflict as 'universal (mis)understanding' contribute to the wider mission of managing the infiltration of extremist rhetoric into the mainstream. That this comes at the expense of permitting a powerful liberal critique of extremism's presence within both youth subculture and official patriotism is part of the 'hegemonic bargain'. *Shkola* epitomizes the potential, but also the considerable dangers, that the serial harbours for Russian television's nation-builders – the reason why, under the conditions we describe in the next part of our book, the prospects for a second series faded so rapidly.

Modalizing prejudice

Three contexts

What, however, of the outer limits of our broadcasting spectrum: those channels, and genres, whose relative distance from the centre of power appears to minimize the nation-building prerogative? Let us turn to our final fictional format – the comedy sketch series – and to a broadcaster at the peripheries of the mainstream. *Nasha Russia* began showing on TNT in 2006. Five series have been made (all available at TNT's website).[37] The show was placed 36th (repeat series) and 48th (current series) in Russian television's top 100 rankings for 2010–11, according to rating data from late November 2013.[38] In 2010 a feature-length film, directed by Gleb Orlov and suggestively entitled *Nasha Russia: Balls of Fate (Iaitsa sud'by)*, was released; it briefly became Russia's biggest box office hit for 2010.[39] *Nasha Russia* (henceforth NR) is made by TNT (Tvoe novoe televidenie). Although one of Russia's top five networks and owned by Gazprom-Media (part of the Putin establishment), it is aimed at a youth market.[40] In contrast to Channel 1 which, in turning to Gai-Germanika, was self-consciously mining the cultural peripheries for fresh material, TNT must perpetually accommodate subcultural discourses at odds with official rhetoric. And unlike *Shkola*, which filters profanity-laden, carnivalesque youth humour through the 'adult' vision of an art-house director, NR aligns itself more closely with its source.

Also important is the fact that NR was adapted from the BBC1 series *Little Britain* (2003–6), whose provocative portrayal of issues of class, gender, sexuality and race generated debate as to whether it had spurned the 'achievements' of political correctness or whether its 'knowing' attitude exonerated it from this heresy.[41] The adoption of *Little Britain* catchphrases like Vicky Pollard's 'Yeah but no but yeah but …' in media assaults on illiterate 'chav' culture certainly indicates a reinforcement rather than a subversion of popular prejudice. 'Political correctness' has yet to establish a firm foothold in Russia, of course, and indeed, judging by the homophobic hysteria and Orthodox-led campaign for 'traditional values' currently sweeping the country, seems as distant as ever.[42]

However, a third factor to consider is that NR's creators originate from the post-1991 version of the Soviet-era show *KVN* (*Klub veselykh i nakhodchivykh*), which achieved cult status among students (the show spawned a nationwide competition in which teams from every Soviet republic participated in 'leagues', outdoing one another in socio-politically biting skits judged by panels of 'experts'). Dating back as far as 1961, KVN was banned under Brezhnev but revived in 1986. It continues to enjoy success among students throughout the former USSR and an annual 'final' is shown on Channel 1, though its satire has recently been diluted.[43] NR is scripted by former KVN player Semion Slepakov and Armenian producer Garik Martirosian, both actors in the series, along with Mikhail Galustian (also Armenian). Although closer to the cultural mainstream than Gai-Germanika, KVN's lingering dissident ambiances and pan-Soviet reach pervade NR.

Cultural (mis)translation

NR begins with a narrated introduction highlighting the quirks of the nation it lampoons. While the equivalent *Little Britain* narration is read by Tom Baker, famous for his appearances in the cult science fiction series *Dr Who*, and drifts into surreal flights of hyperbolic nonsense, NR's anonymous narration hones in on the stupidities of Russian everyday life. It begins with the statement 'We live in the most wonderful country in the world and all other countries envy us', followed by a series of facts contravening that statement, for example:

> It is in our country that women drink beer through straws in order to look like ladies; only our girls put cosmetics on before going swimming; it is we who have so much oil that nobody can possibly steal it all – even us!

The narration always ends with the sentence: 'We call our country 'Rossiia', whilst foreigners enviously refer to us as 'Russia', but still it is ours – *nasha* Russia.'

Like *Little Britain*, NR consists of unrelated sketches featuring recurring comic heroes, each exposing different aspects of Russian reality. The use of men in drag is also borrowed from the British series, as is the grotesque style and the (locally adapted) focus on class, gender, ethnicity and sexuality. The homosexual character is a foreman in a Cheliabinsk factory rather than an effeminate frequenter of a Welsh village pub. Class surfaces in the depiction of two *bomzhi* (tramps) bedecked with designer suits, smartphones and laptops acquired from the refuse bins at Rublevka, favoured residence of Russia's new bourgeoisie. Gender relations are satirized in the interactions between an uncouth male from Taganrog and his television screen and overweight wife. Certain themes are absent from *Little Britain*: corruption (a Voronezh teacher constantly tricks her students into giving her money; two Duma deputies use their privileged positions to plot their own advancement) and the iniquities of private healthcare (a state-funded patient is repeatedly insulted by her consultant, who fawns upon her private counterpart).

The 'stars', however, are Ravshan and Djumshud, two *Gastarbeiter* workers whose widespread notoriety has, as Roland Oliphant suggests, established them in the Russian cultural imagination as 'archetypes of the guest worker'.[44] They were assumed to be Tajik – the nationality which, because its members constitute the overwhelming majority of migrant workers, many Russians associate with a generically conceived Central Asia – but, in response to Tajik protests that NR represented the 'cultural genocide of the Tajik people', TNT argued that the *gastarbeiters*' ethnicity is never specified; in the feature film, they are from the fictitious 'Chuchukistan'.[45] The sketches portray them carrying out building work in, alternatively, a Moscow *novostroika*, the mansion of media celebrity Kseniia Sobchak and the site of the 2014 Sochi Winter Olympics. One of them speaks broken Russian and the other has no Russian at all. Their stupidity and their disastrous inability to follow instructions frustrate their Russian supervisor.

Later series introduce Zhorik Vartanov, the presenter of Sev-Kav TV, a fictitious television channel based in Piatigorsk. He forever harangues his interfering cameraman in a heavy North Caucasus accent for restricting his freedom to promote his own concerns. In one sketch (episode 85), Zhorik's interview with a local police chief about a cannabis haul descends into a self-interested effort to discover how cannabis crops can be hidden from the authorities. Elsewhere (episode 98), he reports on a local beauty competition, pressuring the organizer to favour his sister in the event, and uses a feature on local food to advertise his family restaurant (episode 86). An interview with a plastic surgeon (episode 96) ends with Vartanov enquiring about a penis enlargement 'for a friend'. Vartanov epitomizes images of the North Caucasian as hot-tempered, venal and over-sexed. His ethnicity is never specified, enhancing his stereotyping function.

Ravshan and Djumshud, however, have the monopoly on criticism from migrant communities. Russians, too, lamented the depressingly negative picture of their nation conveyed by NR.[46] The associated feature film aggravated the controversy. As well as depicting the *gastarbeiters* as incompetent, it adds a crude sexual dimension: the plot centres on the disappearance of two golden balls (the testicles of Genghis Khan) from their current owner – a scheming oligarch who owes his fortune to their magic properties. The dénouement turns on the revelation that there exists a third golden ball which Ravshan, a descendant of the Khan (a fact he proves by displaying his anatomical affinity with his ancestor), wears around his neck. Ravshan is eventually reunited with all three balls and the entire population of Moscow *gastarbeiters* gather around him, celebrating the fortune the balls will bring them.

NR's British progenitor was no stranger to scandal. However, it featured only two non-white characters – Desiree DeVere (David Walliams), a black, obese former Miss Botswana prone to displaying herself naked, and Ting-Tong Macadangdang (Matt Lucas), a scheming mail-order Thai bride. The humour is hyper-grotesque and there is little connection between the surreal excesses displayed by the characters and the behavioural stereotypes associated with their ethnic groups. By contrast, ethnicity defines NR's place in Russian cultural discourse, and shapes our discussion of the programme's complex relationship with prejudice.

Frames as modalizers

The title *Little Britain* conflates the imperialistic 'Great Britain' with the anti-nationalist taunt 'Little Englanders' (it is also the name of a street in the City of London). NR refers not to Russia, but to (mis)perceptions of Russia common to the English-speaking 'other'. Yet the title is '*nasha*' Russia. In Russian, the concept of 'our guys' (*nashi*) connotes a strong sense of national self-identification, as reflected in the name of the pro-Putin youth movement. In *Nasha Russia*, the perspective of the anti-western patriot is combined with that of the naïve western admirer. Yet the 'Russia' which foreigners 'enviously' imagine is not that depicted in the succeeding sketches, but one inflected with the mordant voice of the KVN player. National pride is mischievously deflated rather than triumphantly reconfirmed. The title and the multi-voiced introductory sequence infiltrate the sketches, distorting their surface meanings. Each edition is likewise framed at the end with a patriotic rap song peppered, incongruously, with anglicisms and with a chiastic refrain whose grandiose claims have been punctured by the preceding skits: *Strashnaia sila – Nasha Rossiia; Nasha Rossiia – strashnaia sila*' ('An awesome power is our Russia; Our Russia is an awesome power').

Anti-migrant sentiments pervade Russian nationalism, as slogans such as 'Russia for the Russians' confirm. Thus, recalling Todorov's and Ducrot's distinction between 'literal content' and 'attitudinal stance', NR's frames *modalize* the anti-migrant prejudice projected within the comic sketches. If Russian patriotism is open to ridicule, then so is the xenophobia central to its construction.

Intratextual modalizers

The framing effect is reinforced by the narrative dynamics of the sketches themselves, and by the intertextual linkages traversing them. Ravshan and Djumshud form part of a trio which also includes Leonid, their foreman, whom Ravshan addresses as '*nasial'nika*' (a corrupted version of '*nachal'nik*'). Leonid, a bigot who despises his subordinates, persists in exploiting them. The three settings between which the sketches alternate underscore societal hypocrisy on a larger scale: the exploitative use of illegal immigrant labour by the privileged, and by the state itself, undermines the anti-migrant discourse to which both are prone.

In one sketch (episode 62), Leonid arrives back at the luxury flat whose repair he is overseeing drunk, following Russia's 'Day of the Airborne Troops', a key date in the Russian patriot's calendar. He creates havoc, inflicting damage on the flat, uttering profanities and shouting at the *gastarbeitery* that they do not belong in *his* country. When he emerges from his drunken stupor he blames his hapless workers for the damage. An earlier sketch (episode 51) revolves around a Commission of Human Rights visit to a building site. Leonid feigns that he is looking after his migrant workers, but beats them once the inspectors have left. In one scene, the inspectors make a surprise return, catching Leonid in the act of manhandling Ravshan. Thinking on his feet, he tells the inspection team that he is merely administering a massage.

The migrants' incompetence is so grossly overstated as to represent a double-voiced commentary on the inevitable hyperbole at the heart of stereotyping itself. Examples include constructing three toilets on Olympics-style podia in the Sochi complex then defecating in one before the flush system has been installed (episode 49), and accidentally setting fire to an intricate model of the Olympic village (episode 46). The *gastarbeitery* gabble to one another using made-up words; the absurdity of rhymed pairings like 'Schengel'm-Bengel'm' would not be lost on a Russian ear. When speaking Russian, Ravshan merely repeats in a high-pitched voice what Leonid asks him, likewise conforming to a classical stereotype of the ignorant foreigner so obvious as to be immediately identifiable.

The migrant is constructed from Leonid's openly bigoted point of view, a position which audiences are encouraged both to identify with (the 'canned laughter' is inserted when the *gastarbeitery* are at are their most inept) and distance themselves from (Leonid is frequently himself ridiculed), and so to experience their own susceptibility to ethnic stereotyping. The migrants' perceived stupidity is further undercut by learned philosophical conversations between them (in the 'Airborne Troops' sketch they quote Cicero to one another in their own 'language' as the inebriated Leonid is temporarily unable to focalize viewer perceptions). Confirmation of their dual construction comes in the film, which ends with the revelation that the action has been simply a Hollywood-style blockbuster viewed on a provincial Russian television screen.

The intertextual persona

NR's self-reflexivity extends to the roles of its key actors: the Armenian Mikhail Galustian (Ravshan) and the Russian Sergei Svetlakov (Leonid). Both have strong KVN connections (Galustian was a KVN 'champion' in 2003) and both contributed to the politically controversial Channel 1 television comedy talk show *ProzhektorPerisHilton*, a KVN offshoot.[47] Galustian's Armenian identity is a strong component of his stage persona and Galustian is responsible for the Armenian ambiance of much of the show's humour, which can be traced to the absurdist 'Radio Erevan' jokes.

The intertextual linkages are reinforced by the shared audience profile of *ProzhektorPerisHilton*, KVN and NR, and by the NR fanclub's website.[48] In the *Gastarbeiter* sketches, Galustian's biographical background is important. The introductory voiceover describes the two 'heroes' as 'our guests from Central Asia'. Combined with his self-evidently generic language, Galustian's prominent Armenian (non-Tajik) persona exposes the reductionist tendencies informing anti-migrant prejudice.

References to the broader cultural environment are complemented by intertextual effects reliant on ironic juxtapositions within the series. Galustian embodies overblown North Caucasian machismo as well as the child-like naiveté of the Central Asian migrant. But he also adopts the rough-edged proletarian persona of Mikhalich, the unsuspecting 'victim' of the homosexual advances of Dulin (Svetlakov). Svetlakov plays the corrupt Voronezh schoolteacher Snezhana Denisovna, as well as the Taganrog couch potato Sergei Beliakov and the discriminatory hospital consultant Vadim Rol'fovich. The prejudicial stereotypes cancel one another out; Svetlakov's camp Ivan undermines his migrant-taunting Leonid. Galustian's comical North Caucasian bravado contradicts the deference of the simpering migrant. Both are negated by Galustian's quintessentially Russian homophobe. His Armenian-ness, meanwhile, places all three ethnic types in quotation marks.

The criss-crossing meanings shape the overarching image of Russian identity projected by NR. In one of the Cheliabinsk sketches (episode 56), Mikhalich dreams that he is a Tolstoyan hero, scything in a mythical Russian field and flirting with a peasant girl in national costume; as the girl (Svetlakov) gently caresses him, he wakes with horror to find that she has mutated into an all-too-real Dulin, whose homosexual advances besmirch the patriotic idyll.

Performative community-building

Galustian's and Svetlakov's performances are a theatrical tour de force: as they slide in and out of their various skins, they underscore less the attitudes of their different personae than their own comic bravura. The irreconcilable differences between a homophobic *muzhik*, an impassioned North Caucasian television presenter and a diminutive migrant are transcended by the virtuoso talents of a single Armenian comedian. This 'equalization' effect contradicts both the misappropriations of the *Gastarbeiter* sketches by Russian xenophobes and the knee-jerk reaction of the Tajik authorities.

The annulment of difference and the consequent undermining of stereotypes at the level of 'character' are motivated by the KVN-inspired collaboration at the level of 'performance'. Despite its 'dissident' heritage, KVN now fulfils a nation-binding function (Putin occasionally attends performances). Karin Sarsenov argues: 'The show creates a communicative space in which people share the same jokes in spite of national and ethnic differences. Those very differences are subjected to kind-hearted ridicule, taking the edge of existing conflicts.'[49] NR's inter-ethnic inspiration subverts current nationalist mythologies by re-invoking KVN's Soviet-era community-building.

Installed at the heart of the UK's flagship broadcaster, *Little Britain* also ultimately projects a positive image of the nation it mocks, revitalizing the offbeat excesses of British comic tradition, performing the epiphany which occurs when some conclude that, with inter-ethnic relations in post-PC Britain now purportedly 'harmonized', a scabrous sketch about a Thai bride is as 'inoffensive' as a good-humoured Irish joke. From the peripheries of the Russian mediascape, NR, by contrast, fails to close off the ambivalent spaces for nationalist bigotry and for anti-xenophobic indignation that its double-voiced sketches open up. Thrust into pre-PC conditions, NR's imported, post-PC format cannot claim the transcendent position to which both *Little Britain* and Gai-Germanika's liberal universalism aspire. KVN's community-binding legacy subsists not at a level *above* the discourses and counterdiscourses targeted by NR, but alongside and in competition with them. More subversively liberal in intent than *Little Britain* or *Shkola*, NR runs a greater risk than either of them of mobilizing rather than modalizing prejudice.

The programmes we have analysed are, in their different ways, all party to the nation-building mission that state-aligned television plays in Russia as elsewhere (the Rossiia dramas by re-grounding Russian identity in popular images of 'gypsiness'; *Shkola* by attempting to dissolve inter-ethnic difference in a transcendent universalism; NR by 're-enacting' the pan-Soviet community). They each, too, complicate stereotypical representations of ethnic minorities; the Rossiia romances deploy the 'threshold' figure of the gypsy to mediate and thus attenuate ethnic 'otherness'; *Shkola* accommodates the North Caucasian object of nationalist bigotry within an all-embracing humanism; NR self-consciously hyperbolizes, and thus modalizes, stereotypes.

The programmes share an impulse to subject ethnonational and racist prejudice to critique. Both *Shkola* and the gypsy dramas expose bigotry by aligning the viewing subject with its target. Conversely, NR places the viewer in the bigot's position but frames it from without. *Shkola* moderates the xenophobic viewpoint by locating its roots in an alienation endemic to youth subculture, and in a resultant susceptibility to deleterious 'adult' influence. All the genres exploit their fictional status as a licence to explore and exceed the limits of issues as treated in the mainstream news media. As part of that process they situate themselves at the nexus of a circulation of discourses emanating from various points on the official–unofficial spectrum. Their genre specificities account for the different ways in which they negotiate that position. The gypsy drama makes available narrative devices which facilitate the transgression of established boundaries between romantic ideals and familiar news representations of the Roma community. The protracted form of the *telenovela* brings hybrid discourses into conflict but, via its 'provocation–pacification' dynamic, also manages audience resentments arising from the encounter. NR's comic sketch format reframes common prejudices and patriotic truisms from the grotesque viewpoint of subversive student subculture.

Subcultural space demonstrably provides tools of renewal as well as sources of threat. Moreover, it is emerging as the site of an as yet inchoate renegotiation of the relationship between majority and minority ethnic communities. That process is uncertain. As the controversies surrounding NR and *Shkola* demonstrate, discursive strategies are open to misappropriation (by nationalist extremists; by hardline anti-westerners) and to misreading (the Tajik response to Ravshan and Djumshud). Nowhere is this truer than in Russia's heterogeneous social reality, which lacks a stable, consensual source of authority.

Indeed, the deeply unconventional Gai-Germanika's plea for a consensual acceptance of radical difference in all its forms, though contaminated by nostalgia for an imagined version of Soviet brotherhood, represents the mirror image of the exclusionary version of unity we will find later in our analysis of coverage of the Pussy Riot scandal. This exclusionary approach and the earlier latitude shown by Channel 1 (and Putin) to Gai-Germanika highlight state television's (and an increasingly desperate ruling elite's) fractured and contradictory response to its acute awareness that it has failed to speak to and for a growing segment of Russia's younger generation: it 'indulges' that segment through risky ventures such as *Shkola*, while at the same time giving the likes of Irina Iarovaia a platform from which to demonize it as treacherous and degenerate.

One might surmise that, by the time of Putin's third presidency, the latter strategy had prevailed. Nonetheless, the very same period saw the habitually populist NTV broadcast a bold example of 'reality TV' in a genre different from but complementary to that of *Shkola*: a spoof documentary mocking NTV's shock-revelatory style and making the patently absurd claim that scientists have discovered a 'fascist gene' accounting for Russia's nationalist movement.[50] Meanwhile, at the peripheries of the Russian media sphere, TNT continued to host the equally subversive comedy *Nasha Russia*. Still more striking is TNT's launch in 2014, at the height of the Ukraine crisis, of a new sitcom mocking xenophobic prejudice and featuring a stable, happy family consisting of a Russian woman, a North Caucasian man and their two children. With its ironic, provocative title *Druzhba narodov*, the sitcom defiantly contradicted the Russocentric, anti-migrant sentiment which, as we will see in Part III, was to sweep across the federal channels after Putin's re-election as president in 2012.[51]

The fact that, as these phenomena indicate – and despite the developments broached at the end of Part I and in Part III – all is still to play for in Russia's internal hegemonic struggle is part of the legacy of *Shkola*, *Nasha Russia* and, in a different sense, the gypsy dramas, with their complex mediations of identity and alterity. The controversies generated by some of these programmes, their mutations into full-scale media events of the endogenous type, perhaps after all marked the birth pangs of a public sphere that is both tolerant and productive of diversity of all kinds, albeit one enduring a dispiritingly prolonged interruption of its development. If the interruption is to be temporary, and given the abject failure of the DNU project, the role of fictional genres located at the margins of official media discourse in challenging and re-working that discourse's worst excesses can only grow in importance.

Notes

1 Dayan Thussu, *News as Entertainment: The Rise of Global Infotainment*, London: Sage, 2007.
2 John Ellis, 'Television as Working-through', in Jostein Gripsrud (ed.), *Television and Common Knowledge*, London: Routledge, 1999, pp. 55–71.
3 For the narrative dimension to news, see Karen S. Johnson-Cartee, *News Narratives and News Framing: Constructing Political Reality*, Oxford: Rowman and Littlefield, 2005, pp. 155–61.
4 Jesus Martin-Barbero, *Communication, Culture and Hegemony: From the Media to Mediations*, London: Sage, 1993, p. 99.
5 See Oswald Ducrot and Tzvetan Todorov, *Encyclopaedic Dictionary of the Science of Language*, Baltimore: John Hopkins University Press, 1979, p. 313.
6 The 2002 Russian census showed that there were 180,000 gypsies in Russia. See 'Gypsy Culture in Russia', 26 August 2005. Available online at http://en.rian.ru/analysis/20050826/41232225.html (accessed 27 March 2008). Between July 2010 and October 2011 there were only two stories about gypsies on *Vremia*, even though this period coincided with the European Roma expulsion crisis and even though Channel 1 normally has a more international outlook than Rossiia, whose *Vesti* dedicated 30 reports to the crisis, but only one to domestic ramifications. For detailed analysis of Russian television news depictions of gypsies see Stephen Hutchings, 'The Gypsy as Vanishing

Mediator in Russian Television Coverage of Interethnic Tension', *Nationalities Papers*, 41/5, 2013, 1–17.

7 The most comprehensive treatment of Russian gypsy mythology's roots in Pushkin is Alaina Lemon, *Between Two Fires: Gypsy Performance and Romani Memory from Pushkin to Post-Socialism*, Durham: Duke University Press, 2000.

8 Susan Layton, *Russian Literature and Empire: Conquest of the Caucasus from Pushkin to Tolstoy*, New York: Cambridge University Press, 1994.

9 Toby Sonneman, 'Dark Mysterious Wanderers: The Migrating Metaphor of the Gypsy', *The Journal of Popular Culture*, 32/4, 1999, 119–39.

10 For Martin-Barbero, mediation goes beyond the media themselves: 'We are placing the media in the field of mediations, that is in a process of cultural transformation which does not start with or flow from the media but in which they play an important role' (*Communication, Culture and Hegemony*, p. 139).

11 For the gypsy as outsider in European culture, see Angus Bancroft, *Roma and Gypsy Travellers in Europe: Modernity, Race, Space and Exclusion*, Aldershot: Ashgate Publishing, 2005.

12 Alaina Lemon argues that 'the Russian romance with gypsies has no equivalent force in most other countries' (*Between Two Fires*, pp. 31–2).

13 Martin-Barbero terms this 'the real mediation, the function of a medium which mass culture fulfills day by day: the communication between the real and the imaginary' (*Communication, Culture and Hegemony,* p. 56).

14 For the gypsy theme in Soviet cinema, see Stephen Hutchings, *Screening Intercultural Dialogue: Russia and its Other(s) on Film*, Basingstoke: Palgrave Macmillan, 2008, pp. 14–15.

15 Available online at www.kino-teatr.ru/kino/movie/ros/89276/content/ (accessed 5 April 2012).

16 Susan Haywood defines 'suture' as 'the effect of certain filmic codes that stitched the spectator into the film text': *Cinema Studies: The Key Concepts*, London: Routledge, 2000, p. 357.

17 Judith Kornblatt, *The Cossack Hero in Russian Literature: A Study in Cultural Mythology*, Madison: University of Wisconsin Press, 1992.

18 Gerard Genette defines 'focalization' as the way that third-person narrative orients itself towards the perspectives of its characters in order to reveal different levels of 'omniscience' regarding plot outcomes. See *Narrative Discourse: An Essay in Method*, Ithaca: Cornell University Press, 1983.

19 Hayden White, 'The Value of Narrativity in the Representation of Reality', in W. J. T. Mitchell (ed.), *On Narrative*, Chicago: Chicago University Press, 1981, pp. 1–23.

20 Martin-Barbero, *Communication, Culture and Hegemony,* p. 130.

21 Ibid., p. 137.

22 'The Serial *Shkola*: Provocation or Truth to Life?', 15 January 2010. Available online at www.aif.ru/culture/article/32072 (accessed 5 April 2012).

23 A. Belitskii, 'The Serial *Shkola* Has Outraged a Duma Deputy', 13 January 2010. Available online at http://vesti.kz/media/36867/ (accessed 5 April 2012).

24 Ibid.

25 'Putin o seriale "Shkola": ne nado delat' obobshchaiushchie vyvody', 25 January 2010. Available online at www.rian.ru/culture/20100125/206194022.html (accessed 5 April 2012).

26 The programme, entitled 'School – Is it Life?', was broadcast on 14 January 2010.

27 Arina Borodina, 'Serial "Shkola": ot massovogo zritelia do prezidenta strany', *Kommersant,* 8/4308, 2010. Available online at /www.kommersant.ru/doc.aspx?DocsID=1306937 (accessed 5 April 2012).

28 Available online at www.shkola-serial.ru/ (accessed 5 April 2012).

29 S. Grachev, 'Valeriia Gai-Germanika: "Moi fil'm ne o shkole! Ia snimaiu kino o chelovecheskikh otnosheniiakh"', 20 January 2010. Available online at www.aif.ru/culture/article/32144 (accessed 5 April 2012).
30 Further references to the serial itself include the episode number in parentheses. The official online version of the programme is available at www.shkola-serial.ru/smotret-serial-shkola-online.html (accessed 5 April 2012).
31 For the myth of the child in Soviet culture see Catriona Kelly, *Children's World: Growing Up in Russia, 1890–1891*, London and New Haven: Yale University Press, 2007.
32 Available online at www.shkola-serial.ru/smotret-serial-shkola-online/smotret-online-serial-shkola-besplatno.-56-seriya.html (accessed 5 April 2012).
33 Available online at www.shkola-serial.ru/smotret-serial-shkola-online/smotret-online-serial-shkola-besplatno.-47-seriya.html (accessed 5 April 2012).
34 Ibid.
35 Valeriia Gai-Germanika, 'Moi fil'm lish' zhalkoe otrazhenie togo, chto proiskhodit', 23 January 2010. Available online at http://yarportal.ru/topic58303s285.html (accessed 5 April 2012).
36 *Sudite sami* (Channel 1), 14 January 2010. Available online at www.1tv.ru/prj/sudsami/vypusk/2813 (accessed 5 April 2012).
37 Available online at http://nasha-russia.ru/ (accessed 5 April 2012).
38 Available online at http://allcharts.org/tv/russia/rating.htm (accessed 14 April 2014).
39 '"Nasha Russia: Iaitsa sud'by" stala liderom otechestvennogo prokata', 26 January 2010. Available online at www.bfm.ru/news/2010/01/26/nasha-russia-jajtsa-sudby-stala-liderom-otechestvennogo-prokata-v-minuvshij-uik-end.html (accessed 5 April 2012).
40 The Gazprom-Media website describes TNT as 'the entertainment channel of the young generation'. Available online at www.gazprom-media.com/en/tv.xml?&company_id=49 (accessed 5 April 2012).
41 Barbara Ellen attacked the show for its condescending mockery of the working class: *The Guardian*, 12 December 2004. Peter Tatchell praised its 'irreverence' but added that its caricaturing of gay people reinforces old-fashioned clichés. Quoted in Sharon Lockyer (ed.), *Reading Little Britain: Comedy Matters on British Television*, London: I.B. Tauris, 2010, p. 12.
42 *Nasha Russia* has no licence from the BBC. For a full scholarly analysis of *Little Britain*, see Lockyer (ed.), *Reading Little Britain.*
43 For an analysis of KVN, see John Dunn, 'Humour and Satire on Post-Soviet Russian Television', in Lesley Milne (ed.), *Reflective Laughter: Aspects of Humour in Russian Culture*, London: Anthem Press, 2004, pp. 181–92.
44 Roland Oliphant, 'Whose Russia', 24 March 2010. Available online at http://russiaprofile.org/international/a1269451394.html (accessed 5 April 2012).
45 For Tajikistan's protest on behalf of its migrant workers in Russia, see 'Etnicheskaia chistka: krizis zhanra', in *Kommersant,* 14/868, 2010. Available online at www.kommersant.ru/doc/1345030. For TNT's rebuttal, see '"Tadzhikskie trudovye migranty" osudeli fil'm "Iaitsa sud'by"', *Kommersant Online*. Available online at www.kommersant.ru/doc/1341516 (both accessed 5 April 2012).
46 Boris Nevzorov, a local government official, complains: 'It is offensive when they say nasty things about us abroad and we can't respond. And we show ourselves to be such idiots in *Nasha Russia.*' See 'A vam za chto stydno?', *Kommersant Vlast'*, 37/891, 2010. Available online at www.kommersant.ru/doc/1506984 (accessed 4 April 2012).
47 The show premiered in 2008 and airs on a Saturday evening on Channel 1.
48 The website features biographies of both Svetlakov and Galustian, whose KVN past is extensively referenced. Available online at http://fanparty.ru/fanclubs/nasha-russia/forum (accessed 5 April 2012).

49 Karin Sarsenov, 'Televising Soviet Tropes: Reforging a Supraethnic Cultural Identity', *Slavica Lundensia*, 24, 2009, 257–77 (264).

50 The programme, *Total Eclipse* (*Polnoe zatmenie*), was the brainchild of controversial journalist Andrei Loshak.

51 The complete series is available online at http://drujba-narodov.tnt-online.ru/s01e01.

6 Transcending marginality

Ethnicity, identity and religion on *Vesti-Buriatiia*

We now switch our attention from the peripheries of the post-Soviet Russian 'mediasphere' to the geo-political margins of the Russian Federation itself. We simultaneously make a second, related shift for, until now, we have focused on federal television channel coverage of ethnic minorities as objects of representation; even when, as in Chapters 4 and 5, characters from minority cultures speak in their own name within news reports, talk shows or television dramas, their subject positions are framed from an ethnically Russian perspective, and oriented towards a primarily Russian audience. Since Soviet times, however, the state has claimed to empower non-Russian minorities by giving them a voice and a space for self-expression. Ethnic republics of the USSR and of the Russian Federation have been viewed as national homelands of the 'titular nationalities' after which they were named. Within these ethnic autonomies local media have been among the prime outlets allowing minorities to act as subjects of representation. Opportunities for local self-expression were, of course, severally limited in the Soviet Union. After a period of unusual freedom in the 1990s, Putin's government again strengthened 'the vertical of power', imposing restrictions on the ability of Russia's ethnic autonomies to challenge the centre. However, the media in the republics and regions of the Russian Federation have their own distinct profiles, which are shaped not only by current local political realities, but also by the historical specificities of the areas and by their own relations with the outside world. This chapter analyses one such example from the Siberian republic of Buriatiia, on the shores of Lake Baikal. We argue that it furnishes a model for negotiating the relationships between unity and difference, and between centre and periphery, that is both surprising and enlightening for our appreciation of the broader issues of representation and nation-building with which the book engages.

Vesti-Buriatiia, our focus here, is the main news programme of the republic of Buriatiia's state television and radio company, and the only local news provider whose potential audience is the entire population of this Siberian republic.[1] The company broadcasts its programmes on the federal state channel Rossiia, and *Vesti-Buriatiia* technically has the status of a regional news programme, subordinate to the federal *Vesti*.[2] Yet Buriatiia state television, despite being a regional affiliate of the Pan-Russian State Television and Radio Company, seems to have considerable autonomy in selecting topics and has developed its own style of coverage.[3] Most programme reporters are ethnic Buriats and they consciously use television to shape a particular image of the Buriats as a national community. Buddhism, which is virtually ignored by the federal channels, is identified in the dominant public discourse in the republic as the national religion of the Buriat people and receives much attention from *Vesti-Buriatiia*. Whereas the coverage of Orthodox Christianity and Islam on Russian television has attracted significant scholarly attention, little research outside academic circles in Buriatiia itself has been conducted on media representations of Buddhism, the issue which we address here.[4]

There are further reasons for our interest in *Vesti-Buriatiia*. These are connected with the fact that Buriatiia is both typical and exceptional. Its typicality lies in the fact that the titular nationality after which it is named constitutes a minority of the local population, as is the case in more than half of Russia's 21 ethnic republics. This demographic profile complicates the task of local journalists who are required both to represent the republic as the national homeland of its titular nationality and to take into account their predominantly ethnic Russian audience. The way in which the tension is handled provides us with the distinctive perspective on the relationship between unity and difference that is one of the two guiding themes in this chapter.

Official pan-Russian discourse also represents Buriatiia as exceptional. The republic is seen by the federal centre as a model for the rest of the country in terms of the region's achievement of societal, including ethnic, cohesion.[5] Indeed, the level of conflict, whether between people of different ethnicity or among local elites, is relatively low. This renders the republican broadcasters' task somewhat different from that of their federal counterparts, who regularly have to negotiate more complex situations. It also means that Moscow leaves the local leadership to manage its own affairs, including those of the media. As one of our interviewees explained: '[This distinctiveness is permitted] because no one [in Moscow] cares about Buriatiia.'[6] The way in which *Vesti-Buriatiia* covers ethnicity-related issues will therefore furnish an instructive comparison with the approach of the federal channels. Finally, it is important to note that Buriat elites imagine their nation in relation not only to Russia, but also to Mongolian and Buddhist communities beyond Russia's borders. *Vesti-Buriatiia* thus offers a crucial perspective on the local/global axis as well as the local/federal one. Together, these two angles generate the centre–periphery theme that is our second overarching concern in this chapter.

In our monitoring period (September 2010–August 2012), *Vesti-Buriatiia* featured numerous news reports on the start of the harvest season, the production of milk by local collective farms, the testing of heating systems before winter and even the local marking of anniversaries of the Komsomol and Young Pioneers communist youth organizations, which *Vesti* did not cover.[7] Routine speeches and visits by republican political leaders open most daily bulletins, and are covered with even greater frequency and detail than that accorded by *Vesti* to similar events at the federal level. The tone of the news is overwhelmingly positive, with a palpable feeling of local pride running through most stories and found also in the general youthful and enthusiastic appearance of the reporters themselves. If social or economic problems in the republic are noted, it is claimed that they are being addressed by the authorities. What is perceived as news by *Vesti-Buriatiia*, and the very reporting format, at first glance seems therefore to have more in common with the Soviet period, in which predictable, ritualized coverage reaffirming the 'normal' state of affairs predominated, than with what we find on *Vesti*, where this traditional approach is constantly challenged by the perceived necessity to conform to global media formats in reporting extraordinary, scandalous and otherwise sensationalized developments. Most of the *Vesti-Buriatiia* broadcasting is in Russian, which reflects the linguistic profile of the republic.[8] Twice a week, in ten-minute slots, selected reports from *Vesti-Buriatiia* – mostly on cultural issues – are re-broadcast in the Buriat language.

In the monitoring period we found no reports of any criticism of the Kremlin. In contrast to its federal equivalent, *Vesti-Buriatiia* told its viewers nothing about public protests held in Moscow and across Russia between December 2011 and the summer of 2012.[9] The first obvious explanations for this are to be found in the tight control exercised over *Vesti-Buriatiia* by the republic's political leadership, whose position is shaped by Buriatiia's dependency on subsidies from Moscow, and in elite and broadcaster perceptions of the channel's audience, which clearly differs from that of the federal channels; in an economically poor republic with a harsh climate, where 40 per cent of the population still lives in the countryside (as opposed to the national average of 26 per cent), stories about harvests, milk production and preparation for winter which would not be deemed newsworthy by Muscovites are likely to be topical for many people. What the channel's stance in relation to Moscow also points to is a paradox by which its very distance from the federal centre renders it a more reliable proponent of that centre's official values. Similarly, the Soviet period, when centripetalism was at its height, is still viewed with nostalgia by many in the region.[10] As one of our local interviewees put it: 'For Buriatiia the Soviet period was fortunate – industrialization, modernization, the formation of the Buriat nation...'[11] Within this context, the Buriatiia Television and Radio Company's virtually unmitigated glorification, on its website, of the establishment of television broadcasting on the republic's territory in the Soviet period comes as no surprise.[12]

Post-Soviet trends are, of course, also prevalent. For instance, unlike in the Soviet period, the so-called traditional religions of Russia (Buddhism and Orthodox Christianity) are depicted as constituting the core of people's identities and as symbols of the historical continuity of national and ethnic communities. This coverage fits with the Kremlin's view of the role of religion in contemporary Russia and prioritizes a limited number of so-called traditional religions, each associated with one particular religious organization which closely co-operates with the government. Thus, despite the existence of various schools of Buddhism in today's Buriatiia, on *Vesti-Buriatiia* only representatives of the Buddhist traditional Sangha of Russia speak in the name of all Buddhists of the republic.

Nevertheless, similarities with the Soviet approach to news reporting, instances when republican broadcasters follow the current Kremlin line and inversions of the centre–periphery relationship cannot on their own adequately account for the nature of *Vesti-Buriatiia*'s coverage. We will demonstrate, also, that in their coverage of identity politics, *Vesti-Buriatiia*'s broadcasters are engaged in a complex task of negotiating multiple ambiguities connected with Buriatiia's position in relation to its main 'external' reference points (Russia, Tibet and Mongolia), as well as with internal boundaries criss-crossing society.

We attempt to make sense of the complex balancing act by analysing *Vesti-Buriatiia*'s broadcasts through the perspective of marginality theory, which explores strategies with which groups and individuals tend to address their real or perceived, symbolic or physical status of being on the border between two or more cultures. The Buriat elites, we suggest, have a long, historically formed tradition of negotiating what could be perceived as their community's marginality. An appreciation of this tradition is necessary to our understanding of current processes. State television broadcasting in Buriatiia combines this tradition with the ideology of Eurasianism. We suggest that, following the demise of the USSR, for the elites in Russia's ethnic republics, Eurasianism has become an alternative to post-colonial interpretations of the relationship between the Russian/Soviet imperial centre and ethnic minorities.[13]

We start with an overview of the broader historical and current political context of the region within which *Vesti-Buriatiia*'s coverage will be analysed. Buriatiia is a relatively little known region, yet local conditions to a large extent shape approaches to broadcasting – hence the need to discuss them in detail. We then outline the theoretical framework we deploy when analysing individual news reports. The remainder of the chapter explores the representational strategies with which *Vesti-Buriatiia* deals with the republic's actual and symbolic borderland position and negotiates its cultural and religious diversity. We link those strategies to a recasting of centre–periphery relations and to the articulation of hybrid identities, whose potential to interrogate and subvert federal models of representation, we argue, is however undermined by the mythologized Eurasianism to which the strategies are ultimately subordinated.

A historical overview

The Buriats are a people of Mongolian language and cultural tradition, who have lived historically around Lake Baikal in southern Siberia and in northern Mongolia. The formation of the Buriats as an ethno-national community is, to a large extent, a result of Russian imperial and Soviet nationality policies. An organized penetration of Russian Cossacks into the region began in the seventeenth century and the establishment of the border between the Russian and Chinese empires in 1729 facilitated the separation of the Mongolian tribes on the Russian side from other Mongolian communities, thus allowing for the consolidation of the former as a separate Buriat people. The process of ethno-national consolidation was further facilitated when, in 1923, the Soviet government established the Buriat-Mongol (from 1958 Buriat) autonomous republic within the RSFSR, which was conceived as the national homeland of the Buriat people.[14]

Religious affiliations among the Buriats have also been influenced by the policies of St Petersburg and Moscow. The systematic penetration of Tibetan Buddhism into the region, where the local population had previously been worshipping shamanist deities, dates to the eighteenth century, and the process was again strongly influenced by the empire-building and foreign goals of the Russian tsars. Orthodox missionaries were active among western Buriats, particularly in the nineteenth century.

As is often the case with the imperial historiography of colonial conquests, the dominant Russian narrative about the peaceful establishment of Tsarist rule among the Mongolian tribes of the Baikal region glosses over instances of active and passive resistance to the Russian imperial expansion. During Soviet times, the local population resisted the brutality of Stalin's collectivization and the 1930s campaign against Buddhism. Yet these events attracted significant media attention only in the first years after the collapse of the communist regime: already by the mid-1990s, as Caroline Humphrey argues, these complex historical issues had been pushed into the background, partly to avoid any complications in relations with Moscow and the republic's Russian majority and partly because of the positive view of the Soviet period among the republican population at large.[15] Significantly, these issues were not explored in *Vesti-Buriatiia*'s coverage in 2011 of celebrations of the 350th anniversary of 'the unification of Buriatiia with Russia'.

At the same time, as Humphrey notes, we cannot reduce the relationship between Russians and Buriats to a conqueror–conquered dichotomy; instead, hierarchical relations have historically often been challenged and subverted in the region.[16] Russian Cossacks and peasants who moved into the areas around Lake Baikal at times also regarded the Russian imperial administration as an enemy, joining the indigenous population in acts of resistance. In turn, some Buriats joined Cossack troops and fought alongside Russians against Khalkha Mongols.[17] Levels of cohabitation and inter-marriage between members of different ethnic groups were high. In the religious sphere syncretic practices were also widespread, and people did not see contradictions between them.[18] Thus, hybrid identities – which, as we will see, *Vesti-Buriatiia* portrays so positively – emerged historically as a common feature of the region.

The contemporary political and media environment

Buriatiia is a relatively calm and stable region, with no major ethnic conflicts, extreme nationalist movements or aggressive commercialization.[19] It is a multi-ethnic and multi-confessional republic in which the titular nationality, the Buriats, constitutes only 24 per cent of the population. The Russians form a majority of over 70 per cent.[20] Political power mainly resides with the head of the republican government: at the time of writing the position is occupied by Viacheslav Nagovitsyn, an ethnic Russian from the Siberian region of Tomsk. Yet despite the authoritarian nature of the regime, political pluralism is not completely absent in Buriatiia, where leaders who come from outside the republic have to take into account the local ways of conducting politics. Opinion polls suggested that Nagovitsyn enjoyed strong popular support; the relationship between key players among the elites demonstrated a relatively high degree of consensus.[21]

The Buriatiia government website reveals the nature of government-sponsored discourse on the history of the Buriat–Russian encounter, as well as the role of religion and inter-ethnic relations in today's Buriatiia. As we will see below, the influence of this governmental discourse is reflected in *Vesti-Buriatiia*'s output. Here, there is no place for considering local resistance to Russian imperial rule and repressions of the Soviet period. The multi-ethnicity of the region is celebrated as a major strength. Orthodoxy and Buddhism appear as the two main religions today, with shamanism described historically as the first dominant 'religious worldview' of the indigenous population.[22]

The website pays a great deal of attention to inter-ethnic relations. The text of the 2007 'Concept of the State Nationalities Policy' of the Republic of Buriatiia is included in full, alongside annual reports on the implementation of the policy. The document proudly states that 'currently, the ethno-political situation in Buriatiia is marked by a well-grounded stability (*stabil'nost'iu i ustoichivost'iu*) and is characterized by a climate which fosters [ethnic harmony] and ensures the absence of serious conflicts and expressions of extremism'. This is attributed to a long-term tradition in the region of solving controversial issues through 'negotiations and a search for compromise'.[23] Yet potential problems are not altogether denied, as the document warns that the situation could easily deteriorate.

The republican nationalities policy ascribes particular importance to the media in the management of inter-ethnic relations. Contrasting inter-ethnic harmony in Buriatiia with the rise of ethnic conflict in Russia as a whole in the past ten years, the 'concept of state nationalities policy' blames the media above all for inflaming ethnic conflict:

> The Republic of Buriatiia is part of a unified information space in the country and is subjected to the destructive influence of this negative information... We have to admit that in terms of its impact, negative information, broadcast on television and disseminated through the print media and Internet, can sometimes be greater than the enlightened and prophylactic work which public and educational institutions of the republic are conducting.[24]

From the context, it seems that the media subjected to criticism here are particularly those based outside the republic of Buriatiia. Even though this is not spelled out, the statement might be read as a criticism of the coverage of ethnic relations on the federal channels, which report instances of conflict more frequently than the state media in Buriatiia, at times interpreting them through the prism of racialized concepts such as 'conflict of cultures' or 'ethnic criminality' (see Chapter 4). There is no place for such terminology in government-sponsored discourse in Buriatiia. In line with this local policy, *Vesti-Buriatiia* also avoids racializing its conceptual apparatus, preferring instead to evoke the Soviet ideal of the 'friendship of the peoples' which appeared – and disappeared – so suddenly in federal coverage of Manezhnaia.[25] In this sense, it is providing from the peripheries a Soviet corrective to the erroneous, de-Sovietized practices of the centre.

Indeed, in our statistical analysis of *Vesti-Buriatiia* reports, and unlike the situation on the federal channels, we coded the majority of relevant *Vesti-Buriatiia* news stories as belonging to the 'ethnic cohesion' category. These included reports celebrating displays of cultural diversity during local holidays and festivals, as well as the artistic and sporting successes of republican representatives in Moscow. Also assigned to this category were reports on the celebrations of the 350th anniversary of the 'unification of Buriatiia with Russia', because they portrayed the republic as a community of citizens who, regardless of their ethnicity, viewed 'the unification' as wholly beneficial. At the same time, in contrast to the federal channels, reports falling into categories capturing challenges to the picture of inter-ethnic harmony ('separatist violence', 'migration' and 'violent conflict') were minimal or completely absent in *Vesti-Buriatiia*'s coverage[26] (see Figure 6.1).

The republican government has the means to exercise tight control over the media, which, in this economically poor region with underdeveloped market relations, rely heavily on government subsidies for their existence.[27] Yet, as is the case with the functioning of the broader public sphere in the republic, this does not contradict the presence of a degree of pluralism within the republic's media output.[28] Overall, the model of media production followed by state television corresponds to what Olessia Koltsova has called 'pact configuration'. This model emerges when one player dominates the situation, yet for various reasons decides to compromise with others.[29] In Buriatiia the adoption of the 'pact configuration' model can be attributed above all to a local style of conducting politics in which the possibilities for the emergence of an organized, outspoken opposition are minimized.[30] Within this context, the most powerful leader is willing to show a degree of tolerance towards alternative voices, as

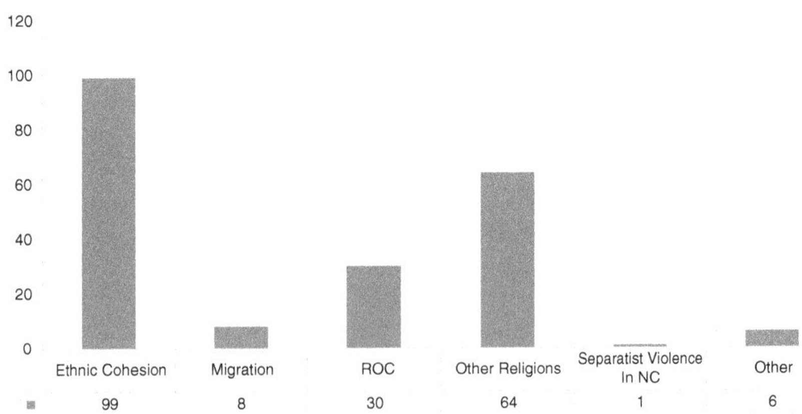

Figure 6.1 Frequency of category-coded news for *Vesti-Buriatiia* over the total recording period (September 2010-May 2012).

long as they do not offer too vigorous a challenge to his position. Buriatiia has six television companies, four of which are commercial. One of these commercial companies, called 'Arig Us', offers viewers alternative, critical perspectives on political issues.[31]

Vesti-Buriatiia's tasks in covering ethnicity and religion-related issues are in some respect straightforward, in others complicated. The programme closely follows the expectations of the republican government, which seems to believe that the best way of ensuring good inter-ethnic tensions is to avoid discussing controversial issues. At first glance, this should simplify the work of *Vesti-Buriatiia*'s team. The fact that the level of inter-ethnic tension is low in Buriatiia compared to many other parts of the Russian Federation allows the broadcaster to follow government recommendations without losing its overall credibility.[32] At the same time, as is the case for all television broadcasting in Russia, more specific regulations about how to handle particular situations in the media are non-existent; journalists are expected to exercise 'their professional judgement', which makes them potentially vulnerable if the decisions they take on this basis fail to coincide with the judgement of their superiors.[33]

While broadly advising the state media to focus on positive examples in the sphere of ethnic relations, the republican government displays ambivalence regarding the very nature of Buriat nationhood. The official position, as outlined on the government website, reflects an unresolved tension between a more inclusive perception of the republic as the homeland of all its residents and an exclusive vision of the region as the national homeland of ethnic Buriats. This situation replays at a local level the pan-Russian dilemma of reconciling civic and ethnic approaches to nation-building.

The confusion created by the tension between the two approaches is evident in the site's definition of the 'indigenous' (*korennoe*) population of the republic. At one point Russians, Buriats and the third main ethnic group, the Evenks (Tungus), are classified as 'the indigenous inhabitants (*korennye zhiteli*) of the republic of Buriatiia'; elsewhere only Buriats are described as the 'indigenous population' (*korennoe naselenie*) of the republic.[34] This inconsistency simultaneously points to the special status of ethnic Buriats and to the fact that the republic is part of a state in which ethnic Russians dominate the political sphere. The duality is reflected in the ethnic composition of the Buriatiia Television and Radio Company staff. The company, like *Vesti-Buriatiia* itself, is led by ethnic Russians, with two Buriats acting as the deputy heads. At the same time, the majority of members of staff are ethnic Buriats (five out of nine members of the leadership team and 12 out of 18 people involved in the production of television programmes).[35] This suggests that, even if *Vesti-Buriatiia* refrains from talking about overt conflicts, the broadcasters cannot avoid making complex choices when covering inter-ethnic relations.

Before looking in detail at how *Vesti-Buriatiia* attempts to influence the ways in which residents of the republic perceive inter-ethnic and inter-confessional relations, we must present the theoretical framework within which we will illuminate the specificity of the programme's approach.

Marginality theory and identity politics in Buriatiia

Buriatiia is quintessentially a borderland zone; it is located on Russia's border with Mongolia, and a flight from Moscow to the Buriat capital of Ulan-Ude lasts six hours. Internally Buriatiia is also divided by numerous – even if normally fluid – religious and ethno-cultural boundaries (particularly between ethnic Russians and Buriats). Scholars have long considered boundary spaces as particularly good loci to study relationships between culture and power and the complexities of identity formation. Especially interesting processes tend to occur when boundaries are (semi-)permeable and can be crossed in both directions without being eradicated. Power relations among people around such boundaries are invariably hierarchical; yet the hierarchies can be unstable. The situation of people positioned, at least implicitly, subordinately around such boundaries has been analysed by scholars since the 1920s through the concept of 'marginality'. Robert E. Park, who first suggested the term, conceived of marginality as 'a state of limbo between at least two cultural life-worlds'.[36]

Since the revival of interest in marginality theory in the 1960s, scholars have offered further insights into how the concept of marginality could be understood and utilized as a tool for analysing particular social and political conditions. In our analysis the 'in-between state' of marginality is crucial, as is the rejection of a simplified perception of power relations in respect to groups whose situation could be described as marginal. Marginality, of course, is predicated on inequalities in access to power and resources between groups separated by different boundaries.[37] At the same time, one should remember that the balance of power between individuals and groups is rarely fixed. Its reduction to a simple dichotomy between a powerful centre, represented by a dominant group, and a weak periphery, represented by minority groups, would gloss over the complexities of local processes.[38]

A form of marginality, at the core of which lies the 'double ambivalence' of a group or individual towards their own and foreign cultures, inevitably provokes differing responses.[39] Each specific case needs to be examined, taking into account the political and social conditions under which different cultures meet. Adam Weisberger has offered a productive general typology of responses to marginality that is helpful to our understanding of the specificity of Buriatiia's case. Weisberger argues that individual and group strategies for dealing with marginality often go beyond remaining in a state of limbo or poise, which Park saw as the only position associated with his definition of marginality. Whereas Park's 'marginal man' did not attempt to resolve ambivalences associated with his status, historically people have developed various strategies to achieve such a resolution. Weisberger suggests three other ways of addressing the condition of marginality. The first is transcendence, entailing attempts to overcome 'the opposition of the two cultures through creating a third way which is supposed to surpass them or reconcile them'.[40] Two further methods are assimilation, when those in the position of marginality strive to integrate into the dominant culture as far as possible, and what Weisberger calls 'return', which entails an overzealous identification with one's own culture and community.[41]

It can, of course, be argued that the term 'borderland zone' can describe not just physical spaces around state borders, but any sites where different cultures interact with each other.[42] In that sense many urban metropolises are borderlands. Yet areas around state borders do have specificities, if only because intensive cross-cultural interactions and actual and symbolic border crossings in such regions tend to have longer histories, dating to a time when most people rarely ventured outside their own communities. Moreover, power relations and hierarchies between different groups, even under conditions of colonialism, are paradoxically likely to be more ambiguous within areas around state borders than around symbolic boundaries in the national centre.

Buriatiia's position as a borderland zone since the seventeenth century has been shaped by the existence of the state border between Russia and China and later Mongolia. Here, too, local people and Russian colonizers interacted closely, meaning that group identities, and not only those among non-Russian minorities, have been historically shaped by conditions of marginality. As our brief historical survey indicates, power relations in the region never neatly mapped onto ethnic divisions. Ordinary Russians have not always enjoyed the dominant position and could find themselves in the condition of not only cultural and social but also political marginality, although the overall balance of power has invariably tilted in favour of the Russian/Soviet centre.

Nonetheless, when, at the turn of the twentieth century, a small number of members of the intelligentsia began to speak in the name of the imagined Buriat nation, they were acutely conscious of their in-between position. Their very desire to speak the language of nationalism reflected their encounter with European culture, which at the time was widely perceived as the yardstick according to which all traditions were to be measured.[43] This perspective inevitably placed Buriats in the position of marginality which these intellectuals attempted to address in the course of their travels between St Petersburg and their homeland in the eastern regions of Baikal, Mongolia and Tibet.

The fact that this period as a whole has sometimes been described as Europe's 'second oriental Renaissance', marked by a fascination among the European elites with non-European cultural traditions and criticism of Eurocentric prejudices, seems to have encouraged these Buriats to develop the strategy of transcendence with which to negotiate their marginal position. In fact, Buddhism, which they claimed to be their community's national religion, provoked particular fascination in Europe and Russia at the time.[44] The Buriats' first-hand knowledge of Buddhist traditions facilitated their creative interaction with Russian imperial scholars, in the course of which, at least occasionally, established power relations and hierarchies were subject to challenge. The Buriat intellectuals developed particular frames through which they interpreted their experiences. These included the centring of Buriat culture and forms of religiosity vis-à-vis other Mongolian and Tibetan communities; partaking in representations of Russia as a space where various cultural traditions mixed and created third-way fusions, of which the Buddhist reform movement (*obnovlencheskoe dvizhenie*) was claimed to be a striking example; and contributing to the development of a view that a strong adherence to the local ethno-cultural tradition, even if distinct from that of the Russians, could facilitate rather than hinder the formation of a strong sense of belonging to the Russian state as a whole. In this vision, a provincial locality within Russia, termed a 'small homeland' (*malaia rodina*), was perceived as a building block, fitting organically within the large construction of the Russian fatherland (*otechestvo*).[45]

In the years immediately following the creation in 1923 of the Buriat ASSR, the politics of culture were firmly in the hands of those Buriats who cultivated transcendence as the way of negotiating between local cultural and religious traditions and Russian/Soviet influences.[46] Although the Buriats who occupied leading positions in the Buriat ASSR in the 1920s fell victim to Stalin's purges, their legacy again began to attract significant attention in Buriatiia from the late 1980s onwards, because the 1920s are seen as a key stage in the Buriat nation-building process. Today elites in Buriatiia draw parallels between current developments and the early twentieth century, arguing that conceptions of the Buriats as a nation and of the relationship of this nation to powerful surrounding cultures, though articulated almost one hundred years ago, remain directly relevant.[47]

Vesti-Buriatiia's representational strategies

Within this context, it comes as no surprise that the strategy of transcendence is consistently at work in *Vesti-Buriatiia* news. As a result, a more creative and varied set of representations of ethnicity-related issues is offered than in the republican government's documentation on the nationalities question.

In *Vesti-Buriatiia*'s output the boundaries around which identities are formed in contemporary Buriatiia are often similar to those highlighted one hundred years earlier. The boundaries and meeting points between Russian culture and Orthodox Christianity on the one hand, and the Buriat ethno-cultural traditions – in which Buddhism occupies a particularly important place – on the other continue to produce the main zones of marginality. At the same time, the importance of Mongolia and Tibet as other reference points for the formation of local identities has only increased since 1991.

Many conceptual frames for addressing the ambiguities arising from Buriatiia's position as a multi-faceted borderland zone are also similar to the ones which have been employed historically: these include the celebration of hybrid identities and the highlighting of instances of successful multi-directional border crossings; the revival of the small homeland (*malaia rodina*) ideology; and the centring of the Buriats' historical, cultural and religious experiences within Russia, as well as within global Mongolian and Buddhist communities. At the same time, the ideology of Eurasianism is also often used to highlight the Buriats' experiences within the pan-Russian context.

Buriatiia – a region of hybrid identities

The governmental 'Concept of the State Nationalities Policies' only briefly addresses the convergence of different cultures on Buriatiia's territory. It argues that one of the aims of 'the preservation of ethnic cultures and traditions of ethnoses' in Buriatiia is 'to create a unified *rossiiskaia* civic (*obshchegrazhdanskaia*) identity'.[48] This statement reflects the current Kremlin line, as well as the legacy of Soviet policies on nationality which, in theory, were supposed to achieve the eventual transcendence of national specificities by means of the initial cultivation and even, in some cases, temporary strengthening of the ethno-cultural identities of the Soviet Union's subject nationalities. As the events of 1991 demonstrated, this utopian project of identity formation failed. Putin's leadership has not been able to develop a practical means of creating a unified '*rossiiskaia* civic identity' while preserving the legacy of Soviet ethno-cultural particularism. The Buriatiia government's outline of its vision of policies on nationality likewise fails to explain how such a goal can be achieved.

Vesti-Buriatiia, however, adopts a consistent discursive strategy in support of the convergence goal by constructing an image of the republic as a place where the successful crossing of boundaries is easily achieved on a daily basis and where people feel entirely comfortable with their hybrid or ambiguous identities. The transcendence of ethnic boundaries and the appearance of creative fusions as a result of multi-directional border crossings are claimed in relation to the three most important points around which, according to Buriatiia's and Russia's elites, identities are formed today – native language (*rodnoi iazyk*), religious affiliation and common historical memory, in which the Second World War is the main episode.

Typically, reports on various government-sponsored activities promoting the Buriat language and culture highlight the fact that they are aimed at non-ethnic Buriat residents of the republic for whom the Buriat language is not expected to be native. Thus, a report of 6 April 2012 on a Buriat-language competition among non-Buriat residents featured a local university student, an ethnic Russian, who successfully negotiated her potential position of marginality vis-à-vis the Buriat culture. Commenting on winning the competition, the student stated:

> For me the Buriat language is a second native language. Sometimes I perceive it as more native (*rodnee*) than Russian. It is easy for me. I grew up in a Buriat village among the Buriats, I am an ethnic Russian, but in my soul I am a Buriat.[49]

The reporter noted that knowledge of the Buriat language could also improve this student's employment opportunities. The student is thus presented as a symbol of the multiple benefits derived from the ability to cross cultural boundaries.

Similarly, the celebration of boundary-crossing is a prominent feature in coverage of religion-related events. Orthodox Christianity and Buddhism (in that order) are officially proclaimed to be the two 'national religions' in Buriatiia, and most coverage is devoted to them. It is not, in fact, immediately obvious what the appropriate balance in the coverage of the religions important to the region should be. From the early twentieth century, the dominant Buriat intellectual tradition represented Tibetan Buddhism of the Gelug school as the *bona fide* national religion of the Buriats, branding shamanism – which was widespread particularly among Western Buriats – as backward, and describing Buriat conversions to Orthodox Christianity as 'artificial'.[50] Yet shamanic beliefs and practices which do not require the existence of an organized religious institution survived better during the Soviet period than Buddhism, whose clergy fell victim to Stalin's repressions and whose institutional base was all but destroyed.[51] Post-1991 shamanism emerged as an important spiritual phenomenon, often described as an element of Buriatiia's 'national tradition', and attracts a following not only of Buriats, but also of local Russians and representatives of other nationalities.[52] At the same time, because ethnic Russians constitute 70 per cent of the republic's population and 70 per cent of the officially registered religious communities (*religioznye ob'edineniia*) are Orthodox, significant attention to Orthodox Christian events on state television is to be expected.[53] *Vesti-Buriatiia* deals with this complex situation by according virtually equal space to Orthodoxy and Buddhism, with shamanism receiving less attention.[54]

Again, the coverage is aimed at facilitating the transcendence of religious boundaries. It underscores the similarities between the two 'main national religions' by identically structuring reports about them and by drawing direct parallels between Orthodox and Buddhist religious holidays and divine figures. The prevalence of syncretic religious rituals in Buriatiia is also highlighted. Finally, the fact that the concept of reincarnation in the Buddhist tradition is based on the perception of crossing temporal, spatial and corporal boundaries also seems to encourage a view of border crossings as an acceptable and even highly desirable condition.

Numerous reports on Orthodox and Buddhist religious festivals and traditions start with a description of their origins in the distant past and a brief overview of a tradition's survival up to the present (a history usually depicted as an unbroken line); interviews with Orthodox priests and Buddhist lamas, as well as believers among lay people, are always included. Considerable attention is paid to various 'miracles' occurring in the context of particular traditions, which both interviewees and reporters invariably take at face value.[55] Indeed, in these reports the voices of the reporters and those of the priests, lamas and believers are indistinguishable.[56]

Reports also tend to draw parallels between Orthodox and Buddhist rituals and belief systems. Thus the Virgin Mary is compared to Green Tara, a female Bodhisattva (enlightened being) in Mahayana Buddhism, with the reporter commenting on how both offer protection to believers in similar ways.[57] Religious syncretism is also presented as a common feature. A report on Shrovetide in 2011 describes it as a holiday marking the end of winter and the beginning of spring and celebrated by most ethnic groups in the region. It then features an event organized by the Ulan-Ude ethnographic museum, during which different nationalities were encouraged to demonstrate their modes of celebrating the occasion.[58] Meanwhile, the coming together of different religious traditions is noted in coverage of rituals featuring both shamans and lamas, such as the *Lusuud takhilga* ritual dedicated to a water spirit.

Last but not least, coverage of the Buddhist traditional belief in the possibility of crossing existing boundaries, even those which in most cultures are perceived as impermeable (such as that between life and death), implicitly endorses that belief.[59] The most prominent manifestation of the belief in contemporary Buriatiia is the phenomenon of the Pandito Khambo-Lama Itigelov. This well-known Buriat Buddhist priest died in 1927. His body was recovered, reportedly intact, in 2002; it is now preserved in the main Buddhist datsan of Buriatiia, and occasionally exhibited to the public. In contrast to the perception of the 'uncorrupted' relics of saints in the Christian tradition, where these 'saints' are understood as being dead, in the Buddhist tradition the 'uncorrupted bodies' are perceived to be in a marginal state, in between life and death. *Vesti-Buriatiia* pays a great deal of attention to the story of Khambo-Lama Itigelov. The language used by *Vesti-Buriatiia* adheres closely to the statements on the issue made by the main governing body of Russia's Buddhists, Buddhist Traditional Sangha, suggesting that Itigelov is more alive than dead and constantly defying one of the most crucial boundaries. As one *Vesti-Buriatiia* report put it:

> The teacher returned on the fourth day of the lunar calendar on 10 September 2002... Retaining his body uncorrupted, Dosha-Dorji Itigelov has reached the highest level of spiritual realization. Living next to us (*zhivia riadom s nami*), he gives advice, helps out, acting out of compassion towards all the living beings (*vsem zhivym*).[60]

Significantly, the federal version of Vesti initiated this merging of the voices of religious authorities and reporters as late as the start of the presidential election campaign in January 2012. On *Vesti-Buriatiia*, by contrast, it was a feature throughout our entire monitoring period, reconfirming Buriatiia as a republic whose very peripheral status enables it to provide an 'exemplary' handling of the 'national question', pre-empting that of broadcasters located at the federal centre.

Finally, in television representations of the collective historical memory of the people of Buriatiia, those individuals who have shown the ability to negotiate cultural boundaries successfully, building new ties across them and encouraging fusion between communities, are particularly celebrated. This, for instance, applies to the selection and representation of Second World War veterans who are interviewed on television. Every year, several days before the Soviet Victory Day of 9 May, *Vesti-Buriatiia* news coverage is dominated by reports about the war, which is perceived as a formative event for the people of Buriatiia, underscoring their membership in the Soviet and now pan-Russian national community. Most interviewed veterans have multiple identities, which they always manage to reconcile. Thus, a veteran interviewed at length on the occasion of Victory Day in 2012 was a 'hereditary' (*potomstvennyi*) Cossack, who was born and raised in Mongolia, outside the Soviet borders, yet reportedly felt equally at home within the Cossack, Buriat-Mongol and Soviet cultures.[61]

In a similar example, in 2011 *Vesti-Buriatiia* reported on a reception in the Buriat Republican Mission in Moscow of war veterans from 'Buriatiia's diaspora community' in Moscow. Whereas in federal television coverage members of Moscow's so-called diaspora communities are invariably identified by their non-Russian ethnicity rather than by their affiliation with an ethnic republic within Russia or a foreign state (for example, they are represented as ethnic Chechens, Dagestanis, Tajiks, etc.), on *Vesti-Buriatiia* the veterans representing the republic at its 'Mission in Moscow' were of multiple nationalities, including Russians. What united them, according to the broadcast, was that when the Second World War started they were living in Buriatiia; only later in their lives did they move to Moscow. The report emphasized these people's ability to transcend ethno-cultural differences by sharing a strong feeling of pan-Russian patriotism, as well as through a shared perception of Buriatiia as their 'small homeland' (*malaia rodina*).[62]

There are other instances of evocation of the notion of '*malaia rodina*', which dates to the late nineteenth century and which contributed significantly to shaping the worldview of the first generation of the Buriat intelligentsia. Even though the notion acquired a second life in the Soviet Union in the 1960s and 1970s and is occasionally used today by elites in the centre, '*malaia rodina*' is not part of the conceptual apparatus with which federal television news reporters attempt to make sense of Russia's diversity. The term foregrounds the ease with which cultural boundaries can be crossed – indicating, however, that the total erosion of these boundaries is unnecessary and could even harm the process by which people achieve a productive membership within the pan-Russian community. It is clearly well matched to *Vesti-Buriatiia*'s representational strategies.

The Eurasian identity

In Russia, Eurasianism is currently championed by many politicians and intellectuals, who interpret it in different ways. Eurasianism's representation of Russia as a special world where different cultures interact, thereby creating a distinct supranational community different from both 'west' and 'east', is very popular among the elites in Russia's ethnic republics.[63] They adapt Eurasianism to local conditions, regarding it as an effective tool with which to overcome the position of marginality through the strategy of transcendence.

The evocation of Eurasianist theory by elites in Russia's ethnic republics in order to emphasize the importance of their local cultural traditions within the pan-Russian (*rossiiskii*) framework can be seen as a specific response by Russia's non-European minorities to the post-imperial situation. Indeed, the original Eurasian movement of the 1920s offered what was arguably at the time the most systematic attack on Europe's cultural and political imperialism. In the context of Russia's ethnic republics, Eurasianism in effect acts as a substitute for the post-colonial perspective. It has a distinct anti-colonial message and it entails the challenging of cultural hierarchies which prevail in Russian intellectual discourses treating Russia as part of Europe. So it offers a framework in which leaders of non-Russian autonomies within the Russian Federation can draw attention to the importance of their local experiences and traditions, which otherwise remain marginalized. Yet Eurasianism treats Russia not as a colonial power, but as a community which is itself under threat of being colonized by the west. In its juxtaposition of Russia's supposedly positive experience of managing multi-ethnicity and the destructive nature of west European colonialism, Eurasianism was and remains an attempt to articulate a new ideology for the Russian empire and does not encourage the questioning of Russia's political practices and cultural attitudes.[64] Today, as in the past, Eurasianism promotes the mythologization of, rather than a critical reflection on, the cultural and political interactions of the peoples of 'Russia-Eurasia'.

The main criticism of the centre which Eurasianism encourages among local elites is in relation to the claims of their federal equivalents regarding the European nature of Russian statehood. This is the case in Buriatiia, where the political programme of the republican branch of the United Russia party represents the republic as 'the cradle of the Eurasian culture'.[65] On important public occasions covered by *Vesti-Buriatiia*, the Eurasian nature of the republic is claimed by Buriatiia's leadership with reference to current times as well as to the distant past. Thus, in September 2010, coverage of celebrations of the 15th anniversary of the establishment of Buriatiia's state university in Ulan-Ude emphasized that various luminaries who spoke on the occasion explored the 'Eurasian nature of Russian (*rossiiskaia*) statehood' and Buriatiia's role in the development of the syncretic Eurasian culture.[66] This role, it was contended, was central and dated back to the period when the great Mongol leader, Genghis Khan, was establishing his vast empire.

The figure of Genghis Khan occupies an important place in the historical mythology of Buriatiia today, since the Khan's birthplace is said to be located on 'Buriat national territory', in the Aginsk Buriat autonomous district (*okrug*).[67] The Buriat perception of Genghis Khan stands in sharp contrast to the Russian historical tradition dominant since the nineteenth century, in which Genghis Khan's image is negative and the Mongol invasion of Rus is portrayed as an event which failed to divert Russia from its necessary development as part of Europe. In Buriatiia, however, the Khan is a 'national' hero, represented as a model leader of a multi-ethnic community, and this is indeed how he is depicted on *Vesti-Buriatiia*. In 2012 the parliament of Mongolia made 14 November, the day on which Genghis Khan is reputed to have been born, a national holiday. The *Vesti-Buriatiia* report covering the occasion depicted the Mongol empire founded by the Khan as a place of unusual ethnic and religious tolerance and 'international cooperation'. A member of the Buriat parliament was quoted as arguing that if the Russians acknowledged fully the 'Eurasian nature' of their statehood, they would have been able to change their assessment 'of this Mongolian military leader from negative to positive'.[68] Buriatiia is thus depicted as being instrumental in revealing to the Russians their true identity. Here, then, the centre–periphery inversion involves not the ideal embodiment by the periphery of the centre's values, but the projection of the values of the periphery onto the centre in order to reveal its 'authentic' core.

Centring Buriatiia's experience

Vesti-Buriatiia regularly attempts to centre the Buriat people's historical experiences not only within Russia, but also within the global Mongolian and Buddhist communities. Yet the challenging of established hierarchies and power relations prevailing in this context never presupposes the perception of Russian and Soviet policies as colonial. Thus, in a report on the 50th anniversary of the Academy of Culture and Arts in Ulan-Ude in September 2010, it was argued that the institution, with its multicultural teaching staff and student body and its location at 'the crossroads of East, West and North', should be seen as 'the centre of a dialogue between the culture of Russia and that of the world'.[69] What this argument challenges is Russia's self-representation vis-à-vis Europe/the west, not the nature of the relationship between ethnic Russians and minority cultures.

Not only Russia but also Tibet and Mongolia constitute important reference points for the Buriats. The proto-Buriat tribes in the regions to the east and south of Lake Baikal received Buddhism from Tibet. The transmission of this religious tradition began relatively late and was further challenged by the activities of Orthodox missionaries in the late imperial period and the anti-religious campaigns of Soviet times. Against this background, members of the Tibetan exile community often see Buriats as in need of assistance through the missionary activities of Tibetan monks.[70] The activities of these monks have become a political issue in Buriatiia, generating public debates.[71] *Vesti-Buriatiia* seems to adopt a particular position in this debate. Based on a rejection of the perception of Buddhism in Buriatiia as a peripheral phenomenon, it challenges the dominant hierarchy in which Tibet emerges as the religious centre in relation to which Buriatiia has an inferior status.

Such coverage reflects the current position of the republican leadership as well as that of Moscow, according to which Russia's Buddhists can revive their traditions and rebuild their monasteries without any help from abroad.[72] This position has a long history, dating back to the Russian imperial policies of offering instrumental support for the development of Buddhism among eastern and southern Buriats at the same time as ensuring their separateness from their co-religionists in Tibet and Mongolia. Within this political context, in collaboration with Russian imperial specialists on Buddhism in St Petersburg, early twentieth-century Buriat intellectuals such as the famous Buriat lama Agvan Dorzhiev and Tsyben Zhamtsarano, a leader of the Buriat nationalist movement during the 1905 revolution, developed a powerful narrative about Buddhism practised among the Buriats. The dominant view of it as an inferior version of the Buddha's teaching was rejected and it was argued instead that it was a form of Buddhism in its own right, which should not be measured against any external yardstick.[73] These developments reveal the complexity of the relationship between the colonial context and the agency of the subalterns who attempt to use this context to achieve their own goals, a relationship that continues to play itself out today.

Dorzhiev's role as advisor to the Dalai Lama helped him develop a discourse about the centrality of the Buriats in the transnational community of Buddhists and among the Mongolian people.[74] Today, Dorzhiev's ideas are popular among Buriat elites. The Agvan Dorzhiev Foundation, set up in 1993, has as its proclaimed aim a 'further development of his [Dorzhiev's] spiritual heritage within the contemporary context'. While noting that this would include the strengthening of ties between Buddhists in Buriatiia and their co-religionists abroad, the foundation argues that in the main Buddhist monasteries outside Russia, 'the memory lives of the great lamas' from Buriatiia. In these foreign places, disciples of the Buriat lamas 'now occupy important positions'.[75] This is in line with the view of the Buddhist Sangha of Russia, which proudly asserts the autocephality of Russia's Buddhist tradition, offering a highly idealized picture of the formation of this tradition as an outcome of the positive benefits of Russian imperial rule.[76]

In our period of analysis, *Vesti-Buriatiia* ran few reports featuring Tibet. Significantly, most of them challenge the perception of Tibet's cultural superiority over Buriatiia and reject the perception of Buriatiia's Buddhist tradition as peripheral. In January 2012, the programme focused on activities organized in the republic by the Buddhist Sangha of Russia to celebrate the lives of two twentieth-century monks who were born in Buriatiia and were eventually to become leaders of important monasteries in Tibet and India.[77] In this coverage, it is not Tibetan Buddhists who offer support for the development of Buddhism among the Buriats, but Buriat Buddhists who exercise leadership in the world centres of Buddhism. Similarly, when the activities of Tibetan monks inside Buriatiia are covered, reports note that participating Buriat Buddhists use the occasion to collect money to support their co-religionists abroad.[78] Again, a reversal of power relations is implied.

The view that the Buriats could become cultural and political leaders in a pan-Mongolian community was first expressed by figures such as Dorzhiev and Zhamtsarano. In the early 1920s, Zhmatsarano's, and other prominent Buriats', roles in the establishment of the People's Party in Mongolia was seen by them as fulfilment of the national destiny of their own community to be leaders among the Mongols.[79] Here again local agency and the imperial context were closely intertwined, as the Buriat leaders attempted to pursue their own vision of their 'national mission' while working for the new Bolshevik regime.

In its coverage of Buriatiia's relationship with Mongolia, *Vesti-Buriatiia* adopts a similar template that simultaneously reflects the Buriats' own historically formed strategy of self-representation, yet also fits with the Kremlin-sponsored line. Buriatiia and Mongolia appear as two separate, competing entities, for instance in the sphere of preserving and publicizing the historical heritage of the Mongolian people and 'the civilization of Central Asia' (*tsivilizatsiia Tsentral'noi Asii*) as a whole.[80] Buriats are expected to take a lead in this competition. The theme emerges in reports regarding the government policy of promoting ethnographic tourism as an important source of revenue for the republic, which suggest that by maintaining various historical sites in good condition and advertising them well to potential tourists, Buriatiia, rather than Mongolia, will emerge as the internationally recognized 'centre of the historical heritage of the Mongolian tribes'.[81]

Russian imperial and broader Eurocentric perspectives seem to have been deeply internalized in Buriatiia, all the criticism of Eurocentrism and attempts to challenge established hierarchies notwithstanding. This introduces further ambiguities in the official discourse about Buriatiia's position within Russia and the wider world. For instance, in *Vesti-Buriatiia*'s reports about ethnographic tourism the main point of reference remains 'the west', whose centrality in world affairs is implied and whose approbation is implicitly sought. The publicity given to travel opportunities in Buriatiia is supposed to attract the particular attention of tourists from Europe, who, according to *Vesti-Buriatiia*, should be encouraged to recognize the vast sight-seeing potentials of the Buriat lands.[82]

In a context where interrogations of Buriat–Russian encounters take place within the ideological framework of Eurasianism rather than post-colonialism, it comes as no surprise that some programmes on republican state-controlled television occasionally lapse into promotion of an unmitigated imperial perspective. The most striking example of this approach was the local television programme *Buriatiia: The Chronicle of the Centuries*, whose launch in March 2011 was advertised by *Vesti-Buriatiia* as a major cultural event.[83] In it, Orientalizing stereotypes are reproduced without being subjected to any scrutiny. From the producers' point of view the history of the region starts with the arrival of Russian Cossacks in the northern areas of today's Buriatiia. Even though the programme acknowledges that the region was populated by various ethnic communities, the Cossacks are described as 'the first comers' (*pervoprokhodtsy*). The presenter's narrative about subsequent developments pieces together lengthy quotes from reports by Cossacks and Russian imperial administrators informing the government in St Petersburg about the distant and exotic peripheries. Again, fully in line with the Orientalizing perspective, the indigenous population of the region – Buriatiia's main 'small minority group' (*malyi narod*), the Evenks – are not given a voice, instead being depicted through an exhibition of their exotic craftsmanship at a local ethnographic museum.[84]

***

Like the output emerging from the peripheries of the post-Soviet Russian mediasphere examined in the previous chapter, but for rather different reasons, *Vesti-Buriatiia*, located at post-Soviet Russia's geographical margins, is no mere 're-transmitter of a Moscow-produced context'. In fact, local elites view the latter critically. The discursive strategies of *Vesti-Buriatiia* reflect not only the local elites' interpretation of the Kremlin line on ethnic diversity as Russia's major strength, but also a long-term indigenous tradition of negotiating the perceived marginal status of Buriatiia and Buriats within Russia and in relation to other important local points of reference, such as Mongolia and Tibet. The republic's experiences are drawn upon in a double inversion of the established centre–periphery axis. Buriatiia emerges as a region which can both epitomize the nation-building strategy emanating from the centre and project its own values and traditions onto that centre, enabling Russia to find its true Eurasian self and realize the ideal of multi-ethnic harmony to which it aspires. These experiences are also depicted as being of crucial importance for the wider communities of the Mongolian people and Buddhists, a strategy facilitated by a globalized environment in which local peripheries can bypass mediation through national centres and operate directly within a transnational context.

Furthermore, by promoting hybrid identities and by cultivating individual and group experiences of multiple forms of boundary-crossing, *Vesti-Buriatiia* draws upon key tactics which the scholarship theorizing the relationship between the media and inter-ethnic relations identifies as conducive to strengthening ethnic cohesion. Its preferred model of 'unity via hybridity' offers an alternative both to the failed 'unity in diversity' model analysed in Part II and the ominous 'unity through division' model that will dominate Part III. In contrast to federal television, *Vesti-Buriatiia* shuns sensationalism and refrains from dramatizing the events it covers. Unusually for a contemporary media outlet, *Vesti-Buriatiia* emphasizes the absence of crisis and conflict and dwells on instances of co-operation and opportunities for consensus-building.[85] It also positions itself as a 'shared medium', equally appealing to Buriats, Russians and other nationalities of the region.[86] In thereby invoking the 'friendship of the peoples' mantra promoted by the former Soviet centre, but in a religious context alien to the latter's atheistic values, it represents a particular kind of post-Soviet experience.

Ultimately, however, the programme lacks the necessary tools to fully interrogate power relations between ethnic minorities and the Russian centre. Its representations of the Buriats' relationship with Mongolia and Tibet likewise fall in line with positions which find Moscow's approval. The reason for this is that the Buriats' own intellectual tradition of transcending marginality originated in the Russian imperial and Soviet contexts, with their limited opportunities for criticism or challenge; moreover, it is now supplemented by the ideology of Eurasianism. While critically assessing west European colonialism, Eurasianism mythologizes Russia's own tradition of managing ethnic diversity. This intellectual framework only increases *Vesti-Buriatiia*'s reluctance to discuss serious challenges to societal cohesion. As a result it lacks both a strategy and a conceptual framework for addressing future internal conflicts which may prove impossible to ignore. In light of the dramatic confrontations that were to sweep the political and media environment at the federal level in 2012, and which we must now address, its intriguing and at times paradoxical mode of negotiating its position on the centre–periphery axis looks all the more anomalous.

Notes

1 Interview with a television journalist in Buriatiia, 14 March 2013. In 2010, only 33 per cent of the population of Buriatiia had internet access.

2 This subordination is visually highlighted by the background image against which *Vesti-Buriatiia's* moderators are filmed. It is identical to that of the federal *Vesti*.

3 Interview, 14 March 2013.

4 On coverage of Islam on Russian television, see Christopher Flood, Stephen Hutchings, Galina Miazhevich and Henri Nickels, *Islam, Security and Television News: A Comparative Study*, Basingstoke: Palgrave Macmillan, 2012. On the Russian Orthodox Church, see Zoe Knox, 'Church, State, and Belief in Post-Soviet Russia', *The Russian Review,* 71/1, 2012, 122–27. On the representation of Buddhism in the Buriat mass media, see Timur B. Badmatsyrenov, 'Sangkha i politika: politicheskie aspekty funktsionirovaniia buddiiskogo dukhovenstva Mongolii i Buriatii', *Vestnik Buriatskogo gosudarstvennogo universiteta*, 14, 2012, 137–43.

5 Vsevolod Chaplin, 'Buriatiia – vdokhnovliaiushchii dlia Rossii primer otnoshenii mezhdu raznymi national'nostiami', 3 August 2012. Available online at http://vtinform.ru/vti/138/59804.php (accessed 21 March 2013).
6 Interview with a Rossiia-24 journalist, 2 April 2013. A similar argument was articulated by another journalist, interviewed on 26 March 2013.
7 *Vesti-Buriatiia*, 'Prazdnovanie 93-letiia VLKSM', 31 October 2011, and 'Reportazh o Dne Pionerii', 19 May 2011.
8 On the predominance of the Russian language in Buriatiia, including among the Buriats, see V. I. Zateev and N. S. Babushkina, 'Dvuiazychie v sovremennoi Buriatii'. Available online at http://ecsocman.hse.ru/data/527/890/1216/011.ZATEEV.pdf (accessed 30 June 2013).
9 Coverage of a pro-Putin meeting in Ulan-Ude on 6 March 2012.
10 Caroline Humphrey, *Marx Went Away but Karl Stayed Behind*, Ann Arbor: The University of Michigan Press, 1998, p. XV.
11 Interview with a television journalist in Buriatiia, 27 February 2013.
12 Available online at 'Istoriia' http://bgtrk.ru/materialyi-o-bgtrk/2010-08-03-09-26-34/istoriya-gtrk-Buriatiia.html (accessed 21 March 2013).
13 On the lack of engagement with post-colonialism in Russia, see 'Postsovetskoe soznanie i Postcolonial Studies', *Novoe literaturnoe obozrenie*, 94, 2008. Availble online at http://nlobooks.ru/sites/default/files/old/nlobooks.ru/rus/magazines/nlo/196/1208/index.html (accessed 1 July 2013); Ilya Gerasimov, Sergey Glebov and Marina Mogilner, 'The Postimperial Meets the Postcolonial: Russian Historical Experience and the Postcolonial Moment', *Ab Imperio* 2, 2013, 97–135, particularly 120–30. See also Sergei Oushakine, 'V poiskakh mesta mezhdu Stalinym i Gitlerom: o postkolonial'nykh istoriiakh sotsializma', *Ab Imperio*, 1, 2011, 209–32.
14 L. R. Pavlinskaia, *Buriaty. Ocherki etnicheskoi istorii*, St Petersburg: Evropeiskii dom, 2008, pp. 171–82; 195.
15 Humphrey, *Marx Went Away but Karl Stayed Behind*, p. XV.
16 Ibid. p. 27.
17 Pavlinskaia, *Buriaty*, pp. 172, 179.
18 Humphrey, *Marx Went Away but Karl Stayed Behind,* p. 30.
19 Ibid., p. X; E. D. Dagbaev, *Institutsional'nyi dizain politicheskikh protsessov v rossiiskikh regionakh Vnutrennei Azii i Mongolii*, Uland-Ude: Izdatel'stvo BNTs SO RAN, 2011.
20 In 1937, as part of the Soviet campaign against 'bourgeois nationalism', the boundaries of the Buriat-Mongol Autonomous Soviet Socialist Republic were cut back. Two areas with substantial Buriat populations were incorporated into the neighbouring Irkutsk and Chita regions.
21 A report, 'Politicheskaia situatsiia i SMI v Buriatii', compiled at the request of the authors by Timur B. Bardmatsyrenov, Professor of Political Science at the Buriatiia State University, Ulan-Ude, March 2013.
22 'Osnovnye religioznye konfessii'. Available online at http://egov-buryatia.ru/index.php?id=272 (accessed 21 March 2013).
23 'O Kontseptsii gosudarstvennoi natsional'noi politki v Respublike Buriatiia'. Available online at http://egov-buryatia.ru/index.php?id=377 (accessed 21 March 2013).
24 Ibid.
25 *Vesti-Buriatiia*, 'Chempionat po national'noi buriatskoi bor'be', 22 March 2011.
26 For statistical analysis of the news coverage of the two main federal channels, see Chapter 2.
27 'Demokraticheskii audit regionov'. Available online at www.freepress.ru/publish/publish043.shtml and 'Respublika Buriatiia', available online at http://freepress.ru/arh/book_2000/014.shtml (accessed 21 March 2013).
28 Dagbaev, *Institutsional'nyi dizain*, p. 126.

29 Olessia Koltsova, *New Media and Power in Russia*, London: Routledge, 2006, pp. 174–6.
30 Interviews with television journalists in Buriatiia, 27 February and 3 March 2013.
31 Lasar Bartunaev, 'Obshchestvo mozhet otstaivat' svoi interesy', 8 June 2006. Available online at http://baikal-media.ru/news/interview/23418/?sphrase_id=4080009 (accessed 21 March 2013).
32 See, for instance, the relatively independent republican newspaper *Inform Polish*, which reported the results of a 2012 opinion poll according to which 61 per cent of the population of the republic positively assessed its inter-ethnic relations. Available online at http://infpol.ru/news/6701/126228.php (accessed 21 July 2012).
33 Interviews with television journalists in Buriatiia, 27 February and 3 March 2013.
34 'Kratkie svedenia o Buriatii', available online at http://egov-buryatia.ru/index.php?id=321, lists three nationalities as 'indigenous' to Buriatiia. In contrast, 'Istoriia Buriatii', available online at http://egov-buryatia.ru/index.php?id=324, refers only to the Buriats as the indigenous population; the webpages offering similar profiles of the Russians and other ethnic groups in Buriatiia do not refer to these nationalities as indigenous to Buriatiia (accessed 8 September 2014).
35 'Kollektiv', available online at http://bgtrk.ru/about/team/list.php?SECTION_ID=18 (accessed 8 September 2014).
36 Robert E. Park, 'Human Migration and the Marginal Man', *The American Journal of Sociology*, XXXIII/6, 1928, pp. 881–93.
37 Rachel Platonov, *Singing the Self: Guitar Poetry, Community and Identity in the Post-Stalin Period,* Evanston: Northwestern University Press, 2012, pp. 40–6.
38 Ibid.
39 Adam Weisberger, 'Marginality and Its Directions', *Sociological Forum*, 7/3, 1992, 429.
40 Ibid., 430.
41 Ibid., 442.
42 R. R. Alvarez and G. A. Collier, 'The Long Haul in Mexican Trucking: Traversing the Borderlands of the North and the South', *American Ethnologist*, 21/3, 1994, 606–27.
43 Vera Tolz, *'Russia's Own Orient'. The Politics of Identity and Oriental Studies in the Late Imperial and Early Soviet Periods*, Oxford: Oxford University Press, 2011, pp. 85–147.
44 Maria Carlson, *'No Religion Higher Than Truth': A History of the Theosophical Movement in Russia, 1875–1922*, Princeton: Princeton University Press, 1993, pp. 193–4.
45 Tolz, *'Russia's Own Orient'*, pp. 37–43.
46 Assimilationists among Buriat leaders, who also became outspoken in the first decade of the twentieth century and were in the minority from the very beginning, were completely sidelined in the first years of the Soviet rule, as were those who favoured the strategy of 'return' – for instance, conservative lamas, who opposed the Buddhist reform movement. See Tolz: *'Russia's Own Orient'*, pp. 121–4 and 140–7.
47 The relevant publications which appeared in the first decade of the new millennium include: E. N. Stroganova, *Buriatskoe national'no-kul'turnoe vozrozhdenie*, Moscow-Irkutsk: Natalis, 2001; B. V. Bazarov and B. V. Zhabaeva, *Buriatskie natsional'nye demokraty i obshchestvenno-politicheskaia mysl' mongol'skikh narodov v pervoi treti XX veka*, Ulan-Ude: Izdatel'stvo BNTs SO RAN, 2008; E. D. Dagbaev, *Agvan Dorzhiev v storii Rossiiskogo gosudarstva*, Ulan-Ude: Belig, 2005; V. A. Khamutaev, *Natsional'nyi vopros v Buriatii. 1980-2000-e gody*, Ula-Ude: Izdatel'stvo BNTs SO RAN, 2008. Numerous works publicizing ideas of the Buriat leaders of the early twentieth century were also published in Buriatiia in the 1990s.
48 'O Kontseptsii gosudarstvennoi natsional'noi politki v Respublike Buriatiia'. Available online at http://egov-buryatia.ru/index.php?id=377 (accessed 8 September 2014).
49 *Vesti-Buriatiia*, 'Festival-olimpiada buriatskogo iazyka', 6 April 2012.

50 Tolz, *'Russia's Own Orient'*, p. 123.
51 Humphrey, *Marx Went Away but Karl Stayed Behind*, pp. 403, 411, 441.
52 Anya Bernstein, 'Remapping Sacred Landscapes: Shamanic Tourism and Cultural Production on Olkhon Island', *Sibirica: Journal of Siberian Studies*, 7/2, 2008, 23–46.
53 'Religioznye ob'edineniia'. Available online at http://egov-Buriatiia.ru/index.php?id=272 (accessed 21 March 2013).
54 In the period between September 2010 and August 2012 we identified 30 reports on issues related to Orthodox Christianity and 26 reports related to Buddhism; there were 14 reports related to shamanism. See Figure 6.1, where reports on Buddhism and shamanism are represented in the category 'other religions'.
55 See, for instance, the following *Vesti-Buriatiia* reports: 'Vosstanovlenie Sartuul-Bulagskogo datsana', 29 September 2010; 'Stroitel'stvo stupy Lkhabab Duisen', 15 October 2010; 'Otkrytie novoi stupy v Ivolginskom datsane', 12 September 2011; 'Moshchi sviatogo Innokentiia Irkutskogo v Ulan-Ude', 25 November 2011.
56 In contrast with such coverage of Orthodox Christianity and Buddhism, in reports about shamanism journalists distance themselves from the position of shamans and those involved in shamanic practices. Reporters clearly indicate that what they describe is what 'shamans themselves say'. See for instance *Vesti-Buriatiia*'s report of 23 May 2012 on the celebration of the shamanic festival of Tailgan.
57 *Vesti-Buriatiia*, 'Otkrytie Sume Zelenoi Tary', 10 September 2011.
58 *Vesti-Buriatiia*, 'Prazdnovanie Maslenitsy v etnograficheskom muzee narodov Zabaikalia goroda Ulan-Ude', 10 March 2011.
59 On the impact of these beliefs on the formation of public discourses and politics in contemporary Buriatiia, see Bernstein, 'Remapping Sacred Landscapes', 23–46.
60 The quote is from *Vesti-Buriatiia*, 'Den' pokloneniia XII Pandito Khambo Lame Dasha-Dorzhi Itigelovu', 19 September 2012. *Vesti-Buriatiia* regularly broadcasts reports on Lama Itigelov.
61 *Vesti-Buriatiia*, 'Reportazh o veterane Velikoi Otechestvennoi voiny', 4 May 2011.
62 *Vesti-Buriatiia*, 'Chestvovanie veteranov-urozhentsev Buriatii v Moskve', 6 May 2011. For other instances of the usage of the notion of *malaia rodina* see, for instance, *Vesti-Buriatiia*, 'Osveshchenie Pandito Khambo Lamoi D. Aiusheevym Shibertuiskogo datsana', 16 October 2010.
63 See Dmitry Shlapentokh (ed.), *Russia between East and West: Scholarly Debates on Eurasianism*, Leiden: Brill, 2007, and Caroline Humphrey, '"Eurasia". Ideology and the Political Imagination in Provincial Russia', in Christopher M. Hann (ed.), *Postsocialism. Ideals, Ideologies and Practices in Eurasia*, London: Routledge, 2002, pp. 258–76.
64 See Marlène Laruelle, *Russian Eurasianism: The Ideology of Empire*, Baltimore: Johns Hopkins University Press, 2008, and Gerasimov *et al.*, 'The Postimperial Meets the Postcolonial', pp. 102–9.
65 *Vesti-Buriatiia* report on the celebration of the 15th anniversary of the foundation of Buriatiia's State University in Ulan-Ude on 28 September 2010. Buriatiia's national orchestra, whose performances are covered on the republican state television, is called 'The Sounds of Eurasia'. See *Vesti-Buriatiia*, 'Piatyi mezhdunarodnyi festival "Zvuki Evrazii"', 20 October 2010.
66 *Vesti-Buriatiia* report on the celebration of the 15th anniversary of the foundation of Buriatiia's State University in Ulan-Ude on 28 September 2010.
67 'Buriatiia ostaiotsia bez litsa', 16 June 2004. Available online at http://pressa.irk.ru/number1/2004/24/005001.html (accessed 21 March, 2013). The *okrug* is now part of the Siberian Chita region. Until 1937, it was part of the Buriat-Mongol ASSR. See also Tatyana Skrynnikova and Darima Amogolonova, 'Symbols of Post-Soviet Buriat Consolidation', *Zhurnal Sibirskogo Federalnogo Universiteta*, 4/6, 2011, 797–8.
68 *Vesti-Buriatiia*, 'Velikii Khural Mongolii priznal 14 noiabria ofitsialnoi datoi rozhdeniia velikogo polkovodtsa', 15 November 2012. It is noteworthy that at the

time of our writing the holiday has not been introduced in Buriatiia. A negative view of Genghis Khan in the Russian tradition is likely to be the main obstacle to establishing it.

69 *Vesti-Buriatiia*, 'Iubilei Vostochno-sibirskoi gosudarstvennoi akademii kul'tury i iskusstv', 23 September 2010.

70 Anya Bernstein, 'More Alive Than All the Living: Sovereign Bodies and Cosmic Politics in Buddhist Siberia', *Cultural Anthropology*, 27/2, 2012, 267.

71 Badmatsyrenov, 'Sangkha i politika', 139.

72 Bernstein, 'More Alive Than All the Living', 267.

73 Tolz, *'Russia's Own Orient'*, pp. 101–9.

74 Ibid., 124.

75 'Buddizm v istorii Buriatii'. Available online at http://cheloveknauka.com/buddizm-v-istorii-Buriatii-i-rol-agvana-dorzhieva-v-ego-rasprostranenii-konets-xix-v-30-e-gg-xx-v#ixzz2Nu5rQEvb (accessed 23 March 2013).

76 Bernstein, 'More Alive Than All the Living', 270, 274.

77 *Vesti-Buriatiia* reports 'Traditsionnyi turnir Buddiiskoi Sangkhi Rossii v chest' Khambo lam Goman-datsana', 20 January 2012 and 23 January 2012.

78 *Vesti-Buriatiia*, 'Tibetskie monakhi prepoodnesli mandalu Buddy mudrosti reke Ude', 13 August 2012.

79 Tolz, *'Russia's Own Orient'*, p. 136.

80 *Vesti-Buriatiia*, 'Sostoianie istoricheskikh pamiatnikov i perspektivy turisticheskoi industrii', 18 April 2011, and 'Vizit Premier Ministra Mongolii v Buriatiiu', 13 June 2011.

81 *Vesti-Buriatiia*, 'Sostoianie istoricheskikh pamiatnikov'.

82 *Vesti-Buriatiia*, 'Prepodavanie Buriatskogo iazyka s ispol'zovaniem iurt', 10 November 2010.

83 *Vesti-Buriatiia*, 'Tsikl peredach "Buriatiia: Khronika vekov"', 25 March 2011.

84 'Buriatiia. Khonika vekhov. Bauntovskii raion'. Available online at www.youtube.com/watch?v=qs6WN205ubc (accessed 23 March 2013).

85 Tarik Jusic, 'Media Discourse and the Politics of Ethnic Conflict: The Case of Yugoslavia', in Pål Kolstø (ed.), *Media Discourse and the Yugoslav Conflicts. Representations of Self and Other*, Farnham: Ashgate, 2009, pp. 24–38.

86 Gadi Wolfsfeld, *Media and the Path to Peace*, Cambridge: Cambridge University Press, 2004, p. 42.

Part III

Difference in question

7 (Un)covering alterity

Television, the 2012 presidential elections and the ethnic underside of Russian political discourse

A powerful new impulse for reform appeared to take Russia by storm in 2012. The explosion of protest following Putin's victory in Russia's latest presidential elections was, in fact, the culmination of a process stretching from the initial outrage at Putin's announcement in September 2011 that he would run for a third term, through protests at the Duma election results later that year, to the remarkable scenes which greeted Putin's re-election and the vigorous campaigns for and against the punk group Pussy Riot, who were imprisoned for 'desecrating' an Orthodox cathedral with an anti-Putin 'prayer' (the focus of our next chapter). During critical historical periods the 'political unconscious' of a society's prevailing discursive regime is revealed. Federal television broadcasts during the period prematurely hailed as the 'Russian Spring' are an ideal source to analyse the key subtexts of the 'authoritative discourse' which dominated state television during the new millennium. In this chapter we examine the initially subdued role of ethnicity in television's approach to consensus management as it struggled to negotiate the competing demands of the democratic and nationalist oppositions. We argue that ethnicity's virtual absence from the surfaces of mediated official discourse during the presidential election campaign belied its presence as a deep underside poised at any moment to disrupt and overturn that discourse's very foundations.

Critical historical periods tend to highlight the tensions underlying a society's prevailing discursive regime. It is consistent with this phenomenon that the journalists whom we interviewed in early 2013 confirmed that federal television coverage had altered in the preceding 18 months, while building on pre-existing framing techniques and analytical templates.[1] Television broadcasts during the first half of 2012 are therefore an ideal source from which to gauge the specificities and assess the effectiveness of the 'authoritative discourse' on political matters which Putin's government has been attempting to articulate in the course of the new millennium. Our approach here is to explore these phenomena by attempting to identify the illocutionary strategies that they reveal: what coverage of the issues at stake is 'doing', or not doing, rather than what it is 'saying', or not saying (we dwell on the latter in more detail in our final chapter). We shift our attention from how federal broadcasts 'stage' a particular 'content' in relation to the management of diversity in the new political environment to how that staging itself 'performs' certain moves, overt and inadvertent, in respect of the ways in which politics and ethnicity articulate with one another.

Recent Kremlin discourse dates to the middle of the first decade of the new millennium, when there was a major turn in the Putin government's policies in response to the colour revolutions in Georgia and Ukraine and the Beslan school hostage crisis. This discourse, emerging during a heightened sense of insecurity, has included representations of the 'west' as Russia's primary 'other', perpetually conspiring to destroy the country,[2] as well as a particular account of Russia's inter-ethnic relations according to which the country's ethnic and confessional diversity, whose origins in Russia's imperial and colonial past remain uninterrogated, have been presented as Russia's main strength. Any challenges to this rhetorical celebration of Russia's multi-ethnicity have been condemned as actions of the country's external and internal enemies. The greatest of these challenges was the Manezhnaia race riots of December 2010. As Chapter 4 demonstrated, the Russian media struggled to explain their significance and the Kremlin was concerned enough to initiate policies designed to deal with them.[3] In their immediate aftermath, President Medvedev singled out the maintenance of 'civic peace and interethnic and inter-faith harmony' as Russia's 'biggest national priority' and announced plans for a new federal programme on 'Strengthening the Unity of the Russian (*rossiiskaia*) Nation and the Ethnic and Cultural Development of the Peoples of Russia'.[4]

Though Manezhnaia remains a milestone in the troubled recent history of Russian inter-ethnic relations, by the time of the 2012 elections it had receded in the public consciousness and did not feature prominently in coverage of the political clashes themselves. Indeed, at first glance, state-aligned news coverage of the presidential election campaign significantly downplayed the importance of the 'national question' as an issue in need of urgent attention. This chapter will argue that it was far more central to the discourses employed by the presidential candidates, as well as by the opposition, than was acknowledged by the Russian media. Western reporting, too, concentrated on suspected irregularities favouring the discredited election winner, and on Putin's prospects for survival in light of popular revulsion at his underhand, anti-democratic practices, his links to corruption, a weakening economy and the increasing confidence that the democratic opposition movement could inflict real damage on his reputation.[5] Yet in the aftermath of Putin's re-election, the importance of inter-ethnic relations to Russia's political and social life was restated with a greater emphasis than ever before, by political leaders of all shades, and was highlighted by state broadcasters.

We begin with the point at which ethnicity and the (pseudo-)democratic process seemed openly to coincide. Prior to the election, Putin wrote seven campaign articles, each tied to a particular theme in his political agenda. One of these articles, published in *Nezavisimaia gazeta* (in theory if not in practice an independent rather than government-controlled outlet), dealt with 'the national question' and provided Putin with a platform for his post-Manezhnaia thinking. The article was Putin's first ever attempt to articulate an authoritative discourse on an issue that hitherto had been mostly debated by journalists and public intellectuals. We first provide a brief account of the article itself, noting (i) explicit connections made between ethnicity and politics within the article and (ii) the article's relationship to the set to which it belongs, and thus to the wider campaign.

We focus primarily on Channel 1's coverage of the article, as it was the most extensive, highlighting six key rhetorical strategies for co-articulating seemingly disparate issues: (i) discursive 'contamination'; (ii) the declaration of false 'equidistance' from issues of different value; (iii) the use of 'code' terms in order to establish relations of equivalence, or of synecdoche, across disparate topics; (iv) the conducting of polemics 'by proxy' (deploying the voice of one's opponents in support of one's own position); (v) the 'performative' enactment of inter-ethnic harmony/disharmony; (vi) the inversion of the logic of the 'universal' and the 'particular' as a means of promoting the specific and ethnic over the abstract and political.

Our analysis enables us to posit the media's central role in a hegemonic process of 'manufacturing consent'[6] in which questions of ethnicity (a priority of Putin's conservative antagonists) are injected into the discourse of democratic debate (the rallying cry of the opposition), but then filtered from it through a focus on the 'struggle against corruption' (an example of our emphasis on illocutionary act over locutionary statement). Drawing on Laclau's theory of populism as the articulation of fragmented demands around a new hegemonic core, we then discuss the figure of Aleksei Naval'nyi, a prominent leader of the opposition movement and the target of a media campaign of vilification and exclusion.[7] We examine the role of Naval'nyi's own populist campaign in constituting such a core and in accounting for federal television's neurotic reaction to Naval'nyi's claim to the anti-corruption territory, pointing, however, to that territory's ethnic substratum, and demonstrating that it is liable to rise to the surface at any point.

Putin's 'national dilemma'

Putin's pre-election campaign articles dealt with the 'national question', economic tasks, democracy and effective government, social policy, military affairs and foreign policy. The national question was the subject of the second of the articles, released in January 2012 in *Nezavisimaia gazeta*, whose associations with free speech gesture toward the insertion of the national question into the realm of democratic debate.[8] From the outset, inter-ethnic cohesion and democracy are linked in the official symbolism of the Kremlin campaign. Indeed, the gesture is largely staged, since despite its title, the newspaper has been frequently used to promote the Kremlin line. Recognizing the dangers posed by a failure to deal with the 'national question', Putin divided his article into five sections, treating: (i) the inadequacy of international models; (ii) Russia's historical experience of multi-ethnic statehood; (iii) ethnic Russian cultural heritage as a 'cultural code' for unifying the nation's diverse ethnic communities; (iv) national politics and the role of strong institutions; (v) the problem of migration.

Rejecting both the American 'melting pot' approach and the separatism fostered by European multiculturalism, Putin is equally forceful in his objections to ethnic purism and nationalist extremism. He emphasizes Russia's multi-ethnic, multi-faith history, invoking Dostoevskii's claims that Russia's very destiny lies in its unique ability to unify disparate peoples. Treading an awkward line between proposing a 'polycultural', civic model and insisting on the primacy of one community over others, he eventually accords ethnic Russians a special position within the national community by referring to them as a 'state-forming' people – the glue which binds the plurality of peoples together.[9] This leads him to advocate adherence to a 'cultural code' based on the Russian language, Russian literature and culture and Russian history, but respectful of the rights of minorities to maintain their own traditions and cultures. The instillation of such a code relies on powerful institutions of state and Putin singles out education as key. He concludes with a proposal to tighten laws to ensure that immigration works in favour of Russia's 'indigenous' (*korennoe*) population. This term introduces further ambiguity, as it can be used either in specific reference to ethnic Russians or Slavs (the dominant definition) or to mean representatives of minorities, in theory including North Caucasians, whose 'ethno-national homelands' are located within the Russian Federation. In his concluding remarks about migration, Putin distinguishes between 'external' and 'internal' migration, acknowledging the rights of non-Russian citizens to move freely around their country without fear of discrimination but, in an echo of the post-Manezhnaia line eventually settled on by the Kremlin, calling on them to respect the behavioural norms of the cities and regions in which they now reside. Given that the dominant flow of migration is from the 'ethnic periphery' to 'Russia proper', the call to migrants to respect the behavioural norms of their new places of residence is one-sided.

State-forming news?

Unsurprisingly, the article received largely reverent coverage on Russia's federal channels. Channel 1 is normally the most reliable guide to Kremlin thinking. The fact that, as we shall see, it unusually included trenchant criticism of Putin's arguments deserves explanation and accounts for the particular attention we devote to it. Two *Vremia* reports in the bulletin of 23 January dealt with the article itself and with Putin's address to a forum held in Kislovodsk, in the North Caucasus. The Sunday edition included a lengthy summary on 29 January. In addition, the weekly discussion programme 'In Context' ('*V kontekste*'), broadcast on 26 January, centred on a debate on the article involving the controversial Eurasianist Aleksandr Dugin, film director Stanislav Govorukhin, the editor of the liberal-leaning Moscow Echo radio station, Aleksei Venediktov, and liberal historian and television journalist Nikolai Svanidze.

Both *Vremia* reports utilized the familiar rhetorical device of eliding direct and free indirect speech in such a way as to blur the boundaries between the authoritative voice of Putin and that adopted by the reports themselves:

> Putin warns against the idea of a mono-national state in our country which has historically gathered together the most varied of peoples around the Russian

> people as its core. Russia is not some kind of immature calque on Western countries which it needs to catch up with. It is a unique civilization ... The article provoked many arguments.[10]

The first report addressed the issues raised in the order in which Putin presented them. However, it illustrated the first section on the failure of European multiculturalism with footage juxtaposing an illegal Central Asian migrant worker and a North African French citizen going about his daily business.[11] The conflation of illegal immigration with minority communities was indicative of the lack of rigour characterizing Russian public debate on inter-ethnic cohesion issues, and of the tendency for state broadcasters to stray from protocol. Likewise, commentary on Putin's rejection of all extremisms (mono-ethnic nationalism; separatism) is accompanied by images of the Manezhnaia, Kondopoga and Sagra conflicts. While the influence of Russian nationalist impulses can be considered as part of the context to these conflicts, 'separatist nationalism' is not an ideological factor. The potential dangers of this elision are reinforced by the fact that one of the points from the article omitted by *Vremia* is Putin's warning about the dangers of political parties established on a regional basis, and conversely by the extent of the attention paid to the section on the 'behaviour' of 'visitors'. Overall, the issue of migration was not central to *Vremia*'s coverage. In contrast, *Vesti*'s 23 January report on the article paid considerably more attention to problems associated with migrants than did the article itself.[12]

Given that the article was presented as part of Putin's electoral manifesto, the inclusion of critical commentaries is noteworthy. *Vremia*'s main report of 23 January incorporates an interview with Geidar Dzhemal (President of the Islamic Committee of Russia), who criticizes Putin for failing to distinguish sufficiently clearly internal and external migration, and for implying that it is always 'visitors' whose norms are at fault:

> The assumption is made that local populations are always more cultured and civilized than migrants, regardless of whether these are internal or external migrants ... And the Soviet experience is completely ignored because Soviet policy was founded precisely on extensive internal migration.

This is followed by an observation from Vitalii Tret'iakov, a well-known journalist, dean of the Faculty of Television Broadcasting at Moscow State University, editor of the journal *Political Class* and a vocal critic of the government's nationalities policies for their alleged failure to accommodate ethnic Russian interests. He claims that Putin overestimates the capacity of education to change ethnic identity. Dissent is also featured in the Sunday report.[13] Oleg Orlov, president of the human rights organization Memorial, argues that tightening registration rules will only increase corruption. Boris Makarenko, president of the Centre for Political Technology, warns that Central Asians forced to study Russian literature may shun Russia altogether. The incorporation of dissenting voices is a reflection of genuine disagreements over inter-ethnic relations, but, above all it performs the 'democratic process' by permitting a dialogue within tightly controlled parameters. Before critics are given a voice on the Sunday *Vremia*, the reporter notes that on 'matters of principle' (*glavnoe*) 'everyone agrees; disagreements relate only to specific points' (*raznoglasiia po chastnostiam*). Critics are in the minority and their remarks are preceded and followed by those of Putin's unequivocal supporters.

The range of voices serves other discursive purposes. A question addressed to Putin at the Kislovodsk forum and aired within the second *Vremia* report consists of an attack on 'liberal forces' which 'reject and hate Russia'. The questioner identifies such forces with 'some of those who aspire to be our leaders' – an indirect reference to the opposition.[14] Putin, naturally, agrees. Yet Orlov's doubts about migration rules are accompanied by an attack on Russian nationalists whose presence at the anti-Kremlin demonstrations Orlov considers 'a mistake'.[15] Following Orlov's intervention, the report cites Putin's own implicit critique of Aleksei Naval'nyi, the popular opposition figure who is not averse to deploying nationalist rhetoric: 'Slogans like "Stop Feeding the Caucasus" appear. If that logic is pursued, Putin writes, ... you bring about the collapse of the entire country.'[16]

Orlov's status as a member of that same opposition authenticates first, the aspersions cast upon Putin's democratic opponents, second, his own rejection of the nationalist right and third, the sense that critiques of the absence of democratic debate in Russia are unjustified. Orlov is used as a proxy weapon, targeting two opponents at once (nationalists and pro-western liberals) and guaranteeing their mutual contamination.[17]

The Sunday *Vremia* report also accommodates the voices of Aleksandr Prokhanov, editor of the nationalist newspaper *Zavtra*, and Aleksandr Belov, ex-leader of the far-right anti-immigration movement DPNI, who reinforces Putin's argument in favour of 'visitors' adhering to the behavioural norms of their places of residence by invoking the familiar stereotype of the Muslim slaughtering a sheep on the streets of Moscow. The pluralizing strategy establishes Putin's position as equidistant from both extremes. Through it, Channel 1 enacts 'democratic pluralism', and constructs the official line as occupying the locus of consensus.

The same strategy is aimed at enacting the 'unity in diversity' which Putin claims to be central to Russian national identity, for the appearance of ideological pluralism is matched by an inter-ethnic equivalent. The voices of a Russian, the Chechen Kadyrov and a Tatar (Dzhemal) are joined by that of Alla Gerber, selected to represent the liberal position within Russia's Jewish community, who endorses Putin's rejection of Naval'nyi's 'separatist' logic. The Sunday *Vremia* report is structured as a collation of different views on Putin's account drawn from across the political spectrum, and from a multitude of nationalities.[18] The purpose of the overblown rhetorical performance of unity can thus be seen as providing a polemical rebuttal to perceptions of disunity. When presenting the report on Putin's contribution to the Kislovodsk forum (an event whose North Caucasian location was itself significant), the newsreader includes in his introductory gloss the claim:

> Old and young, Orthodox and Muslims, people of different experience and convictions, are united in their love for their common motherland. In terms of its national, religious and social make-up, this is perhaps the most complex region in our country and it has survived more than one historical drama. But in spite of these problems, when we are united we are undefeatable, Putin recalled.[19]

But the unanimity expressed in the Kislovodsk forum barely confounds the image of a region torn by conflict. In this sense, the assertion of harmony uncovers the disharmony against which it is directed. Putin's own attempt to set the tone for the forum illustrates the perils of the polemical strategy: 'One of the most impressive examples of our strength of spirit and capacity for self-sacrifice was the legendary defence of the Brest Fortress. Its immortal defenders consisted of people from different nationalities.'[20]

Not only is the Brest Fortress not in Russia, but the state which defended it (the USSR) fragmented in 1991, precipitating the very conflict that brought Putin to Kislovodsk at that time. Moreover, there is more to the hegemonizing strategy than the mere achievement of ideological equidistance. It must be accompanied by an effort to articulate the common values underpinning that unity. Gerber's endorsement of Putin in the Sunday report is followed by the reporter's suggestion that 'Supporters and critics of Putin's proposals agreed on one thing: that we all need to begin with ourselves – to behave with respect toward visitors, no matter where they are from.'

Implicit here is one of the main themes to emerge from Putin's article: the call for adherence to behavioural norms. The Sunday report deploys the voice of the Russian writer, Zakhar Prilepin, to convey the message that the call is a general one, reflecting a universal value:

> We give each other bribes, steal from one another, scratch one another's backs and the Caucasians who come here and do the same irritate us. But let's resolve our own problems and the Caucasians and other people living in Russia will come on board.

Prilepin's comments undermine the objectivity claimed for the appeal for adherence to common norms. His superficially 'balanced' summary is preceded by the rather more partial commentary of a politologist, Leonid Davidov, who confirms that Putin's call for a universal commitment to cultural norms on all sides is a euphemistic mask for an assault on the specificities of migrant behaviour:

> In the article the accent was placed on the fact that migrants must not break local customs and traditions. This applies to everyone and the most vulgar crude and flagrant behaviour ought to be harshly, though legally, dealt with. It is an especially good signal for the peoples of Russia's South.[21]

In effect, Davidov highlights the suspicion of all sub-cultural diversity inherent in Putin's 'traditionalist' model of multiculturalism, which is based not on a shared liberal adherence to individual rights but on the recognition of historical and cultural commonalities of citizens of Russia. Channel 1's use of proxy voices enables the different sub-textual currents in Putin's ideological hybrid to be recalibrated so that the universal terms are coded with strong ethnic undertones. Analogously, the use of neutral terms such as 'welfare recipient' serves as a right-wing code for America's black underclass.[22]

Closely tied to the 'behavioural norms' theme is that of corruption and 'the equality of all before the law'. The rhetorical strategy is identical; an issue is extracted from the Manezhnaia 'archive', detached from its North Caucasian context, then transformed via quasi-direct discourse into a universal principle authenticated by Putin, and finally adopted as a code term:

> The disturbances on Manezhnaia Square, the shoot-out in Sagra and the earlier riots in Kondopoga: these events are … unified by one thing – corruption and the incompetence of the law enforcement organs. 'We need to raise the level of competence of our law enforcement organs … So that the local population feels protected … When all are equal before the law – then there will be no problems'.

These gestures short-circuit the universal/particular relationship. Rather than the universal ('the equality of all before the law') subordinating to it all particularities (Russian, Caucasian, Tajik, etc.), the meaning of that universal is covertly altered to apply to one term (Caucasians) exclusively. This is the reverse of what happens with what Putin terms the 'cultural code'. Here, a singularity ('Russian language, culture and tradition') is overtly accorded universal value.

In the *V kontekste* edition of 26 January dedicated to 'the national question' and focused on the discussion of Putin's article, Svanidze acknowledges the danger posed by the insidious ethnic coding of terms such as 'corruption',[23] offering his own socio-economic account of the phenomenon. Yet, despite the fact that Svanidze's view has support from Putin's own statements, the host, Maksim Shevchenko, is sceptical of the idea that corruption arises solely from a generalized inequality, and he challenges Svanidze twice on the point. Shortly after Svanidze's intervention, Shevchenko attempts a third time to 'bring the discussion back to concrete issues': 'Can we talk about concrete things? … What about the specifics of when a person from a Tajik village arrives? …'

Venediktov supports Svanidze's liberal position on the implications of Putin's article. They are opposed by Dugin and Govorukhin. The selection of participants in this talk show is an attempt to give voice to 'extremes' – Svanidze and Venediktov on the left, versus Dugin and Govorukhin on the right. Putin's hybrid position thus appears as consensus-building and stability-endorsing. Indeed, every main point of his article is enthusiastically supported by one or another speaker. The programme ends with Shevchenko's own description of Putin's article as a clear guide for how to manage inter-ethnic relations in Russia by ensuring everyone's 'equality before the law' and by accepting and respecting cultural differences, yet maintaining the primacy of the Russian language and of Russians as the 'state-forming people'.

Compared with post-election talk show coverage of the 'national question', the participants and the moderator of the 26 January edition of *V kontekste* are careful not to present the situation in Russia as alarming and out of control, nor to disagree openly with any of Putin's statements. When an edition of the same programme broadcast just three days before the election discussed the key issues that the new president would need to address, the 'national question' did not feature at all.[24] This downplaying of problems in the sphere of inter-ethnic relations other than in connection with the debate around Putin's article reveals an acute awareness of the potentially explosive nature of the issue, and of the fact that it tends to polarize rather than unite society. It also points to broadcasters' uncertainty concerning their ability to maintain the superiority of Kremlin-sponsored discourse on the 'national question' over competing discourses.

Rather than offering more opportunities for disagreements, other *Vremia* reports during the presidential election campaign further narrowed down coverage of the 'national question' to construct an image of Putin as the supreme guarantor of inter-ethnic harmony. In a highly staged representation of Putin's meeting with his supporters at Moscow's Luzhniki stadium on 23 February 2012, the Sunday *Vremia* report foregrounded Putin's capacity in this regard in its account of how his supporters viewed him. (Putin's own address to the crowd did not feature any discussion of inter-ethnic relations.) The first pro-Putin voter to be given a voice was from Dagestan. Referring to the Russian army's repulsion of the invasion of Dagestan by the Chechen separatist leader Shamil Basaev's forces in 1999, he hailed Putin for bringing peace to the region at a time of crisis.[25] The choice is unlikely to be accidental. The 1999 invasion of Dagestan has acquired a symbolic meaning in anti-western conspiracy theories, which have flourished in Putin's Russia and were frequently evoked on state-controlled television during the election season. The Basaev-led invasion is claimed to be precisely the point at which unspecified enemies of Russia, acting in consort with western intelligence services, attempted to trigger the country's disintegration. These accounts represent Putin as the politician who saved Russia from being partitioned and put under the direct control of western powers.[26]

State television coverage of the presidential election campaign made no reference to how the candidates dealt in their public speeches with the so-called 'Russian question' or of whether the ethnic Russian majority should enjoy special status or rights, despite the fact that the issue is of significant public interest. Opinion polls indicate that the constitutionally acknowledged equality of all the nationalities of the Russian Federation is viewed critically by large segments of the population. Of particular concern are the attitudes of young ethnic Russians (aged between 18 and 25). A 2011 Levada Centre poll on the state of inter-ethnic relations found that only 37 per cent of respondents from that group supported the equality of all the nationalities of the country, whereas 55 per cent were either in favour of the notion 'Russia for the Russians' or favoured greater rights for ethnic Russians.[27]

To one degree or another, every candidate exploited such sentiments in their public statements. Putin's evocation of the notion of the 'state-forming' people was among the more cautious responses.[28] Zhirinovskii was, predictably, rather more vocal in addressing the 'Russian question'. The broadcasters' decision not to dwell on this aspect of the electoral debate represents an acknowledgement of its divisive nature and of the Kremlin's inability to articulate a convincing consensual position on it. Similarly, the presence in opposition rallies of Russian ethno-nationalists, including those at the extreme pole of the spectrum, tended to be mentioned only in passing. Neo-Nazi groups' participation in anti-Putin rallies was ignored. In order to discredit the opposition, by contrast, state broadcasters continually underscored its heterogeneity and lack of effective leadership.

Yet it is clear that Putin's electoral team was aware that inter-ethnic relations were perceived in the popular imagination as being in acute crisis, and that violent incidents interpreted by those affected as having an ethnic basis provoked widespread fear. New media were used by the Putin campaign's organizers to exploit this fear. It was reportedly Putin's electoral headquarters (*vybornyi shtab*) that placed on the Internet a pro-Putin advert warning that unless he was elected, Russia would disintegrate along ethnic lines amid a bloody civil war. Alternative media, such as REN TV's *Nedelia*, discussed the clip, but its appearance and the public controversy around it were ignored by the state-controlled channels.

The agenda 'beyond' ethnicity

The performative strategy adopted by the state-aligned media during the presidential election campaign relied, then, on managing consent by both manipulating and suppressing the national question. In a parliamentary democracy, the consensual centre ground is the point from which hegemony operates. In Gramsci's theory, consensus and hegemony are mutually dependent.[29] The question of the extent to which the hegemonic process is at work in Russia is contested. According to Andrew Wilson, the restricted pluralism accorded to Russian state-controlled television is far from hegemonic in the Gramscian sense and amounts to no more than a cynically manufactured, superficial, 'imitative' democracy imposed by a fully subservient media.[30] While not rejecting Wilson's linear account, the position we adopt throughout this book is one for which that account oversimplifies a relationship between Kremlin, media and populace which is more dynamic, 'hegemonic' and complex than the 'imitative democracy' model allows. Why, for example, should an all-powerful Kremlin need to create such an elaborate artifice if not to respond to a popular demand for pluralistic dialogue? The question also arises of how well this enacted pluralism fulfils its main function in ensuring that Kremlin-sponsored discourse enjoys authoritative status.

Channel 1's reporting strategies are more consistent with a reading of Russia's wider political environment in which the media are ideologically aligned with, but do not merely regurgitate, a Kremlin position that is itself fragmented and sufficiently vague to generate multiple interpretations. For example, the deployment of proxy voices to articulate different emphases within the official stance on the national question facilitates the recalibration of the stance from within that of opposing forces, enhancing its pre-eminence within the hegemonic process. Inversions of the relationship of particular to universal indicate a struggle to inflect definitions of values such as 'equality before the law' and 'the rejection of corruption' with meaning favourable to one political or social force. Yet the manner in which the polemical subtext to the performance of inter-ethnic harmony exposes the disharmony against which that performance is directed confirms the fragility of any position of hegemonic dominance.[31]

In exploring how ethnicity informs the hegemonic process beyond the topic's seemingly limited presence in Putin's overt election campaign agenda, we might first touch upon the broader fate of axiological principles emerging from discourse specific to the 'national question' article. On 6 February 2012, Putin published in *Kommersant* (another newspaper with a reputation for independence) his fourth campaign article, dealing with 'Democracy and the Quality of Our State'. It includes one whole section on 'corruption' and another on 'Russian federalism'. In the latter, Putin refers to 'countering blatant separatism as well as its latent, subtle varieties, along with breaking up the convergence between certain regional authorities and organized crime or nationalist groups.' He grounds his defence of 'a strong federal centre' in the claim that

> People's lifestyles are determined by their specific traditions, customs and models of behaviour. Therefore, we have certain assets of unquestionable value, namely the powerful consolidating factors such as the Russian language, Russian culture, the Russian Orthodox Church as well as Russia's other traditional religions.

Here Putin is clearly developing the 'cultural code' concept from his earlier article. The *Vremia* and *Vesti* reports on this article broadcast the same evening single out the passages on corruption, ensuring that the longest direct quote from the 'Democracy' article is drawn from them.[32]

The initial 'overview' article, published in *Izvestiia* on 16 January 2012, likewise references corruption and the 'fight against separatism'. *Vremia*'s coverage of the third article (published in *Vedomosti* on 30 January 2012), on the economy, highlights Putin's identification of the key issue facing Russia's domestic economy, connecting it with the legal context familiar from his approach to the 'national question': 'The main problem is a lack of transparency … in the workings of state representatives, from customs and tax services to the courts and law enforcement system. To call a spade a spade, we are talking about systemic corruption.'[33] In the reports broadcast by *Vremia* and *Vesti* on Putin's fifth article, published in *Komsomol'skaia pravda* on 13 February 2012, the theme of corruption surfaces again, this time in relation to housing issues: 'The recipe is familiar: fighting corruption and monopolism; greater transparency and private investment.'[34]

Naval'nyi and the populist 'chain'

Ethnicity's centrality to the hegemonic process is indicated by the subliminal manner in which it colours the corruption theme so ubiquitous in Russian political discourse. The corruption theme is most closely associated with Aleksei Naval'nyi, who first rose to fame in the western media following the 2011 Duma elections, when he brazenly dubbed the Kremlin-backed United Russia 'the Party of Swindlers and Thieves'. In July 2013 the charismatic blogger was sentenced to five years in prison, having been convicted of a specious embezzlement charge. Released in a surprise move before he had served any of his term, he went on to provide an unexpectedly robust challenge to Sergei Sobianin in the Moscow mayoral election of September 2013, and remains a popular force. In February 2014 he was again placed under house arrest for participating in an opposition demonstration against the sentencing of the Bolotnaia protesters.

Naval'nyi has used his specialist legal knowledge to conduct a relentless campaign against corruption at the heart of the Russian body politic, a phenomenon he links directly to Putin. Revulsion at corruption is widespread and the authorities attempt to use it as their own tool for political mobilization. Putin and his most serious opponent are engaged in a battle to inscribe themselves into this popular sentiment, and to externalize one another as its objects. The official counter-strike against Naval'nyi is based on a strategy of tarring the campaigner with his own brush and undermining his popular reputation by associating him with the criminality that he, in turn, attributes to his pursuers. Here, all pretence at pluralism is abandoned. Channel 1 turned to Mikhail Leont'ev, presenter of the virulently pro-Kremlin opinion piece, *Odnako* (However):

> That our 'Bolotnaia' opposition is so painfully concerned about corruption … and that Comrade-Mister Naval'nyi is at the forefront of these concerns, every child knows. Yet the fact that Mister Naval'nyi is himself a highly paid, hired entrepreneur in the services of that very same corruption is known to a much narrower circle.[35]

Though Leont'ev's accusation lacks credibility, Naval'nyi is a more complex figure than is suggested by the western media's image of him.[36] His anti-corruption fight aligns him with the movement for political change in Russia and has enabled him to form allegiances with other leading liberal opponents of the Putin regime.

Naval'nyi was a prime mover in the protest against electoral fraud on Bolotnaia Square between December 2011 and February 2012, and in demonstrations on Pushkin Square that followed the March 2012 presidential elections. He has, along with the likes of Grigorii Iavlinskii of the liberal Iabloko Party, called vociferously for 'honest' elections and democratic reforms. Yet Naval'nyi often underscores his Orthodox faith. He was initially ambivalent in his support for Pussy Riot, rejecting their 'desecration' of the Cathedral of Christ the Saviour but vocally opposing their incarceration and trial.[37] Furthermore, he played a leading role in the Russian nationalists' 'Stop Feeding the Caucasus' campaign. In his LiveJournal blog he rebutted those accusing him of fomenting extremism, suggesting that they had 'fallen into the traditional trap of the liberal and are describing chimeras and prejudices' when they claimed dramatically that 'the virus of nationalism is very dangerous' and that 'a packet of yeast and we all die. Pogroms will begin'. He went on to contend:

> I don't see in the discussion of questions of budgetary equality, fairness and the rationale for government grants any threats of universal catastrophe ... The Kremlin swindlers are paying for the 120% vote extracted for the Party of Swindlers and Thieves by ploughing unheard of amounts of money into the North Caucasus, not forgetting to take their 30% cut.[38]

Naval'nyi grounds his participation in the campaign through references to financial transparency, electoral propriety and equality. Like Putin, though from a different angle, he frames his position from a universalist stance.

Still more controversial is Naval'nyi's participation in the 'Russian March', whose slogan 'Russia for the Russians' has become a nationalist battle cry. Interrogated about his political allegiances during a Moscow Echo radio interview, he described himself as a 'national democrat': 'I am in favour of a democratic form of government but I consider that Russia has particular national qualities. We differ significantly from European countries, but must definitely seek a European path of development. Nonetheless, Russia is no Holland.'[39]

When pressed on his apparent support for the 'Russia for Russians' slogan, and on his participation in the 'Russian March', Naval'nyi negotiated the tension by attempting to combine a commitment to a civic notion of Russian identity with an antagonism toward illegal migration:

> I mean Russia for Russian citizens, of course ... It is complicated to explain the meanings that people put into the concept of 'Russia for Russians', but I think that it is definitely not Nazism. We know that Russia has the highest number of illegal migrants in Europe and ... those [who support the slogan] are really saying 'Kick migrants out of our country'... When I attend events like 'Russian March', I see that the overwhelming majority of people are quite normal, and they consider themselves nationalists or national democrats ... If you ask them what they mean by 'nationalism' they will ... say that they are against corruption, against migration, and against our oligarchic economy. According to that terminology, you and I are both straightforward nationalists.[40]

Naval'nyi is a signatory to the manifesto of a small political grouping known as 'Narod' (The People), which confirms that 'corruption' is the biggest problem facing the Russian state and which envisages its first task as:

> to stop the process of the degradation of Russian civilization and the creation of the conditions for the development of the Russian people, their culture and language and their historical territory. The indigenous (*korennye*) peoples of Russia are firmly inscribed into Russian civilization and have long since tied their fate to that of the Russian people. National minorities have all the opportunities for successful assimilation, and for the preservation of their national specificity.[41]

Acknowledging its nationalist orientation, the manifesto attacks not nationalism itself but extremist xenophobia, for giving nationalism a 'bad name': 'National provocateurs foist xenophobia on the state with their calls for violence toward 'aliens' (*inorodtsam*), thus creating a negative image for nationalists.'[42]

Naval'nyi's meteoric rise to fame, combined with first his tortured efforts to steer a path between the liberal democratic wing of the anti-Putin protest movement and its nationalist right; second the Kremlin's desperate efforts, via the media, to marginalize and discredit him; and third the vicious competition between the opposing forces to achieve 'ownership' of the corruption issue, are all indicative of a set of conditions favouring the rise of a populist movement.[43] When Naval'nyi's accusations of corruption became intolerable, and in an example of an old media 'reframing' of a new media subversion, *Vesti* broadcast a report on the discovery of Naval'nyi blogs containing on one hand falsified charges of vote-rigging against United Russia, and on the other 'extremist statements inciting hatred toward Dagestan, Mordovia and Ingushetiia'.[44] Unsurprisingly, no space was found for Naval'nyi's rebuttal. The report disrupts what Laclau calls the 'chain of equivalent demands' on which populism relies, and which threatened to coalesce around Naval'nyi.[45]

For Laclau, populism is not a fixed ideological constellation 'but a series of discursive resources which can be put to very different uses', and which 'proceeds by articulating fragmented and dislocated demands around a new core'.[46] He stresses the need for heterogeneity as a precondition for populism, but argues that 'heterogeneity … has as a defining feature a dimension of deficient being or failed unicity'. The plurality of demands must remain discrete, yet contain the shared, unfulfilled urge toward unity which constitutes them as a 'chain'. [47] When one demand becomes core, it acquires the capacity to occupy the position of 'unity', ensuring that the chain is precisely one of equivalences, without erasing either its own distinctiveness or the antagonism of competing demands: the particular 'can also have a non-partitive meaning: not a part of a whole, but a part that is the whole'. It 'transforms its very partiality' in the name of a 'transcendent universality' whose impermanence guarantees the continuation of the hegemonic struggle.[48]

We have already observed the instability of relations between universal and particular in our analysis of *Vremia*'s coverage of Putin's election campaign. Moreover, the 'transcendent universality' can, as Laclau insists, never be permanent, for 'this moment of fusion between partial object and totality represents, at one point in time, the ultimate [and unattainable] historical horizon'. Driven by the persistent encounter of mutual antagonisms, history is in truth 'a discontinuous succession of hegemonic formations that cannot be ordered by a script transcending their contingent historicity'.[49]

It is in keeping with Laclau's account that anti-corruption protests, calls for free speech, pleas for economic reform and campaigns around slogans such as 'Stop Feeding the Caucasus' and 'Russia for the Russians' should remain discrete, even mutually 'antagonistic', demands. The antagonism of representatives of Central Asian migrant workers toward the liberally inclined makers of the subtly anti-xenophobic television comedy *Nasha Russia* (see Chapter 5) illustrates how difficult it is for chains of equivalent demand to form in political environments lacking the basis for hegemony, such as Russia's. In the populist process, one demand is required to constitute the whole as an equivalence chain and fill the 'empty signifier' that is the claim to universality with its particular meanings. Despite the anti-hegemonic environment, several such impulses converged in the Manezhnaia crisis. Naval'nyi, meanwhile, as absent from Russia's state-controlled mediasphere as he is omnipresent in its blogosphere, juggles them with consummate skill. State-controlled television provides a plane on which the state enacts its part in the process.

Rather than colouring the entire political landscape, ethnicity features within the chain of demands from which a universalist claim to populist 'transcendence' might emerge. In themselves, anti-corruption protests may or may not carry hidden xenophobic motivations. But the effectiveness of the anti-corruption protest as a mobilization tool depends on whether it can establish itself as a core demand capable of investing universality with its meanings by establishing equivalence between disparate demands, several of which have ethnic undertones. Little surprise, then, that Naval'nyi, who seems best placed to accomplish the elusive feat of transcendence, is *persona non grata* on Russian state television, nor that Vladimir Pozner, who has survived as one of the few relatively independent voices left on Channel 1 – at the expense of numerous compromises with his principles – issued a threat to cancel his own show after a discussion of Naval'nyi, which took place on one edition, was cut (Pozner has prided himself on the fact that his show is never censored).[50]

Naval'nyi's intuitive ability to re-contextualize popular prejudices promulgated by his state media opponents, including those relating to the ill-mannered, unintegrated migrant worker, adds a further tool to his armoury.[51] He was, for example, closely associated not only with drafting the anti-migrant statement issued by Russia's Opposition Coordination Council, formed in 2012 following the 2013 murder of a half-Russian, half-Tatar soldier by a Chechen youth in Pugachev, but also of persuading the usually liberal journalist Oleg Kashin, to sign it (most liberal oppositionists did not follow suit, though the statement attracted some support from *Moscow Echo* listeners).[52] During the Moscow mayoral campaign of September 2013, he became embroiled in a scandal relating to an inflammatory entry in his *Live Journal* blog – an open letter to Sergei Sobianin, his Kremlin-backed rival, calling on him to desist from ordering his illegal Uzbek migrant worker employees to tear down banners from the balconies of Naval'nyi supporters, claiming that the migrants 'would tell their compatriots how to break into Moscow apartments', leading to 'an increase in burglaries'.[53]

The image of the migrant in Russian culture, and the related mythology surrounding the *Gastarbeiter*, is, as Naval'nyi well understood, redolent with significance, and in the fast-changing, febrile atmosphere following Putin's re-election it was repeatedly, and often with deeply contradictory intent, deployed by those with rather more access to the broadcast waves than Naval'nyi. We will explore the full extent of the contradictions in our final chapter.

Naval'nyi's knack for reconciling his tactical endorsement of elements of the far-right, anti-migrant agenda with his status as a figurehead of the democratic opposition to Putin illuminates the suggestive but problematic distinction that Tomasz Zarycki makes between core (western) and periphery (eastern) nationalisms in Europe. Zarycki argues:

> For the core-countries nationalism the key significant other is the migrant; for peripheral nationalism the other is the cosmopolitan liberal, who is betraying his country by assisting the core countries in imposition of their symbolic and material rule over peripheries.[54]

He concludes:

> In such a context Russia seems to be a case of both a peripheral and a lower-level core country at the same time. As a regional centre for part of Eurasia it attracts migrants, mostly from the poorer countries of the former Soviet Union. At the same time it is increasingly peripheral in relation to the western core and resents its pressure in acute forms. These two forces may mutually reinforce themselves producing a specific kind of nationalism.[55]

Zarycki thus accounts for the co-presence within mainstream Russian nationalism of both a powerful anti-migrant tendency and the virulent anti-liberalism that emerged in television coverage of the Pussy Riot scandal (the subject of our next chapter). However, he understands nationalism in a narrow sense and solely as the domain of inward-looking, isolationist and xenophobic movements and ideologies. Nationalism of all shades has in fact from its inception engaged in a transnational trading of ideas and iconographies. It has invariably staked universalist claims (Dostoevskii's Russia as the 'Christ among nations' representing just one example), and it has frequently sought out cross-border allegiances (hence the professed admiration of far-right US politicians such as Pat Buchanan for Putin's social conservatism). Binarism of the sort embraced by Zarycki inevitably tends toward collapse. Naval'nyi highlights a further complication in that, as his demonization by Russia's official media demonstrates, Russia's internal 'liberal other' is as capable nowadays of presenting an anti-migrant nationalist front as it is of promoting European multiculturalist tolerance. Ultimately, perhaps, the Laclauan approach to hegemonic 'chains of equivalence', not without blind spots and limitations of its own, provides the better apparatus with which to interpret the Naval'nyi phenomenon.

Federal television's neurotically exclusionary approach to the figure of Naval'nyi went hand in hand with its pseudo-pluralistic coverage of Putin's pre-election article. Each phenomenon pointed to the restricted hegemony characterizing the environment in which it operated. In each case, ethnicity operates as a disturbing, intractable underside; it is rarely foregrounded as such. In the maelstrom at the centre of the Russian media prior to and in the aftermath of Putin's re-election, it is displaced onto the theme of corruption around which, as we began to see in our analysis of the Manezhnaia riots, the battle for populist hegemony raged.

In the Laclauan model, ethnicity remains beyond the limit of the chain of equivalences, restricting its scope for expansion and limiting its power. But it remains as a trace, tainting and disrupting the hegemonic process. The situation is fluid, unpredictable and with any number of conflicting outcomes possible. And lest we forget, Gramscian hegemony of the sort which prevails in mature democracies only reflects a deeper level of repression; as the 2011 summer riots in the UK revealed, mass disorder sparked by ethnic tension is liable at any moment to shatter the brittle edifices holding together those societies, too. The 2010 Manezhnaia riots (when ethnicity leapt to the forefront of the movement for change in Russia) and the Pussy Riot scandal of 2012 (when, by contrast, the call for freedom of expression was pitted against the aggressive promotion of a virulently Church-state symbiosis directed against western liberalism in all its manifestations) appear to represent alternative scenarios. As we shall show in the next chapter, however, the spectre of the ethnicized 'other' remained a constant presence lurking beneath the shallow surface of that scandal, too.

Notes

1 Interview with a Rossiia journalist, 29 March 2013; interview with a newspaper journalist, 26 March 2013.

2 Ilya Yablokov, 'Conspiracy Discourse in Post-Soviet Russia: Political Strategies of Capture of the Public Sphere', PhD thesis, Russian and East European Studies, University of Manchester, UK, 2014.

3 In June 2012, Putin issued a decree on the establishment of the 'Presidential Council for Inter-Ethnic Relations' as the culmination of the process. Available online at http://eng.kremlin.ru/news/3997 (accessed 15 December 2013).

4 Dmitrii Medvedev, 'Opening Remarks at Joint Meeting of State Council and Commission for Implementation of Priority National Projects and Demographic Policy', 27 December 2010. Available online at http://eng.kremlin.ru/transcripts/1531 (accessed 18 November 2014).

5 See, for example, Andrei Piontkovsky's article in the 14 December 2011 issue of the *Wall Street Journal*: 'The Russian Spring has Begun'. Available online at http://online.wsj.com/article/SB10001424052970203518404577094780010444466.html (accessed 5 May 2012).

6 Edward Herman and Noam Chomsky, *Manufacturing Consent: The Political Economy of the Mass Media*, New York: Pantheon Books, 1988.

7 Ernesto Laclau, *On Populist Reason*, London: Verso, 2005.

8 V. V. Putin, 'Rossiia: Natsional'nyi vopros', *Nezavisimaia gazeta*, 23 January 2012.

9 The notion of 'polyculturalism' which Putin uses several times is derived from one of the unacknowledged sources for his article: a book by liberal, western-inspired thinkers on inter-ethnic relations. In the book the term 'polyculturalism' is virtually synonymous with 'multiculturalism' , which, of course, Putin rejects. This is one example of the article's awkward ideological hybridism. See A. Iu. Daniliuk, A. M. Kondakov and

V. A. Tishkov, *Kontseptsii dukhovno-nravstvennogo razvitiia i vospitaniia lichnosti grazhdanina Rossii*, Moscow: Prosveshchenie, 2009.

10 *Vremia*, 23 January 2012. Available online at www.1tv.ru/news/polit/197193 (accessed 27 January 2012).

11 Ibid.

12 *Vesti*, 23 January 2012.

13 *Vremia*, 29 January 2012. Available online at www.1tv.ru/news/social/197703 (accessed 30 January 2012).

14 *Vremia*, 23 January 2012. Available online at www.1tv.ru/news/polit/197197 (accessed 30 January 2012).

15 *Vremia*, 23 January 2012. Available online at www.1tv.ru/news/polit/197193> (accessed 30 January 2012).

16 Ibid.

17 *Vesti nedeli* of 29 January 2012 explicitly links the opposition to the alleged threat of Russia's breakup.

18 *Vremia*, 29 January 2012. Available online at www.1tv.ru/news/social/197703 (accessed 30 January 2012).

19 *Vremia*, 23 January 2012. Available online at www.1tv.ru/news/polit/197197 (accessed 30 January 2012).

20 Ibid.

21 *Vremia*, 23 January 2012. Available online at www.1tv.ru/news/polit/197193 (accessed 30 January 2012).

22 Herman Gray, *Watching Race: Television and the Struggle for Blackness*, Minneapolis: University of Minnesota Press, 1995.

23 *V kontekste* (Channel 1), 26 January 2012. Available online at www.1tv.ru/sprojects_edition/si5851/fi13481 (accessed 30 January 2012).

24 *V kontekste* (Channel 1), 1 March 2012. Available online at www.1tv.ru/sprojects_edition/si5851/fi14097 (accessed 19 December 2012).

25 *Voskresnoe Vremia*, 26 February 2012. Available online at www.1tv.ru/news/election/199993 (accessed 11 December 2012).

26 Nikolai Starikov, 'Sviaz' vremen', 18 July 2011, blog. Available online at http://nstarikov.ru/blog/10548 (accessed 11 December 2012).

27 L. M. Drobizheva, 'Rossiiskaia identichnost' i mezhetnicheskie otnosheniia'. Available online at http://rodnaya-istoriya.ru/index.php/mejnacionalnie-otnosheniya/mejnacionalnie-otnosheniya/rossiieskaya-identichnost-i-mejetnicheskie-otnosheniya.html (accessed 22 November 2012).

28 Russia's draft state strategy on inter-ethnic relations adopted in October 2012 did not include such a definition of ethnic Russians, even though it was present in the previous strategy document adopted under Boris El'tsin in 1996: 'Russia Drafts New Ethnic Policy of "Multi-People Nation" – Report', 18 October 2012. Available online at http://rt.com/politics/russia-drafts-nation-report-689/ (accessed 14 December 2012).

29 For Gramsci, hegemony amounts to 'the achievement of social consensus'. See Antonio Gramsci, *Selections from the Prison Notebooks*, New York: International Publishers, 1971, p. 151.

30 Andrew Wilson, *Virtual Politics: Faking Democracy in the Post-Soviet World*, New Haven: Yale University Press, 2005.

31 Gramsci refers to hegemony as 'unstable equilibria': *Selections from the Prison Notebooks*, p. 182.

32 *Vesti* and *Vremia*, 6 February 2012. Available online at www.1tv.ru/news/election/198419 (accessed on 15 February 2012).

33 *Vremia*, 30 January 2012. Available online at www.1tv.ru/news/election/197829 (accessed 30 January 2012).

34 *Vesti* and *Vremia*, 13 February 2012. Available online at www.1tv.ru/news/election/198956 (accessed 15 February 2012).
35 *Odnako* (Channel 1), 20 July 2012. Available online at www.1tv.ru/news/print/211705 1/2 (accessed 21 July 2012).
36 The BBC has frequently referred to him as de facto leader of the opposition movement.
37 In an interview with *Der Spiegel*, for instance, Naval'nyi said of the Pussy Riot 'punk prayer': 'I find the action itself despicable. I would be outraged if my daughter, who is 11, were to take part in something like this in a few years' time.' In the same interview, and using his Orthodox credentials to support his stance, he claimed of the trial: 'It was clearly a political show trial. As an Orthodox Christian, I can only see this as a repugnant, downright heathen act of revenge.' See 'Pussy Riot Trial Reminds Me of the Inquisition', *Der Spiegel*, 20 August 201. Available online at www.spiegel.de/international/world/spiegel-interview-with-russian-activist-alexei-navalny-a-850914.html (accessed 25 August 2012).
38 Aleksei Naval'nyi, 'Obshchenatsional'naia kampaniia "Khvatit kormit' Kavkaz"', 30 September 2011. Available online at http://navalny.livejournal.com/627082.html (accessed 15 January 2012).
39 For a transcription of the interview, which took place on 26 December 2011, see http://echo.msk.ru/programs/albac/842708-echo/ (accessed 15 January 2012).
40 Ibid.
41 The manifesto was broadcast on multiple websites on 27 June 2007. See, for example, www.apn.ru/publications/article17321.html (accessed 15 January 2012).
42 Ibid.
43 For further analysis of how Naval'nyi navigates between liberalism and nationalism, see Marlène Laruelle, 'Alexei Navalny and Challenges in Reconciling "Nationalism" and "Liberalism"', *Post-Soviet Affairs*, 30/4, 2014, 176–97.
44 *Vesti*, 9 August 2012.
45 Laclau, *On Populist Reason*, 2005, pp. 93–5.
46 Ibid., pp. 76–7.
47 Ibid., p. 225.
48 Ibid., p. 226.
49 Ibid., pp. 225–6.
50 'Posner Threatens to Cancel Show Over Censorship', 8 February 2012. Available online at www.theotherrussia.org/2012/02/08/posner-threatens-to-cancel-show-over-censorship/ (accessed 15 December 2012).
51 Peter Beaumont, 'Alexei Navalny: Firebrand Bidding for Russia's Soul', *Observer*, 11 July 2013, for a discussion of how Naval'nyi's attitude to migrant workers undermines his liberal credentials.
52 'Russian Nationalists Score Victory in Opposition Council', 18 July 2013. Available online at http://globalvoicesonline.org/2013/07/18/pugachev-the-opposition-weighs-in/ (accessed 30 July 2013).
53 'Sobyanin Is Stealing a "Navalny" Banner From The Balcony. Video'. Available online at http://navalny-en.livejournal.com/96879.html (accessed 15 December 2013).
54 Presentation by Tomasz Zarycki at the international workshop 'Nationalism in Russia and Europe: Understanding the Past, Exploring the Present', London School of Economics, 14 March 2014.
55 Ibid.

8 An unholy scandal

Profanity, abjection and the production of Russian-ness in the 'punk prayer' affair

The making of an (un)orthodox scandal

On 21 February 2012, Aleksei Naval'nyi's meteoric rise in the firmament of Russian opposition 'stars' was overshadowed by an event that was to become the lodestone of recent Russian cultural history. Four women from the Pussy Riot feminist collective entered Moscow's Cathedral of Christ the Saviour and began performing a 'punk prayer' in the sacred altar area (the musical accompaniment was added to the soundtrack later).[1] After 40 seconds they were escorted from the church. The 'prayer' called upon the Mother of God (*bogoroditsa*) to 'evict Putin' (*progoni Putina*), and attacked the Church hierarchy.[2] Its chorus was a variant on the English phrase 'holy shit' (*Sran' gospodnia*). After the YouTube clip went viral, days before the presidential election, three of the women – Nadezhda Tolokonnikova, Mariia Alekhina and Ekaterina Samutsevich (one evaded capture) – were charged with hooliganism, intruding on a sacred part of the cathedral, wearing clothing 'manifestly contradicting common church rules' and 'openly expressing disrespect to the Christian world'.[3] The charges were pressed on the grounds that the performers acted out of 'hatred towards Orthodox believers as a social group'. On 30 July their two-week trial began amid accusations that their performance was the devil's work. On 18 August the women were sentenced to two years' imprisonment. On 10 October Samutsevich was released on probation after appeal. In December 2013, Tolokonnikova and Alekhina were also released three months short of their full term, thanks to an amnesty of political prisoners which also saw the dissident oligarch Mikhail Khodorkovskii freed. (The amnesty was widely believed to have been a publicity exercise prior to the Sochi Winter Olympics).

During the legal investigation into the case, law enforcement organs systematically used archaic lexicon referring to Christian dogmas and rules of conduct in church dating to the medieval period. The terms were echoed throughout state-aligned media coverage of the scandal. 'Experts' supported their conclusion that the women were motivated by religious hatred by reference to a seventh-century Church Council. The preliminary investigation's indictment stated that the women 'denigrated the Russian Orthodox Church's centuries-old principles and basic guidelines in a blasphemous way' and accused them of 'opposing themselves to the Orthodox world'.[4]

The scandal took place on the eve of the presidential elections, and in the context of the liberal opposition's criticism of the Moscow Patriarchate for offering its overt support to Putin. It is likely that political leaders were involved in the decision to mete out exemplary punishment to the performers. But as our interviewees stressed, in implementing the decision, the media had considerable agency[5] in turning the incident into what Simone Gigliotti calls a 'limit event': 'an event or practice of such magnitude and profound violence that its effects rupture ... so-called civilising tendencies that underlie the constitution of political and moral community'.[6] According to self-appointed 'moral entrepreneurs' among Russian media personalities, the performance was part of a larger plot to destroy Russia's key values and norms. In the words of one influential journalist, it forced the entire society to ask a fundamental question about national identity and values: 'Who are we?' In his view, society's overwhelming response was 'We are Orthodox (*pravoslavnye*)'.[7] Liberal critics of the current regime responded by warning about the danger of Russia turning into a clerical state on a par with Khomeini's Iran.[8]

In fomenting the furore surrounding the 'punk prayer', the discourse of federal broadcasters was systematically co-ordinated with that of the Orthodox Church hierarchy and of Russia's political leadership, constituting the scandal as a 'media event' of the 'collusive' variety identified in our Introduction. The careful orchestration of the television coverage was evident from the fact that a particularly large number of talk shows and other feature programmes depicting Pussy Riot in a negative light were broadcast on the very days when the Moscow court controversially decided to place the punk performers under arrest during the preliminary investigation, reconfirming this decision despite appeals from defence lawyers. Until it began to attract the opprobrium of prominent television personalities and to incite dispute throughout the Internet, the incident, no more shocking in essence than other less notorious acts of *épatage* performed by Pussy Riot and its affiliate group, *Voina*, was known and discussed only at the fringes of Russian society.[9]

The mediatization of the 'punk prayer' centred on representations of the performance in the cathedral as a historical turning point which triggered 'the madness of vandalism' against the Orthodox Church and unexpectedly put the Church at the centre of 'passionate battles' between the opposition forces (depicted as an unrepresentative minority) and law-abiding, loyal Russian citizens. In the months to follow, news programmes and talk shows on the two main television channels, as well as on NTV, widely covered the Pussy Riot case. Before the presidential elections, there had been sharp differences between the more measured approach typical of news reports and the approach of talk shows, where extreme opinions were rife. The Pussy Riot coverage eroded this distinction. Now, apocalyptic visions of moral breakdown based on bizarre conspiracy theories and a strikingly Manichean representation of the world pervaded programmes of both genres. They invariably linked the incident to a recurring set of issues: a moral crisis in society stemming from the violation of boundaries between the permissible and the impermissible and the sacred and the profane; threats to Russia and the Orthodox Church from opposition plots masterminded by internal and foreign enemies;

societal demands for tougher actions against transgressors of public morals and order; and Russian citizens' overwhelming adherence to traditional values. These issues were directly linked to key elements in the emergent official discourse characterizing Putin's third presidency.

Once the media had performed its work, Russian society found itself sundered, with liberals expressing despair at the abandonment of secularism and Orthodox believers divided between those who called for Christian mercy and those inflamed with distorted rage at the 'blasphemy' of the acts. In a distortion of another kind, western media sources meanwhile depicted a supine court system enacting its masters' wishes, glossing over the complex relations between the Kremlin and its intermediaries. On state television (itself an intermediary institution), these complexities are foregrounded and will form one of the issues we treat in this chapter. Our eventual goal is to show how, via the 'space of opinion'[10] provided by television coverage of the scandal, the 'ethnic residue' that we discussed in the context of Naval'nyi's incipiently hegemonic anti-corruption campaign rose dramatically to the surface.

In addition to our usual analysis of news reports, we focus on eight shows devoted to Pussy Riot that aired between March and October 2012: Channel 1's *V kontekste* and *Pust' govoriat* (*Let Them Speak*), three special editions of Rossiia's *Spetsial'nyi korrespondent* (*Special Correspondent*), Leont'ev's *Odnako* opinion piece on Pussy Riot presented within the format of Channel 1's *Vremia* and two editions of Rossiia's *Poedinok* (Duel) debate show. Although not all these programmes are talk shows *strictu sensu*, they each feature discussion and rely on oral polemic for their effects. In media event mode, Pussy Riot function as a mechanism for renegotiating the boundaries of consensus in a society divided between conservative forces loyal to Putin's embattled regime and an opposition identified with the forces of 'otherness'. Within the rapidly changing Russian media environment, the precarious balancing act of the immediate post-Manezhnaia period is abandoned as the new boundaries provide a gateway through which sub-official and non-official voices of a distinctly uncompromising tone infiltrate the mainstream. Moreover, official discourse fails to re-authenticate itself by internalizing those voices, and thus fails to occupy the locus of authority. We demonstrate how, by boldly erasing key temporo-spatial distinctions and by permitting associative displacements from the main narrative line, scandalized talk show participants further undermine the authority of that discourse.

Prime among the discursive displacements from the Pussy Riot scandal was the paranoid vitriol launched by the Orthodox Church and the hyperbolic professions of piety. Viewed through the prism of Kristeva's 'theory of abjection', these phenomena are linked to central issues of consensus, identity and inclusion/exclusion,[11] for both the visceral disgust and the religious indignation relate to the 'abject' status of the three women as 'of the national community' yet 'not of it', an ambiguity Kristeva associates with religious rituals of defilement of the sacred. In re-making consensus through abjection, and in filtering populist voices from the peripheries into the mainstream, the Pussy Riot scandal produces a new, radically exclusionary form of Russian-ness, but it also exposes contradictions deep within the consensus-making process that we examined in Chapter 7. The instability to which they point, and which penetrates the discussion show format, foreshadows an alarming transposition of the mark of 'absolute difference' from Pussy Riot to 'the ethnic other' as embodied in the *Gastarbeiter* – the topic of our final chapter.

Context: 'Orthodox nation' to the fore

Before we trace the manufacturing of the Pussy Riot affair in detail, we should set it in the context of a change in television coverage of religious affairs during the preceding period. Until near the end of the Medvedev presidency, federal media approaches to religion were characterized by a managed pluralism of faiths; Orthodoxy retained a privileged position in a hierarchy of coverage, albeit that this gave limited space to Islam, and less still to Russia's other faiths. Most notably, Rossiia has since 2004 broadcast a ten-minute weekly programme entitled *Musul'mane* (*The Muslims*). While presenting a positive image of Islam as it has been historically practised in Russia, in every edition the programme re-enacted Russia's multi-confessional and multi-ethnic ideal, according to which the peaceful coexistence of the four traditional religions have made the Russian state powerful. It did so, however, with a consistent emphasis on the exceptional importance of Orthodox Christianity.[12]

The same carefully restricted pluralism was reflected in the representation of opinions. An instructive example is to be found in the reporting of the controversial law of 19 November 2010 on the transfer of state property to religious organizations. Coverage of debates around the law was undoubtedly selective. For instance, unlike alternative broadcasters such as Dozhd' and REN TV's *Nedelia*, Rossiia and Channel 1 failed to report scandals around the Moscow Patriarchate's attempts to use the law in order to take possession of buildings which, prior to the region's annexation by the Soviet Union, had belonged to the Catholic or Lutheran Churches.[13] The news reports there were on the subject did not overtly marginalize the law's critics, even though the Patriarch branded this criticism 'blasphemy' (*koshchunstvo*).[14] Channel 1 journalist Maksim Shevchenko, who was to become the Moscow Patriarchate's self-appointed campaigner on the Pussy Riot scandal, noted that society was split in its assessment of the law, pointing out that the Church was not the only institution which had suffered injustices under communist rule.[15]

On 7 January 2012, Orthodox Christmas, there was a subtle but significant change. In a seemingly innocuous news story, *Vesti* reported Patriarch Kirill's comments on the display across Russia in October–November 2011 of the cincture of the Theotokos, a medieval relic whose story is described in a sixth-century Byzantine legend. The display of the relic in different Russian cities was widely covered by state-aligned broadcasters in late 2011. When recounting the story of the cincture in October–November 2011, news reporters introduced their accounts with the words 'according to legend' (*po predaniiu*). While noting a great interest in the relic among Orthodox believers, it was acknowledged that far from everyone was happy about its arrival in their city. The complaints were recognized by *Vesti* as a legitimate position for a Russian citizen to hold. Church leaders' comments on the importance of the relic were distinguished from the newsreader's reporting speech and did not dominate the coverage.[16]

The *Vesti* report of 7 January 2012 was very different, consisting entirely of Patriarch Kirill's comments. According to the Patriarch, 'not a single event, either political or sport-related, had ever mobilized so many people'. 'This astonishing event', he claimed, 'has stirred the whole of society'. He concluded that such strength of religious feelings 'shocked those who hate our faith'. In the report the Patriarch's direct speech and the reported speech of the newsreader merged seamlessly.[17] This story signalled a departure from hitherto dominant techniques of representing the role of Orthodoxy in society. Now the Church and the Orthodox faith were Russia's key binding force, the primary marker of Russian identity and the very foundation of Russian nationhood. Reflecting a Manichean and conspiratorial view of the world, those holding alternative views of the exhibition of the cincture were depicted as enemies of Russia, whose only possible motivation was hatred.

This shift must be understood within the context of the political situation. The widespread falsification of the results of the 4 December 2011 parliamentary elections provoked mass public protests across the country. On 10 December 2011, less than three months before the March 2012 presidential elections, tens of thousands of people demonstrated in Moscow; the same day protests took place in 98 other Russian cities and in 42 cities abroad.[18] Oppositional media presented the 10 December demonstration as the largest political protest in Russian history.

The developments could not but create major concerns for the Kremlin, as the official discourse drew parallels between protest demonstrations in Russia and the Arab Spring. As early as 8 December, Putin's prominent ideological adviser, Vladislav Surkov – who would be officially entrusted with oversight of the mass media and of the relationship between the state and religious organizations in the course of the next few months – met with a group of journalists and intellectuals in the Presidential Administration headquarters to discuss how to react to growing public activism. For our understanding of the relationship between the media and the Kremlin, it is important to note that, according to Shevchenko and the well-known writer and journalist Viktor Shenderovich, both of whom reported on the meeting in their blogs, this was not an occasion at which the Kremlin's wishes were merely 'handed down'. Instead, the two sides engaged in a collective brainstorming.[19] The participants reportedly included influential television journalists in addition to Shevchenko, as well as the political commentator Sergei Kurginian, who became a key figure organizing pro-Putin public events in 2012.

Following this meeting, representatives of the Church, commencing with the Patriarch's confessor (*dukhovnik*), and later including the Patriarch himself, began to urge Russian citizens not to participate in opposition protests.[20] This first intervention by the Patriarch's confessor signalled a new turn in the relationship between the Russian Orthodox Church and the state. From this point the Church was to participate in politics to a far greater degree than had hitherto been the case, with the Patriarch playing a particularly important role. It would not be too far-fetched to speculate that, in December 2011, the conclusion was reached that the Moscow Patriarchate and Orthodoxy more broadly could be a vital tool in mobilizing the population against the opposition.

Opinion poll data might help explain why the interpretation of Russian Orthodox Christianity offered above all by Patriarch Kirill was now framing the official discourse so overtly, and why state-aligned media began adopting particular representation techniques in their religious coverage. According to Levada Centre surveys, by the end of the first decade of the new millennium over 70 per cent of respondents identified themselves as Orthodox, even though 40 per cent of those said they did not believe in God. While the polls indicated poor Church attendance, they also pointed to high levels of superstition. A poll of August 2010 reported that 61 per cent and 53 per cent of the respondents admitted their belief in the evil eye and miracles, respectively.[21] In 2012 the Church and the media began exploiting the popular fascination with miracles. The state has a dual reputation, emphasizing corruption and aggression but also unity and protection; the latter is strengthened by the state's association with the Church.[22] With an understanding that the Church enjoys considerable 'symbolic capital', and amid concerns about the weakening of its position, the Kremlin appeared to have decided to draw more actively on this capital in order to authenticate its own enfeebled discourse.

Anathema, talk and consensus

From the outset, the Church's symbolic capital was at the centre of the strategy to constitute the Pussy Riot scandal as a media event. For this reason, reporting on the incident amounted to more than mere demonization and had a 'positive' aspect to it, based on the opportunity for active consensus-building that the 'punk prayer' performance offered. The Channel 1 talk shows *Pust' govoriat* and *V kontekste* provided complementary contributions to this process. The *V kontekste* edition of 15 March 2012 quoted in the opening pages of this book was hosted by Shevchenko, whose guests included Church representative Maksim Kozlov, human rights group leader Ol'ga Kostina, art critic Marat Gel'man and liberal campaigner Kseniia Sobchak.[23]

Shevchenko introduced the discussion by dismissing portrayals of 'these maidens (*devits*)' as 'tragic heroines'. Exploiting his position as moderator to expel them beyond the boundaries of accepted opinion, he intervened constantly in the debate, for example characterizing what they did as 'spitting into the souls of believers'. The use of sarcastic lexicon (*devits*) and the deployment of hyperbole offer a vernacular embodiment of the act of expulsion, but with an anachronistic religious colouring. The trajectory of the discussion served the dual roles of authenticating the expulsion as something accomplished by actors at odds with one another with regard to all other things, and tracing the progress of the expulsion across the timespan of the broadcast.

The discussion concluded with unanimity that the women must be punished. The fact that the preceding debate included sharp exchanges merely reinforced the authenticity of the eventual congruence. Indeed, amid a vigorous exchange of views held by people from opposing ends of the political spectrum, the intervention from Kozlov quoted in our Introduction, in which he refers to 'the slowly maturing consensus' reached via 'maximal disagreement', simultaneously states *the fact* of the consensus, *displays* it 'in action', *inserts* itself within the 'maturing' process and *names* it. This illustrates televised discussion's centrality to the state-led consensus-building process advocated in Putin's pre-election article, exemplifying how its pseudo-Socratic format facilitates its ideological work. Consensus-building is always bi-aspectual and must exclude in order to include. As we shall see, the balance was later to shift toward exclusion, with a correspondingly diminished performative effect.

The structure of a *Pust' govoriat* edition dedicated to the scandal bears superficial comparison with that of *V kontekste*. Like Shevchenko, the host, Andrei Malakhov, set the parameters for the debate, adopting Orthodox lexicon to frame it: 'We are going to … get to the bottom of what … punishment those who … defaced holy Orthodox sacraments deserve, and how to exorcize the devils which have insinuated themselves into our souls'.[24] Initially undercutting the consensus-building mechanisms, such lexicon was soon to sweep across federal media discourse on the 'punk prayer', eventually establishing itself at the heart of a new, highly exclusionary and aggressively oriented consensus. The performance was regularly described as an example of 'outrageous blasphemy' (*bezobraznoe koshchunstvo*), 'an act of demonic possession' (*besnovanie*), 'a witches' Sabbath' (*shabash*), 'a satanic ritual', 'a demonic dance' and 'spitting in the soul of believers'. The performers were described as 'Macbeth's witches' and 'the possessed' (*oderzhimye* and *besy*). A Rossiia journalist who interviewed Tolokonnikova in prison claimed that he saw 'sparkles of devilry (*nechistoi sily*) in her eyes'.[25]

However, rather than lending this medieval lexicon popular authenticity, Malakhov's informal demeanour, the raucous atmosphere cultivated in his studio discussions and the show's 'permissive' title, *Let Them Talk*, undermine it. Having seemingly resolved the tensions permeating the discourse of *V kontekste*'s community of 'moderate' intellectuals, official rhetoric fails to embed itself within the vernacular discourse dominating *Pust' govoriat*. Like Shevchenko, Malakhov positions himself as a believer; in what became a leitmotif of television coverage of the scandal, his first question to the father of Tolokonnikova, the only Pussy Riot sympathizer present, was 'Are you a believer?' Tolokonnikov cut a despondent figure facing a hostile studio audience, his voice silenced by a chorus of condemnations from a panel of guests whose very selection consolidated the boundaries of national consensus,[26] including representatives of the Church and Duma deputies, a Muslim hierarch, an actress and writer, a group of Cossacks and Maksim Shevchenko, present as a guest rather than a host, with Jewish activist Alla Gerber the only discordant voice. The range of guests thus models the 'multi-cultural, multi-faith Russia' from which Pussy Riot excludes itself.

There was, however, robust argument over the Orthodox value of forgiveness. Again exploiting the informal setting to deploy non-standard lexicon, Shevchenko suggested that these 'females' (*baby*) be made to sweep the streets. The Cossacks went further, proposing a good whipping. Others in the *Pust' govoriat* discussion, however, argued for leniency, recalling Christ's meek nature. A lively exchange also took place over the role of parental responsibility: Tolokonnikova had left a four-year old daughter behind after her arrest and Malakhov blamed Tolokonnikova's own degeneracy on her upbringing. Her father's silent isolation within the studio symbolized his position beyond the space within which 'free speech' can occur.

The Cossack call for physical retribution was not isolated. In *Odnako*, Leont'ev's vitriolic diatribe concludes with the proposal that the women be spanked – a suggestion accompanied by Leont'ev's claim that such a punishment would 're-establish a link with reality through tactile contact'.[27] *Odnako*'s position at the heart of, yet separate from, the *Vremia* bulletin epitomizes vernacular talk's problematic role in elaborating official policy by inflecting it with the colourful language in which popular discourse abounds. Vernacular speech is also more permissive than official news discourse with regard to the use of hyperbole and false equivalence. Thus, Leont'ev deploys a crude rhetorical ruse when presenting Pussy Riot's protests as the metaphorical equivalent of abuse directed against the Patriarch, referring to his 'persecution' by a witches' Sabbath' and to 'a bestial rage' indicated by the 'abundance of foul language'.[28]

False equivalences abound. In the *V kontekste* debate, Kostina lists Pussy Riot's actions alongside a notorious, unrelated incident of a policeman in Tatarstan who sodomized a detainee with a champagne bottle.[29] Its status as a subplot deriving from the peripheries of the news realm is what, in the logic of popular discourse, enables it to reinvigorate the mainstream account with the vitality of public scandal, to be transmitted from person to person and eventually to insinuate itself at the core of the 'approved' Pussy Riot narrative.

The conflation of violent and non-violent law-breaking is not restricted to the popular sphere with its relative laxity with respect to linguistic and conceptual categories. It is also inherent in the hooliganism charge: as the defence lawyers pointed out, their clients caused no physical damage to person or property. In an instance of official–sub-official symbiosis, the outpouring of media vitriol, with its breaches of rhetorical convention, provides a discursive underpinning for the conflation. But it is also associated with a process by which the three women are progressively 'othered'. Leont'ev's use of the phrase 'bestial rage' is not coincidental. It is reinforced visually by the main news bulletins covering the Pussy Riot trial, whose photographs of the women huddled together as objects of curiosity in the glass prison cage, known as the 'aquarium', in which they stood trial came to be iconic images. In a curious instance of discourse appearing to generate reality, subsequent reports of food and sleep deprivation barely raised an eyebrow outside the most opposition-minded circles.

The animal imagery was closely connected with Pussy Riot's 'bestial' act of 'defilement'. This enabled the scandal to be drawn into the orbit of inter-ethnic relations, with commentators taunting liberal opponents with questions about how Muslims might have reacted if their own sacraments had been defiled. The singer Elena Vaenga generated a new media subplot around this theme in the verbal web of scandal when she misspelled the word 'mosque' (*mechet'*) in her Twitter diatribe, eliciting scorn from Sobchak. Retrieving this 'meme' from the shadowy hinterlands of official Kremlin thinking, Putin himself proposed that, had the controversy occurred in the Caucasus, 'the police even would not have had time to arrest the women' (in other words, they would have been lynched), implying that their alterity exceeds that even of the less-than-civilized Muslim world. With Putin's approval, this line gained momentum and was re-circulated throughout the state media, eventually eliciting a rebuttal from the Russian Council of Muftis, which noted that the Quran cites an incident in which Mohammed demanded forgiveness for a man who urinated in a mosque.[30] The business channel, RBK, also included an acerbic rejoinder from imam Nail Mustafi, who branded the Pussy Riot controversy implicitly anti-Semitic for its propagation of conspiracy theories centring on a 'Russophobic' intelligentsia, as well as Islamophobic in the assumptions made by the likes of Putin.[31]

Meanings circulated between the realms of the official (the Kremlin line), the sub-official (the state media; the Orthodox Church) and the unofficial (popular culture). Because the Putin regime's legitimacy was draining away, and because its ideological framework had not rooted itself in popular consciousness, its pronouncements were authoritarian rather than authoritative, and relied upon the credibility of spheres over which its power was less than absolute (primarily the Church). Putin's own comments on the Pussy Riot episode in his 3 August 2012 interview in London, his subsequent interview with *Russia Today* on 5 September and his October interview at a meeting of the Valdai club confirm this reliance.[32] His definition of the 'punk prayer' as a 'witches' Sabbath' (an expression used earlier in Leont'ev's *Odnako* and elsewhere[33]) and his lynching claim point to an unreflective dependency on lexicon and rhetoric developed by media personalities in the preceding months.

The Moscow Times singled out from among the prosecution's far-fetched accusations the charge that the 'punk prayer' inspired a terrorist attack on Muslim leaders in Tatarstan.[34] The *Moscow Echo* highlighted another interview in which the prosecutor at the Pussy Riot trial drew parallels between the 'punk prayer' performance and 9/11.[35] The lexicon of 'religious hatred' became a leitmotif of conservative interventions into studio discussions of the trial, the talk show functioning as a filter for controlling the entry of sub-extreme rhetoric across the boundaries of official space whose construction it has, in a complementary gesture, modelled. In the *V kontekste* debate, a variant on the theme emerges in Kozlov's invocation of Huntington's 'civilizational clash' rhetoric during the exchange between him and Sobchak quoted in our Introduction. By the end of the Pussy Riot controversy, the talk show's dual mechanisms had served their purpose and re-foregrounded that rhetoric within the official mainstream, a phenomenon whose earlier prominence in coverage of the Manezhnaia riots confirms the linkage between these two seminal events, and between issues of ethnicity and the related form of alterity represented by Pussy Riot.

The blasphemer as conspirator

In beginning to explicate the construction of that alterity, let us first recall Kozlov's reference to the alluvial 'dregs' contaminating 'traditional Russian culture' from the exchange quoted in our Introduction. Kozlov's visceral image captures the disgust Pussy Riot elicited in its opponents. While Putin was more restrained, Deputy Prime Minister Rogozin used Twitter to abuse the American pop star Madonna for offering her support to Pussy Riot, calling her a 'clapped out, moralizing slut' and advising her to 'either take off your cross, or put on your knickers'.[36] Nonetheless, Rogozin's vulgar insult corresponded with the genuine revulsion felt by sections of Russian society.

The gendered aspect to the discourse of disgust emerges clearly in Arkadii Mamontov's trilogy of programmes all bearing the title *Provokatory* (Provocateurs), broadcast in Rossiia's investigative *Spetsial'nyi korrespondent* series at the height of the scandal – on 24 April (*Provokatory-1*), 11 September (*Provokatory-2*) and 16 September (*Provokatory-3*). The first consisted of a documentary film purporting to uncover the 'truth' behind what the mainstream media referred to routinely as the 'debauchery' (*debosh*) in the cathedral. The film was framed by animations of black snakes writhing lasciviously, condensing sexual revulsion, animal imagery, religious obscurantism and the notion that the women serve a poisonous external force. It included a confrontation with an Amnesty International representative whom the Rossiia reporter presented with a photograph of a pregnant Tolokonnikova copulating in public, asking if these are truly 'prisoners of conscience'.[37] The camera cuts to the studio audience, wearing expressions of repulsion, as they model the disgust that the film provokes in viewers.

A reporter interviews the women in prison. Tolokonnikova receives most attention, with the camera lingering in extreme close-up on her striking features, demonstrating the reverse mirroring of desire and revulsion. Samutsevich is asked why she has no children: her response – that she rejects conventional gender roles – meets with predictable scorn. The themes of motherhood and anti-motherhood infiltrate the discursive space around Pussy Riot. We confront the 'revolting' image of the pregnant, copulating Tolokonnikova and the anti-maternal views of the feminist Samutsevich. Later, in a further proliferation of the verbal promiscuity that scandal generates, Sobchak is attacked, via *Vesti*, by the husband of the singer whom she had ridiculed for criticizing Pussy Riot: 'She is proposing that Valeriia, a mother of three, should take a rope and commit suicide. It would be better if she got herself a husband and had even one child before telling people how to live.'[38] The implied counter-images to the anti-mother are, of course, the Mother of God, defiled by the punk performance, and – beyond the epicentre of the 'punk prayer' controversy – the maternal image of the Buranovsk Grannies, Russia's entry in the 2012 Eurovision Song Contest, which conveniently coincided with the scandal. The fact that the Grannies sang in both Udmurt and English enabled them to negotiate the links between Russian-ness, (ethnic) difference and modernity in a configuration which, as Putin revealed in a presidential address, was altogether more reassuring for him and his media acolytes.[39]

Provokatory attempted to rationalize the visceral antagonism it incited against the anti-mothers of Pussy Riot by answering the question, 'who lies behind the 'outrage?'. Predictably, the female blasphemers are for Mamontov and his guests merely the 'puppets' of a 'global government' in which Berezovskii, the exiled

Jewish oligarch, plays a prominent role. Footage of the 9/11 attack is replayed within the same sequence in which Berezovskii is shown writing to Naval'nyi, who is also drawn into the conspiracy. Conservative bloggers noted that the international campaign for Pussy Riot was led by Aleksandr Gol'dfarb, a New-York based Jewish publicist. Pro-Putin writer Eduard Bagirov suggested that Pussy Riot's support was limited to 'emigrants, kikes and queers'.[40] In his tract on the clash between 'universal liberal values' and 'national cultures', Patriarch Kirill traced the source of these values to 'protestant theologians and Jewish philosophers'.[41] Anti-Semitism, Islamophobia and fear of the democratic movement are synthesized in an irrational, paranoid narrative, Pussy Riot's anonymous balaclavas and shapeless, unfeminine clothing confirming the conspiratorial logic.

The conspiracy was portrayed as having two sources: one historical and specifically Russian; the other contemporary and purportedly emanating from beyond Russia. The historical origins of the crisis were found in the anti-Church activities of the pre-revolutionary liberal Russian intelligentsia, and in the Soviet regime's anti-religious campaigns. A direct link was then made between these past events and the liberal opposition's current criticism of the Moscow Patriarchate and alleged hostility to Christianity. For instance, a member of the *Provokatory* audience, well-known nationalist author Egor Kholmogorov, claimed that the liberal intelligentsia had been manifesting 'its hatred of the country, the people (*narod*) and religion since the Crimean war [of 1853–6]'.[42] For Kholmogorov and Mamontov, the Pussy Riot performance was a continuation of that historical trend. Similar comparisons between past and present were repeated in many other programmes, where the 'punk prayer' and acts of vandalism against churches and religious objects reportedly inspired by the performance were referred to as 'a wave of crimes' or even 'a war' against the Church on a par with Soviet anti-religious campaigns.[43] Frequent references to Soviet repressions against the Church reflected a shift in the current official representation of Soviet history as compared to the previous decade, in which the official line often amounted to a whitewashing of Stalinism.[44] In 2012, Stalin's terror and other repressive Soviet policies were systematically condemned and the need to suppress the opposition was justified by references to the government's determination not to allow a repetition of the bloodshed which followed the Bolshevik revolution.

The second, contemporary source of Russia's purported 'moral crisis' was located in the west. In 2012, federal broadcasters began placing developments in Russia within the broader context of a broader religious crisis in the world, enabling them to universalize the 'moral transgressions' identified in Russia. Echoing a claim made by Patriarch Kirill,[45] Shevchenko maintained in the Rossiia talk show, *Poedinok*, that the west had initiated 'a war' against religion, excluding it from the public sphere. According to Shevchenko, the assault on religion was part of a larger campaign to destroy the traditional values of non-western societies in order better to control them. The *Poedinok* broadcast was devoted to the discussion of a controversial draft law which proposed introducing a punishment of up to five years in prison for 'offending the religious feelings of believers'.[46] A staunch supporter of the law, Shevchenko further argued that the new law would act as a shield between Russia 'and the godless, anti-Christian civilization of the west' whence threats to 'our religion and traditional values' emanated.[47]

According to Shevchenko, Mamontov and other federal broadcasters, then, the Pussy Riot performance was an episode in an international war in which Russia was on the front line. It was a conspiracy or a provocation (*provokatsiia*) by internal and external forces who aimed to undermine the Church, which, according to the Patriarch, and now also leading television journalists, had historically been Russia's 'main binding force' (*skrepa*).[48] Sometimes plotters were vaguely hinted at; at other times their identities were revealed. Most prominent among them were members of the liberal opposition allegedly funded by western agents and by Berezovskii, who had been working for regime change in Russia. The Church was their prime target because, according to the Church hierarchy (in words echoed by state broadcasters), for centuries it had 'constituted the foundation of Russian statehood'. When interviewed on *Vesti nedeli* on 9 September 2012, the Patriarch stated 'I cannot help thinking that [the Pussy Riot performance] was an act of reconnaissance [*razvedka boem*]', preceding a more sustained attack on Russian statehood.[49]

In keeping with the conspiracy narrative, every speaker in the 'studio discussion' which dominated *Provokatory* began by declaring his religious faith and sympathy for the Patriarch's wounded sensibilities. Mamontov concluded the discussion with a sentimental anecdote about how in Soviet times, when swimming in the pool built on the site of the pre-revolutionary Cathedral of Christ the Saviour, he had dreamed of a time when cupolas would rise again from the spot, and how this dream had been sullied by the 'demons' of Pussy Riot.

Sacred boundaries and political apostasy

The Pussy Riot conspiracy narrative was frequently shaped by an obsession with delineating the boundaries separating the sacred from the profane, the lawful (*dozvolennoe*) from the lawless (*bespredel*) and, thus, the authentic Russian 'self' from its deviant, conspiring 'others'. Widespread violations of these boundaries were persistently reported during the course of the 'punk prayer' scandal, fostering a state of moral panic capable of delegitimizing attempts to defend the 'transgressors' and geared toward soliciting public support for tougher state control over public order.

The sacred/profane boundary gradually became central to the media event, displacing the more inclusive boundary of the societal consensus articulated in the *V kontekste* discussion mentioned earlier. Unlike the latter, with its abstract, political connotations, the sacred/profane boundary goes to the heart of a community's sense of selfhood and has for this reason played a key role in the history of most cultures; even secular societies today retain a strong awareness of the taboos which violate what is 'sacred' to them and are liable to prove unacceptably provocative to large sections of the public. As Emil Durkheim puts it, 'there is no other example of two categories of things as profoundly differentiated or as radically opposed to one another' as the categories of the sacred and the profane.

'The sacred thing is, par excellence, that which the profane must not and cannot touch with impunity.'[50]

In constructing the 'punk prayer' incident as a media event, state-aligned broadcasters went to great lengths to depict the Church of Christ the Saviour as a supremely sacred space, therefore rendering the Pussy Riot performance the ultimate act of desecration. The strategy proved effective. As we began to see even in the *V kontekste* discussion, members of the liberal opposition who had criticized the authorities tended nonetheless to accept the view that a religious space had been violated. Marat Gel'man framed his contributions to various television discussions of the Pussy Riot incident by invoking the sacred/profane boundary.[51] Speaking on the *Poedinok* programme on 27 September 2012, the liberally inclined Rossiia journalist Nikolai Svanidze argued that spaces and objects relating to the memory of the Second World War could also be regarded as sacred, and that their violations should not be left unpunished.[52]

The sacred/profane boundary is often called upon to construct a particular religious tradition and to bind it into the history and identity of a nation in such a way as to free it from the bureaucratic abstractions of the state.[53] During the Pussy Riot campaign, broadcasters projected an image of the Russian Orthodox Church not as the state's powerful ally, but as a vulnerable, long-suffering institution in need of protection from attacks by those whose aim was to weaken Russia. Countering representations of the Moscow Patriarchate as the dominant state-aligned religious institution in the country, the state-controlled channels claimed that in Russia and elsewhere, 'Christianity and Orthodoxy are subjected to particular hostility'.[54]

From the period of witch hunts in late medieval and early modern Europe, the violation of the sacred/profane boundary has been associated with sexual transgressions and the violation of gender roles.[55] In line with this tradition, talk shows dedicated to the incident incorporated regular attacks on gays and lesbians as the people 'who corrupt our youth'.[56] A programme about Pussy Riot broadcast by the populist NTV channel on 6 April 2012 was a particularly striking example linking religious, sexual and gender transgressions.[57]

The overt political goals behind the emphasis on boundary-drawing via an assault on 'deviance' are revealed in the repeated claim that it is these 'transgressors' who lead the criticism of the government. Alongside liberal critics of Putin's regime, television viewers are told, the opposition included gay and lesbian activists and those who present acts of asocial behaviour as modern art. Despite their heterogeneity, these groups are purportedly united by their disdain for Russia and for the Orthodox Church. Strikingly, such reprehensible attitudes are also linked to the ethnically specific transgressors. Drawing a direct parallel between the current opposition and liberal parties in late imperial Russia, Arkadii Mamontov claimed in *Provokatory-2* that 'At the turn of the twentieth century... so-called progressive-liberal newspapers supported all religions apart from Christianity... Then the aim was to destroy Russian statehood.' While he spoke, the screen displayed a caricature of a Jewish rabbi sacrificing a cock, along with the head of the last Russian emperor.

As we saw with Manezhnaia, a stable Pussy Riot narrative took time to crystallize. In March 2012, programmes like *V kontekste* acknowledged, even if hesitantly, that society was far from united in its position on Pussy Riot. In the 15 March edition, Marat Gel'man was permitted to argue that the Russian Orthodox hierarchy provoked the Pussy Riot performance.[58] By April 2012, society was being portrayed both in talk shows and in news coverage as homogenous, with the overwhelming majority of people belonging to 'a social group of Orthodox believers', 'out of hatred' for whom Pussy Riot had performed its act of hooliganism. The possibility that there could be a variety of opinions even among Orthodox believers had rapidly diminished. Believers could now profess only one view. In the words of Kozlov, who regularly represented the Church in television debates on Pussy Riot, anyone with a 'religious mentality' would understand that the 'punk prayer' performance was 'an act of blasphemy'.[59]

Official television accounts of the scandal reflected a clear order of voices, regardless of genre. The opinions of those perceiving the Pussy Riot performance as causing deliberate offence to religious feelings, a transgression which the authorities were obliged to prevent, invariably featured in the introductions and conclusions to talk shows and news reports, with those holding dissenting views given a brief slot in the middle. The dissenters themselves were located beyond the bounds of normal, mainstream, Orthodox society.

In 2012, two events were covered particularly intensively on state television as part of an effort to perform society's unity through its adherence to 'traditional Orthodox values'. One was the display across Russia of the cincture of the Theotokos. The Patriarch's claim that it had solicited a level of public participation unprecedented in Russian history was aired repeatedly in television news reports and talk shows, as a means of countering the opposition's claim that the 10 December 2011 anti-government demonstration in Moscow had been the largest public gathering ever.[60] Another such event was a series of communal prayers organized by the Patriarch across Russia on 22 April 2012 in response to what was claimed to be a wave of violence against the Church in the aftermath of the Pussy Riot performance. *Vremia*'s report on the event framed it in the context of a 'mighty force' loyal to the state and a scheming, apostate opposition which regarded the state as of questionable value.[61]

Also in common with Manezhnaia, the dominant narrative eventually infiltrated alternative media accounts propagated by the very liberal groupings that it targeted. Dozhd' and other non-official media were undoubtedly alert to disunity among prominent spokespeople for the Patriarchate, and also highlighted the facts that one of the Pussy Riot artists, Mariia Alekhina, was a member of an Orthodox youth movement and that some priests and lay people active within the Church has condemned the arrest of the performers and were generally critical of the Church hierarchy.[62] At the same time, like their state-aligned rivals, alternative channels structured much of their reporting of the scandal around the sacred/profane boundary, though often in inverted form.[63] By focusing on material rather than spiritual issues, they argued, the Church itself transgresses – for instance, by renting out space in the Cathedral of Christ the Saviour for secular events.[64]

They shared the official view of the scandal as an important turning point, but they defined the nature of the societal change differently and suggested that the case signalled a turn toward 'growing aggression and intolerance'. They also resorted, albeit less consistently than the state channels, to conspiratorial speculation. Speaking on Dozhd' on 20 June 2012, biblical scholar Andrei Desnitskii suggested that the authorities deliberately manufactured the Pussy Riot incident to enable a split between the Church and the liberal intelligentsia.[65]

Furthermore, paradoxically, the alternative media at times endorsed state television's representation of society as Orthodox. REN TV's *Nedelia* broadcast of 29 September 2012 included a report on the impact of the Pussy Riot case on society at large, arguing that one could now talk about 'two Russias': one was represented by a young church warden prone to violent religious activism; another by a businessman who collected religious art.[66] Significantly, the businessman also emphasized his Orthodox belief, placing both Russias on the same side of the sacred/profane boundary as the Russia portrayed in the dominant narrative.

Abjection and the imaginary

How, though, do we link the media's orchestrated narrative of sacred and profane with the flights of sheer conspiratorial fantasy and visceral disgust exemplified in Mamontov's *Provokatory*? The question means we must return from the realm of apostasy and alterity to that of identity. In order to better understand the identification process at work, we must revisit the authentic revulsion felt by sections of Russian society in reaction to the 'punk prayer'.

Offering an account of identity formation inflected by, but not reducible to, psychoanalytical models, Julia Kristeva centres her ideas on the disgust evoked in us by bodily excretions, gaping wounds, the taste of certain foods and, above all, the smell of human corpses. The automatic gagging response produced by such experiences indicates the body's impulse to reject what it finds alien (note Pussy Riot's disgustingly Kristevan refrain '*Sran' gospodnia*', or 'excrement of the Lord', and the fact that the women made Medvedev 'physically sick').[67] But the underlying aetiology of disgust is attributable to the fact that the phenomena evoking it are 'of the body', yet separate from it:

> These body fluids, this defilement, this shit, are what life withstands ... on the part of death. I am at the border of my condition as a living being ... The corpse, the most sickening of wastes, is a border that has encroached upon everything ... the utmost of abjection. It is death infecting life.[68]

While she writes in an uncompromising personal mode, for Kristeva the principle of abjection clearly operates also at the level of culture. Indeed, as Kristeva recognizes, to engage with issues of identity (whether national, community, group, personal or other) is *unavoidably* to deal with affinities in the relationships between self and other, internal and external, individual and collective; with the liminal and with acts of transgression:

> It is thus not lack of cleanliness ... that causes abjection but what disturbs identity, system, order. What does not respect borders, positions, rules. The in-between, the ambiguous, the composite. The traitor, the liar, the criminal with a good conscience, the shameless rapist, the killer who claims he is a saviour ... By abjection, primitive societies have marked out a precise area of their culture in order to remove it from the threatening world of animals.[69]

Importantly, rather than a secondary process, abjection is the act by which the subject, collective or individual, is constituted:

> I expel myself, I spit myself out, I abject myself within the same motion through which "I" claim to establish myself; it is thus that they see that "I" am in the process of becoming an other at the expense of my own death.[70]

Equally crucially, the 'constitutive other' common to identity theories in multiple disciplinary domains (political, anthropological, sociological, or cultural) dwells not beyond but *within* the borders of the subject: 'I experience abjection only if an Other has settled in place and stead of what will be "me" ... an Other who precedes and possesses me, and through such possession causes me to be.'[71] Kristeva's theory can help us explicate the associations linking several features that we have identified in our analysis so far: the visceral disgust expressed toward the women of Pussy Riot by sections of Russian society, the persistent animal imagery accompanying their media representation and cultural perceptions of their 'treacherous ambiguity': they are ethnic Russians, but aligned with alien forces (the 'global government'; Islamist extremism); they are women, yet somehow not 'feminine'; they are child-bearing, but non-maternal. Also consistent with the Pussy Riot episode is Kristeva's account of the links between abjection and the sacred, according to which 'a whole facet of the sacred, true lining of the sacrificial, compulsive, and paranoid side of religions' has to do with invoking 'rituals of defilement' in order to 'ward off the danger' that the emergent community might yet slide back into an earlier state in which its boundaries lack distinction.[72] Paradoxically, the very aggressive certainties spouted by Mamontov and others betray a deep-seated fear of the loss of a community identity that has yet to acquire enduring authenticity.

In Kristeva's account, moreover, phobia is merely the 'opposite correlative' of narcissism.[73] Thus, the disturbing liminality of the abject underlies the mirror-like close-up of the desirable yet repulsive features of Tolokonnikova, the anti-mother, in Mamontov's *Provokatory*, for 'The more or less beautiful image in which I behold ... myself rests upon an abjection that sunders it as soon as repression, the constant watchman, is relaxed ... Abjection is therefore a kind of narcissistic crisis.'[74] The phobic obsession with 'global governments' corresponds closely to the ambivalence which Kristeva locates in the image of the Jew, which she links in turn to the liminality of the woman:

> He is ... the one we do not differentiate ... a conjunction of waste and object of desire ...The Jew becomes the feminine exalted to the point of mastery ...

> the border where exact limits between same and other, subject and object, inside and outside, disappear – hence an Object of fear and fascination. *Abjection itself.* He is abject: dirty, rotten.[75]

In a parallel reading, Slavoj Žižek associates Jewishness with Lacanian 'excess': 'that already within us which is more than us', and which can stand for an infinite range of incompatible 'others':

> Jews are … not another subject that I encounter in front of me but an alien, a foreign intruder within me …. The amorphous intruder of infinite plasticity and an un-dead 'alien' monster who cannot ever be pinned down to a determinate form.[76]

In the Jew, capitalism and communism, Islamist and western liberal, can co-exist without contradiction. The overt anti-Semitism which peppered Russian nationalist discourse in earlier decades had been superseded by anti-Caucasian and anti-migrant sentiment as the dominant form of prejudice in Russia;[77] however, it is now re-emerging in a covert form as the unspoken referent of liberalism, the west, homosexuality, non-conformism and terrorism.

What, though, of offended Orthodox sensibilities? Here, the temporal logic of the scandal is inverted. Rather than a pre-existent Orthodox community, repelling the incursion across its borders, that community is constituted *by the very act of repulsion* – for abjection, and the accompanying repression, coincides with the constitution of the 'social dimension of man'.[78] In Kristeva's schema, full sociality is preceded logically by immersion in an imaginary phase, in which the emergent community suddenly recognizes itself in the other, and as other, and is thus able to constitute itself as whole and integral. For Fredric Jameson, the imaginary, 'whose logic is essentially visual', marks 'a fundamental gap between the subject and its own self or *imago* which can never be bridged'.[79] It is characterized by obsessive narcissism, and by a transitional phase in which, struggling to come to terms with the realm of alterity, the emergent subject misattributes its own actions to others. Jameson refers to this as the 'indifferentiation of subject and object', when 'the child who hits says he has been hit, the child who sees another child fall begins to cry … slave [is] identified with despot, actor with spectator, victim with seducer'.[80] This is precisely the logic of the Pussy Riot scandal, in which the extreme hostility displayed *toward* the 'blasphemers' is transferred back *onto* them, and a mythologized Russian people must 'defend themselves' from the conspiratorial forces of liberalism. 'The state is not defending us' was a recurring refrain in the *Provokatory* series. *Provokatory-2*, meanwhile, features an activist who, mimicking his nemesis, 'defiles' a t-shirt worn by a Pussy Riot supporter, an illegal act met with approval by the purportedly law-upholding audience.

The notion that Russian society remains in a pre-social, imaginary phase is strengthened by Dmitrii Kiselev's startling claim in his interview with us that Russian television's core audience is so primitive that it needs to be 'colonized';[81] in the imperial discourse which Kiselev consciously invokes, colonial subjects were regularly compared to (and viewed as) children. For Kristeva and Jameson, that pre-social phase is also characterized by category confusion and the childlike inability to distinguish subject from object, metaphor from reality. The accusation that Pussy Riot metaphorically 'spat in the soul of believers' is thus equated with the reality of violent aggression.

The category confusion was by no means restricted to a small number of television talk shows, nor even to the 'punk prayer' scandal. It manifested itself too as a conflation of legend and fact in a co-ordinated campaign of exhibiting relics across Russia which began in early 2012. From that point, even news reports ceased to include explanations that the description of the relics' nature derived from medieval legends; instead they were now presented as real and miracle-performing objects, belonging unquestionably to Mary, Jesus and various saints. On 2 April 2012, *Vesti* reported on the display in a church in Birobidzhan of a shroud, robe and belt supposedly belonging to the Virgin Mary. Without acknowledging that legends surrounding these objects date to the fifth century, the reporter commented: 'Many people regard it as a miracle that things used by the Blessed Virgin Mary have survived until today'.[82] From June to September 2012, in the wake of the 'punk prayer' scandal, a series of 'miracle-performing' relics, some brought from abroad, was displayed in churches across Russia. In highlighting these events, television effectively enacted a 'collective psychosis', in which the logical categories of the figurative and the referential, the imaginary and the real, broke down.[83]

As Jameson points out, within the realm of the imaginary, the confusion of subject and object often precipitates an inversion of roles. In the *Provokatory* trilogy, blasphemy's rhetoric of inversion begins in part 1, where Mamontov defends his patronizing use of the term 'girls' in reference to Pussy Riot by citing his characteristically Russian 'over-generosity of spirit' (transferring his own aggression to the threatening other). In part 3, he uses the term 'blasphemers' (*koshchunitsy*) as a fixed epithet (his aggression toward the other is now reassumed by the self). Rather than a weapon to repel the threat posed by an increasingly vocal opposition, Mamontov's trilogy is, in Kristevan terms, a ritualistic enactment of the process by which the community of Orthodox Russians guards against the threat of a loss of distinctness and identity.

It is in accordance with the Kristevan account, too, that in the first and second instalments speakers line up to agonize over how 'we' have allowed 'our' young people to descend into ungodliness. Throughout the third episode centring on the figure of Tolokonnikova's partner, Petr Verzilov, who lived in Canada as a boy and has Canadian residency rights, part of the collective 'we' separates off to become 'they'. The circles in which Verzilov moves belong to a virulent strain of an anti-Russian disease introduced into the collective body from abroad. Elena Vaenga, meanwhile, refers specifically to Pussy Riot as 'traitors'. In the first episode, the dominant appellation was 'puppets' (*mannekeny*). In *Provokatory-1* and *2*, as an expression of their outrage at the acts of blasphemy spreading across Russia, several speakers ask pointedly 'Who are we?' Their question is echoed in Pussy Riot's masks of anonymity, indicating that the processes of boundary (re)definition and (threatened) collapse cut across multiple communities.

The scandal chronotope as anti-media gesture

If televisual anathematizing is akin to religious rite, how do the media formats in which the process takes place facilitate it? Let us return to *V kontekste* and *Provokatory*. Broadcast at either 'end' of the scandal, they represent two poles of a narrow spectrum. Shevchenko's programme follows the outline of the democratic forum in which people holding different opinions challenge one another and the moderator, and genuinely compete for pre-eminence in the discussion. Despite his brazen efforts to impose his own views, the questions that Shevchenko poses are open-ended and the participants reflect the main strands of opinion about Pussy Riot.

By contrast, with the exception of a single, token 'atheist', Mamontov's guests all express the same, conservative views. Representatives of the liberal wing are absent. The speakers reinforce rather than challenge one another, in a form of what Iurii Lotman calls 'auto-communication'.[84] Mamontov poses leading questions with pre-scripted answers. The very thematic question motivating the trilogy – 'Who is behind them?' – does not allow for competing responses. The only acceptable answer is provided by the 'documentary' films screened before each discussion, which leave little open to debate. While the trilogy is conceived as a sequence, memories are wiped clean between the films, and the argument they project requires constant recapitulation. At the beginning of *Provokatory-3*, Mamontov plays the role of a teacher dubious that his 'class' has assimilated previously taught material. Re-presenting the 'facts' of the case, he 'speculates' that there may be a plot behind them, before turning to his guests to ask 'but what do you think?' and receiving their dutiful response.

Rather than dismissing the *Provokatory* trilogy as mere paranoid propaganda, we must ask why, when state television is clearly capable of presenting its case more subtly, it should resort to the antics of the playground. The playground metaphor is apposite, not only in connoting a pseudo-infantile regression but also because Mamontov's shows put one in mind of the rabble of the playground, in which a victim is subjected to conspiratorial gossip, ritual abuse and mockery, and where the purpose is not to debate but to re-confirm group identities through an act of exclusion. The inclusion of the token atheist enables the scapegoating hitherto reliant on an absent victim, to test the sentiment that it has generated in an 'encounter' with the victim himself. In *Provokatory-3*, Mamontov intimidates this victim in the manner of the playground bully, infringing his personal space and addressing him with sarcastic menace as 'my dear fellow' (*dorogoi moi*).

The logic of *Provokatory* is that of Bakhtin's carnivalized public scandal: the shock of the exposed, sexualized body (the women's obscene dancing), the intrusion of the profane into sacred space and the inversion of hierarchies (the despicable punk displacing the high priest). Such breaches of propriety provoke hyperbolic outrage ('this God-defiling action') rabble-rousing abuse (‘*koshchunitsy*’; ‘*izvrashchentsy*’), gossip concerning their origin (Berezovskii; the global government; Gol’dfarb), scapegoating and mockery. By showing the documentaries in a continual loop as the discussions proceed, the programmes re-enact both the original scandal and the shocked reaction of its witnesses.

The space of the studio is a model for scandalized society. The rabble-rouser-in-chief occupies the central position in front of the ‘crowd’. A privileged front row is occupied by Orthodox dignitaries and intellectuals loyal to the state (the antithesis to the art collective, *Voina*), journalists and representatives of official youth movements (mediators between the centre of power and broader society). Behind them sit ‘the people’, whose role is limited to clapping and hissing their disapproval.

Scandal space is accompanied by a collapsed temporality in which events from different periods (the ‘punk prayer’, the 2006 Litvinenko scandal, 9/11, the 1917 revolution, the pre-revolutionary period) can be edited into an undifferentiated, scandalized ‘present’. The waiving of the requirement for logical sequence affects the relationship between the three parts of the trilogy, and between different narrative lines within each. The fact that the focus on Berezovskii in part 2 is replaced in part 3 by a new emphasis on Verzilov – the Canadian citizen and ‘new Lenin’, as he is described contemptuously – is contradictory only within conventional time–space. In the scandal chronotope, stories, rumours, gossip and conspiracies circulate in parallel, and in sequence, without any requirement for causal links.

Nonetheless, Mamontov’s trilogy is a radical departure from even Russian television’s distorted approach to balance. Dismissing accusations that he was skewing the discussion in his favour, Mamontov claimed in an interview with a popular Russian newspaper that he was countering the media sympathy that the ‘blasphemers’ had elicited from liberals:

> The presenters of current affairs programmes seemed to be reporting the information neutrally, but with the subtext that these girls were in the right. I … decided that I would not take part in this devil’s choir … I invited people representing that part of society which has no access to the media, which is not so eloquent.[85]

Mamontov’s trilogy was a counter-strike against a western-influenced media biased against the traditional values of those who felt the ‘punk prayer’ episode was an outrage.[86] When pressed on the lack of evidence for his claims, Mamontov abandoned all pretence at reason:

> So who is Berezovskii, the Pope or what? I was told he was involved once and that’s all I need. … And why have you decided that I must reveal everything exactly when you want me to. I will explain when I want.[87]

Mamontov’s refusal to follow the conventions of televised discussion is itself a provocation against preferred liberal media formats. ‘Form’ and ‘content’ converge: the calculatedly ‘anti-liberal’ format of *Provokatory* coincides with the assault on liberal sentiments expressed within them. We have turned full circle from Kozlov's intervention in *V kontekste* signalling the correspondence between the 'idea' of a consensus in favour of traditional values and free-flowing, 'western-style' discussion. In *Provokatory*, there is again a meta-level correspondence between format and 'ideas', but as an 'anti-media' gesture in which, by adopting the scandal chronotope, the very principle of open-ended discussion is rejected as an imposition from a media in hock to western powers.

When the media via which anti-western sentiments are expressed are themselves subjected to those sentiments, the very notion of 'representing otherness' becomes problematic, since the represented 'other' inhabits the tool of representation. Other areas of the Russian establishment experience similar meta-level transformations. The law itself was rejected as alien to Russian tradition even as it was being practised. When Mamontov is reminded in the interview that blasphemy is not illegal, he replies ominously: 'Well, it will be from now on! Our criminal code was created by atheists, who in 1917 began to destroy our church.'[88]

From *Provokator* to *Gastarbeiter*

As Mamontov's anathemas (and those of other leading television personalities) broadened to include multiple assortments of the marginalized and the excluded, the indirect implications for ethnic cohesion became apparent. However, the theme of ethnicity had already infiltrated Mamontov's discourse in a more direct sense. In *Provokatory-2*, reference was made to a *Voina* performance in which three dummies (representing a *Gastarbeiter*, a Jew and a homosexual) were 'hanged' in a Moscow supermarket. Mamontov criticizes the performance, pointing out that it soon became known in the Internet community as 'The Execution of the Homo and the *Gastarbeiter*'.[89] A member of Pussy Riot, claims Mamontov, had participated in the 'dummies' performance, undermining her claims to speak for enlightened, 'liberal' Russia. Here, the conspirators not only scheme to destroy Russia, but unbeknown to their western admirers, align themselves with the forces of unenlightened intolerance with which these same admirers identify Pussy Riot's antagonists.

Mamontov's interpretation of the 'puppets' performance wilfully misreads it as not a *satire* on intolerance but an *endorsement* of it. The misreading was repeated by Putin and by Dmitrii Chizhov, Russia's representative in Brussels, who was reported by *Vesti* as asking:

> How would the European Parliament react to one of their [*Voina*'s] escapades when they organized a public burning (*sic*) of the dummies of a Jew, a homosexual and a *Gastarbeiter*? How would these opponents of anti-Semitism, and defenders of the rights of minorities and *Gastarbeiters*, square these things?[90]

Chizhov is guilty of no fewer than four errors (replicated by Putin and Mamontov): the performance involved people rather than dummies; there were five rather than three (the performance was dedicated to the memory of the five Decembrists who, in the Soviet historical canon, initiated the movement which led to the 1917 revolution); they were 'hanged', not 'burned'; the performance was directed against *intolerance*, not liberalism. Deliberate deception aside, such distortions occur when an incident circulates across the multiple platforms of a sustained media event. Putin and Chizhov are repeating stories generated by that media event helpful to their position. The logic of rumour, gossip, subplot, pre-history, misinterpretation, over-interpretation, tirade and counter-tirade surrounding Pussy Riot tends toward displacement and obfuscation and the scandal's status as media event drives that process.[91]

Whether or not the disingenuity of Mamontov and others involves calculation or immersion in the web of obfuscation, they momentarily occupy the subject position of the tolerant liberal, affronted by incitement of anti-minority hatred. Their stance is facilitated by the fact that the Decembrist performers temporarily adopted the viewpoint of the intolerant extremist precisely to expose the inhumanity that is incited by an atmosphere of hostility to minorities. *Voina*'s discursive strategy – appropriating the mentality of their target in order to mock it – is evident, too, in the accompanying paraphernalia: a banner reading 'Siberia for the Siberians', a parody of the anti-migrant 'Russia for the Russians'.

Rather than appropriating the other's discourse to ridicule it, the tactical occupation of the 'liberal' position transforms anti-Kremlin antagonists into extreme versions of the intolerant self, while recasting the latter as a moderate version of the other. The fact that *Voina* adopted the same strategy in reverse indicates a mutual entanglement of discourses, rather than a mere hybrid or a crude instrumentalization.

However, the knowing misattribution of referential meaning to a metaphorical hanging occurs at a time when a rampant, and all-too-sincere, new literalism was thrusting the Pussy Riot scandal into the orbit of issues of ethnicity, thereby highlighting the erosion of the notions of tolerance and multiculturalism within official discourse. Mamontov and Putin misread the *Voina* performance from a pseudo-multiculturalist viewpoint mindful of the evils of anti-Semitism. Yet much of the official media vitriol derived from a visceral hatred of the very tolerance of minorities with which multiculturalism goes hand in hand – a hatred which, as the currency of the pun term '*tolerast*' in Russian xenophobic slang highlights, invokes the sexualized disgust we noted earlier. Putin would not have succeeded in donning the multiculturalist mask had the rhetoric of tolerance been rooted within public discourse (or, in Laclauan terms, had Putin succeeded in inserting his reading of *Voina* within a hegemonic–populist chain of equivalence).

This veering between pro and anti-tolerance stances illustrates the dynamic of identity and abjection. It also points, as we demonstrated earlier, to a consensus management strategy on the brink of disintegration. It is significant that as anti-Pussy Riot sentiment was peaking, the Kremlin began to intimate that the new law on offending religious believers was, perhaps, excessive (though the law has since been adopted[92]), and that the amnesty granted to Tolokonnikova and Alekhina on 23 December 2013 coincided with an intensification of the anti-western paranoia which provided the background to Russia's widely condemned intervention in Ukraine two months later. Such abortive half-gestures appear to have been motivated by a contradictory cocktail of fear and desire that the 'absolute outsidedness' accumulated by Pussy Riot in the course of its 'trial by media event' might transfer permanently to other forms of alterity, producing an exclusionary form of Holy Russian-ness bearing within it catastrophic potential. Chapter 9 will demonstrate that the contradiction was to be resolved decidedly in favour of desire.

Notes

1 'Pank-moleben "Bogorodista, Putina progoni" Pussy Riot v Khrame'. Available online at: <http://www.youtube.com/watch?v=GCasuaAczKY> (accessed 15 December 2013).
2 'Punk prayer'. Available online at: <http://andmand.livejournal.com/63504.html> (accessed 15 December 2013).
3 Accusation against Pussy Riot. Available online at http://pravaya.ru/news/22696 (accessed 15 December 2013).
4 SOVA report, 7 June 2012. Available online at http://www.sova-center.ru/en/religion/news-releases/2012/06/d24590/ (accessed 1 November 2012).
5 Interviews with Rossiia journalists, 29 March and 24 April 2013; and with a journalist working for a variety of media outlets, including NTV, 26 March 2013.
6 Simone Gigliotti, 'Unspeakable Pasts as Limit Events: The Holocaust, Genocide, and the Stolen Generations', *Australian Journal of Politics and History*, 49/2, 2003, 164–81.
7 Rossiia's talk show *Spetsial'nyi korrespondent*, the broadcast titled *Provokatory-2,* 11 September 2012. Available online at www.youtube.com/watch?v=NrqpbuItPB0 (accessed 1 November 2012).
8 See, for instance, Irina Levinskaia, 'My stanovimsia pravoslavnym Iranom'. Available online at www.fontanka.ru/2012/08/22/098/ (accessed 1 November 2012); Irina Iasina's interview on the channel Dozhd', 20 April 2012: 'Nam grozit Iran 1979 goda – KGB v tserkovnykh riasakh'. Available online at www.youtube.com/watch?v=Qu10lIDbgog&feature=youtu.be (accessed 1 November 2012).
9 *Vremia* first mentioned the episode only in connection with the public prayer at the Cathedral of Christ the Saviour organized by Patriarch Kirill on 22 April 2012.
10 Ronald Jacobs and Eleanor Townsley, *The Space of Opinion: Media Intellectuals and the Public Sphere*, New York: Oxford University Press, 2011.
11 Julia Kristeva, *Powers of Horror: An Essay on Abjection*, New York: Columbia University Press, 1982.
12 See, for instance, a report on North Ossetia-Alania as a model region of Russia in terms of the relationship between Orthodoxy and Islam in the *Musul'mane* broadcast of 29 October 2010.
13 'Interviu mitropolita Volokolamskogo ilariona telekanalu "Dozhd"', 15 March 2012. Available online at www.patriarchia.ru/db/text/2075242.html
14 'Sviateishii Patriarch Kirill: ne mozhet byt' spravedlivogo obshchestva, osnovannogo na popranii spravedlivosti', 7 November 2010. Available online at www.patriarchia.ru/db/text/1313788.html (accessed 1 November 2012).
15 'Strategiia s Maksimom Shevchenko', 20 November 2010. Available online at http://radio.vesti.ru/fm/doc.html?id=408078 (accessed 1 November 2012).
16 See, for instance, 'Sviatynia - v Krasnoiarske', *Vesti*, 2 November 2011.
17 *Vesti*, 7 January 2012.
18 '10 dekabria. Protesti. Rossiia i mir', 11 December 2011. Available online at: http://newtimes.ru/articles/detail/47588/; 'Vlastiam dali dve nedeli', 10 December 2011, available

online at www.gazeta.ru/politics/elections2011/2011/12/10_a_3922210.shtml (both accessed 2 November 2012).

19 Blog, 10 December 2011. Available online at <http://echo.msk.ru/blog/shenderovich/837971-echo/. 'Otvet durakam', 10 December 2011. Available online at http://shevchenko-ml.livejournal.com/3284.html (both accessed 3 October 2013).

20 'Dukhovnik Patriarcha Kirilla, starets Ilii (Nozdrin) ne blagoslovliaet miting oppozitsii', 24 December 2011. Available online at www.o-megas.livejournal.com/374604.html (accessed 15 October 2012).

21 'Verite li vy v religioznye chudesa'. Available online at: www.levada.ru/archive/religiya/verite-li-vy-v-religioznye-chudesa>; 'Verite li vy v sglaz i porchu', available online at: www.levada.ru/archive/religiya/verite-li-vy-v- sglaz-porchu> (both accessed 15 October 2012).

22 Interview with Levada Centre director, Lev Gudkov, *gazeta.ru,* 20 August 2012. Available online at www.levada.ru/20...2012/devushki-iz-pussy-riot-p (accessed 1 December 2012).

23 *V kontekste* (Channel 1), 15 March 2012. Available online at www.1tv.ru/sprojects_edition/si=5851&d=15.03.2012 (accessed 1 December 2012).

24 *Pust' govoriat* (Channel 1), 15 March 2012. Available online at www.1tv.ru/sprojects_edition/si5685/fi14384 (accessed 1 December 2012).

25 '24.04.2012. Spetsial'nyi Korrespondent "Provokatory"'. Available online at www.youtube.com/watch?v=g_QkRPkTWzk (accessed 15 December 2013).

26 See Jacobs and Townsley, *The Space of Opinion*, for how the talk show establishes the limits of the 'space' within which consensual opinion is negotiated.

27 *Odnako* (Channel 1), 3 April 2012. Available online at www.1tv.ru/news/leontiev/203292 (accessed 1 December 2012).

28 Ibid.

29 *V kontekste* (Channel 1), 15 March 2012.

30 'Musul'mane vstupilis' za Pussy Riot po predpisaniu propoka Mukhammeda'. Available online at: <http://top.rbc.ru/society/10/08/2012/664049.shtml> (accessed 1 December 2012).

31 Ibid. For a full discussion of anti-Semitism in the Pussy Riot affair, see Mike Agapov, 'Proiavleniia anti-semitizma v obshchestvennoi diskussii po delu Pussy Riot'. Available online at www.academia.edu/3372441/_Pussy_Riot_Manifestations_of_anti-Semitism_in_the_public_discussion_on_Pussy_Riot (accessed 15 December 2013).

32 'Putin vystupil protiv strogogo prigovora za pank-moleben', 3 August 2012. Available online at http://ru.reuters.com/article/topNews/idRUMSE87200F20120803; 'Interviu telekanalu Russia Today', 6 September 2012. Available online at www.kremlin.ru/news/16393 (accessed 1 December 2012).

33 *Odnako* (Channel 1), 3 April 2012. Available online at www.1tv.ru/news/leontiev/203292 (accessed 1 November 2012).

34 'Court Extends Detention of Pussy Riot Women Until January', 20 July 2012. Available online at www.themoscowtimes.com/news/article/court-extends-detention-of-pussy-riot-women-until-january/462391.html#ixzz21GzogmdW (accessed 1 November 2012).

35 Interview online, 20 July 2012. Available online at http://echo.msk.ru/programs/beseda/911106-echo/comments.html (accessed 1 November 2012).

36 NTV, 9 August 2012.

37 The photo was taken at a *Voina* performance, entitled 'Fuck for the Little Bear' (*Ebis' za medvezhonka*) and mocking then-President Medvedev. It took place in February 2008.

38 *Vesti*, 10 August 2012.

39 The Eurovision final took place on 26 May 2012. While addressing a new Presidential Council on Inter-Ethnic Relations, Putin referred to the 'great enthusiasm with which

these grannies represented our multinational country at an international cultural event'. Available online at http://eng.kremlin.ru/news/3997 (accessed 15 December 2013).

40 'Pussy Riot verdict greeted with defiance', 17 August 2012. Available online at www.guardian.co.uk/world/2012/aug/17/pussy-riot-verdict-defiance (accessed 1 December 2012).

41 Patriarch Kirill, *Freedom and Responsibility: A Search for Harmony*, London: Darton, Longman and Todd, 2011.

42 *Provokatory-1*, 24 April 2012. Available online at www.youtube.com/watch?v=g_QkRPkTWzk (accessed 30 April 2012).

43 *Novosti* (Channel 1), 22 April 2012. Available online at www.1tv.ru/news/social/205068>; *Vesti nedeli*, 9 September 2012; *Poedinok* (Rossiia), 15 March 2012. Available online at www.youtube.com/watch?v=G6p3auva4Gc (accessed 1 December 2013).

44 Aleksei Miller, 'Istoricheskaia politika v Rossii: novyi povorot?', in A. Miller and M. Lipman (eds.), *Istoricheskaia politika v XXI veke: sbornik statei*, Moscow: Novoe literaturnoe obozrenie, 2012, pp. 334–7.

45 Kirill, *Freedom and Responsibility*.

46 'V Gosdumu vnesli zakon o nakazaniiakh za oskorblenie chuvstv veruiushchikh', 26 September 2012. Available online at www.rg.ru/2012/09/26/veruiushie-anons.html (accessed 1 November 2012).

47 'Poedinok s Vladimirom Solov'evym vypusk No 63: Nikolai Svanidze protiv Maksima Shevchenko', 27 September 2012. Available online at http://poedinoktv.net/poedinok-s-vladimirom-solovevym-nikolaj-svanidze-protiv-maksima-shevchenko/> (accessed 1 November 2012).

48 Available online at http://poedinoktv.net/poedinok-s-vladimirom-solovevym-nikolaj-svanidze-protiv-maksima-shevchenko/ (accessed 1 November 2013).

49 *Vesti nedeli*, 9 November 2012.

50 Emile Durkheim, *The Elementary Forms of Religious Life*, New York: The Free Press, 1995 [1912], pp. 1–3, 35–44, 419–23; see in particular pp. 41–2.

51 *V kontekste* (Channel 1), 15 March 2012. Available online at www.1tv.ru/videoarchive/44706 (accessed 1 November 2012), and debate about boundaries in *Poedinok* (Rossiia), 15 March 2012. Available online atwww.youtube.com/watch?v=G6p3auva4Gc (accessed 15 December 2013).

52 'Poedinok s Vladimirom Solov'evym Vypusk No 63: Nikolai Svanidze protiv Maksima Shevchenko'. Available online at http://poedinoktv.net/poedinok-s-vladimirom-solovevym-nikolaj-svanidze-protiv-maksima-shevchenko/ (accessed 13 December 2013).

53 Talal Asad, 'Anthropological Conceptions of Religion: Reflections on Geertz', *Man*, New Series, 18/2, 1983, 237–59.

54 This was the CPRF leader Gennadii Ziuganov's comment on *Vesti,* 26 September 2012.

55 Raymond Kelly, 'Witchcraft and Sexual Relations: An Exploration in the Social and Semantic Implications of the Structure of Belief', in Paula Brown and Georgeda Buchbinder (eds.), *Man and Woman in the New Guinea Highlands*, Washington DC: American Anthropological Association, 1976, pp. 36–53.

56 'ChP. Rassledovanie' (NTV), 6 April 2012. Available online at http://poedinoktv.net/poedinok-s-vladimirom-solovevym-nikolaj-svanidze-protiv-maksima-shevchenko/ (accessed 1 November 2012).

57 Ibid.

58 'Chto stoit za aktsiei "Pussy Riot" v Khrame Khrista Spastitelia', 15 March 2012. Available online at www.1tv.ru/sprojects_edition/si5851/fi14392 (accessed 1 November 2012).

59 Ibid.

60 See, for instance, Egor Kholmogorov's comments in *Provokatory-1*, 24 April 2012; and *Vesti nedeli*, 9 September 2012 (a report by Dmitrii Kiselev, *Ataka na tserkov'*).

61 'Sotni tychiarch pravoslavniykh po vsei Rossii molilis vmeste vo imia very i v zashchitu tserkvi'. Available online at www.1tv.ru/news/social/205068 (accessed 1 November 2012).
62 Available online at http://tvrain.ru/articles/svjaschennik_pavel_adelgejm_o_pussy_riot_rpts_i_borbe_za_prava_polnaja_versija-331600/ (accessed 1 November 2012).
63 'Marat Gel'man o dele Pussy Riot: ne tol'ko RPTs teriaet'. Available online at http://youtu.be/1EmZQY8ybnw; 'Posledstviya aktsii "Pussy Riot"'. Available online at www.snob.ru/selected/entry/47121 (both accessed 1 November 2012).
64 See for instance the programme *Govorite* (Dozhd'): www.youtube.com/watch?v=1EmZQY8ybnw&feature=youtu.be (accessed 1 November 2012).
65 'Delo Pussy Riot prevrashchaet RPTs v KPSS', 20 July 2012. Available online at http://tvrain.ru/articles/delo_pussy_riot_prevrashchaet_rpts_v_kpss-311777/ (accessed 1 November 2012).
66 *Nedelia* (REN TV), 29 September 2012. Available online at www.nedelya.ren-tv.com/index.php?limitstart=10 (accessed 1 November 2012).
67 *Vesti*, 12 September 2012.
68 Julia Kristeva, *Powers of Horror,* p. 3.
69 Ibid. pp. 4, 12.
70 Ibid., p. 3.
71 Ibid., p. 10.
72 Ibid., p. 66.
73 Ibid., p. 43.
74 Ibid., p. 13.
75 Ibid., p. 185.
76 Slavoj Zizek, *Living in the End Times*, London: Verso, 2010, p. 135.
77 Shireen Hunter, *Islam in Russia: The Politics of Identity and Security*, New York: M.E. Sharpe, 2004.
78 Kristeva, *Powers of Horror,* p. 68.
79 Fredric Jameson, 'Imaginary and Symbolic in Lacan: Marxism, Psychoanalytic Criticism, and the Problem of the Subject', *Yale French Studies*, 55/56, 1977, 338–95 (353).
80 Ibid., 354.
81 Interview with Kiselev, 27 March 2013.
82 *Vesti*, 2 April 2012.
83 Reports on the display on 28 April 2012 of the relic of Saint Matrona of Moscow; the display on 11 June 2012 of the relic of Saint Lazar, brought from Cyprus; the display on 19 and 20 June 2012 in Ivanovo of a relic described as part of the robe of the Virgin Mary.
84 In auto-communication 'a message is transmitted by the collective "I" of humanity to itself' and serves the purpose of re-confirming societal ties. See Yurii Lotman, *Universe of the Mind: A Semiotic Theory of Culture*, Indianopolis: Indiana University Press, 2001, p. 33.
85 'Arkadii Mamontov, Sud i prigovor Pussy Riot spravedlivy', *Komsomol'skaia pravda*, 14 September 2012. Available online at www.kp.ru/daily/25949/2892800/ (accessed 1 November 2012).
86 'Ia angazhirovan Bogom', *Lenta.ru*, 18 October 2012. Available online at: http://lenta.ru/articles/2012/10/18/mamontov/ (accessed 1 November 2012).
87 Ibid.
88 Ibid.
89 'Provokatory. Fil'm Arkadiia Mamontova ob aktsii Pussy Riot'. Available online at www.youtube.com/watch?v=aeT0dZbGkzc (accessed 1 January 2014).

90 *Vesti*, 28 September 2012.

91 For the role of gossip in television news, see Andrew Tolson, *Media Talk: Spoken Discourse on TV and Radio*, Edinburgh: Edinburgh University Press, 2006, pp. 64–73.

92 'Russian drafts new ethnic policy of "multi-people nation" – report', 18 October 2012. Available online at http://rt.com/politics/russia-drafts-nation-report-689/ (accessed 15 December 2013).

9 'There is war on our streets...'

The 'national question' and migration on state-aligned television after the 2012 presidential elections

Our previous two chapters appeared to indicate a progressive hardening of the authoritarian features of Russia's political system since late 2011–early 2012. During the unfolding of the Pussy Riot scandal examined in Chapter 8, the convergence of Kremlin, Church and state-aligned media in a co-ordinated campaign to construct a societal consensus by claiming that the liberal wing of Russian society was alien to the nation would appear to offer clear evidence of this phenomenon. In Chapter 7, the struggle to consolidate the consensus by seizing ownership of the ideological territory linking corruption and ethnicity from populist opponents such as Naval'nyi offered further corroboration of the trend.

The Kremlin did not, however, entirely vacate the territory carved out in Putin's pre-election article and, given the alarming rise in hostilities perceived by the public as inter-ethnic, could not afford to do so. Paradoxically, given media–state collusion over the Pussy Riot affair and the rise of Naval'nyi, rather than continuing to endorse this line, since 2012 Rossiia (and to a lesser extent Channel 1) began systematically to challenge it, departing, often radically, from previous interpretations of ethnicity-related issues and going beyond the far more cautious recalibration undertaken by the Kremlin. We witnessed this to an extent in Chapter 8 in the scandalizing discourse of Arkadii Mamontov in relation to Pussy Riot. Through the prism of television coverage of migration issues, the current chapter focuses in more detail on how these challenges have been manifested, on why such a development might occur and on what it tells us about the relationship between the Kremlin and federal broadcasters.

During the first 18 months of Putin's third presidency, one of the main issues around which the disjunction between the Kremlin's pronouncements and televisual representations of the issues they address became particularly apparent was migration. Coverage of this topic, particularly on Rossiia, changed drastically in the post-election period compared to the earlier period. Further analysis of the change will enable us to explore the role of the third element in the triangular process of circulation within which both state-aligned broadcasters and Kremlin officials operate – the public. Over the past 15 years migration has been of increasing concern to citizens of most European states, and the European media have both reflected and shaped this concern. The same has been true in Russia, and this development began to impact significantly on its state-aligned television output in 2012.

Following a discussion of the political context of post-election Russia, the chapter will consider how television's overall treatment of the 'national question' has shifted following Putin's re-election. We will argue that heightened post-election concern with the issue is not reflected in the actual amount of coverage, which remains small within the news agenda as a whole, but is manifested in the regular inclusion in news and non-news programmes of accounts of the state of inter-ethnic relations in Russia characterizing it as catastrophic. Kremlin-sponsored discourse generally emphasizes positive changes in domestic affairs and foreign policy under Putin relative to the 1990s but in the post-election period official media sources, which normally reproduce this line of argument, began to make an exception for inter-ethnic relations. Here, the focus is on the significant deterioration since 2000. The chapter will then examine the orchestrated anti-immigration campaign on state television which began in the spring of 2012 and intensified further in the spring of 2013, eventually subsiding after a series of ethnic riots across Russia in the summer of that year. It will be argued that this campaign has been accompanied by a shift in representations of the role of Islam in Russia and Europe. Analysing the campaign within the broader context of European media reporting on migration, we will consider what this campaign tells us about the increasingly contradictory public discourse on Russian identity. As a lead-in to the conclusions to our book, the chapter ends with discussion of the reasons for the remarkable shift in the broadcasters' approach and its potential impact on ethnic cohesion in the country.

A new context

Some liberal Russian observers regard the latest political trends and changes in the media discourse during Putin's third presidency as yet another indication of the country's reverting to the past (*vozvrat k proshlomu*).[1] In contrast, we will argue that the three factors shaping state-aligned broadcasting today have developed after the election in such a way that differences with the Soviet period have, if anything, become more pronounced. These factors, which we have been highlighting throughout the book, are: the nature of post-Soviet media culture; the relationship between the Kremlin, broadcasters and the public; and the regime's lack of consistent ideological underpinnings.

Despite superficial similarities with Soviet propaganda rhetoric about the ever-conspiring west and multiple enemies of the state, post-election state-aligned television coverage and the official discourse of Putin's third presidency have little in common with their Soviet predecessors. The socialist realist approach, representing life as it should be, is rarely adopted. Broadcasters are acutely aware that, in contrast to the past, the public now needs to be entertained and that popular concerns, as well as debates on the Internet and in alternative media, must be taken into account. The media first have to win viewers' trust, even if only to exploit it subsequently in order to strengthen the legitimacy of the existing regime. This legitimacy is no longer based on a denial of the existence of problems, as had particularly been the case during the final 30 years of communism; instead it is founded on reverse tactics, which were further radicalized during the 2012 presidential election campaign and its aftermath. Today, portrayals of the country as following a straight path leading to a bright future have been superseded by depictions of a society in deep crisis, faced with multiple and ever growing external and internal threats. The television viewer is not placated and comforted, but systematically horrified and frightened. If, during the election campaign, the main goal of the state-aligned media was to perform societal unity based around Putin, in the post-election period the strategy of governing the country, which is inevitably reflected on television, has entailed legitimization of Putin's regime through the constant identification, marginalization and demonization of various groups claimed to embrace values conflicting with traditional Russian ones.[2] Against the backdrop of this discursively constructed reality, the president is depicted as the only politician able to protect Russia from imminent collapse. The crudeness of these representations has burgeoned during Putin's third presidency. The shift can be partly attributed to the Kremlin's decision to reject the approach of the former chief ideologist, Vladislav Surkov, who promoted a discourse accommodating considerable ambiguity about Russia's relationship with its 'others'.[3]

As we have consistently argued, the relationship between the Kremlin, broadcasters and the public is quite distinct from that of the Soviet era, when the public was unable to influence what was broadcast and journalists were clearly subordinated to political power. Today, while responding to the broader political goals of the regime, state television journalists have far more leeway when deciding what to cover and how to cover it; and one of the essential features of post-1991 media, namely the staged pluralism of views, can never be fully controlled.[4]

The interviews we conducted demonstrate that influential broadcasters, both those loyal to the regime and those who expressed sympathy for the opposition, frequently articulate xenophobic prejudices. Most of our interviewees manifested a hostile attitude towards migrants and certain ethnic minorities. A Rossiia journalist who was open about his sympathies towards the opposition avowed: 'One can be triple-tolerant towards visitors (*priezzhie*), towards differences, towards Islam, towards Muslims, but this is clearly going too far.'[5] In an interview with a member of our project team, a pro-Kremlin celebrity journalist from Rossiia argued: 'Would you have wanted Turks to populate the cities of [the Austrian provinces] of Carinthia and Tyrol, so that there were only mosques in Innsbruck?... Who are educated? The Turks?...'[6] In the absence of clear guidelines about how to address sensitive and complex issues, journalists have ample space within which to indulge their prejudices.[7]

At the same time, opposition protests of the electoral period have made the political establishment and the state-aligned media more sensitive to public opinion – particularly, as we saw in relation to corruption in Chapter 7, regarding those issues which are exploited by the opposition. Inter-ethnic relations and migration are the primary case in point. The fact that opposition leaders such as Naval'nyi were deploying anti-immigration rhetoric undoubtedly began to influence official discourse in the post-election period.[8] Opinion polls demonstrate that the Russian public, whose opinions are influenced by the discourses of media and politicians, increasingly views migration as a major concern.[9] According to Levada Centre surveys conducted in 1996, 2003 and 2012, attitudes towards migrants among Russia's citizens have been getting progressively more hostile. According to the October 2012 survey, 71 per cent of those polled believed that migrants contributed to the rise of crime in the country, 46 per cent believed that they were 'destroying Russian culture' (as opposed to 19 per cent who rejected such a view) and 65 per cent supported the proposition that migration should be reduced.[10] It is in this context that the increased attention given by federal broadcasters to migration issues must be understood.

As we have argued in earlier chapters, Putin's own pronouncements on the 'national question' during the election campaign remained fluid, as did the broader Kremlin-sponsored Russian identity discourse, and both are open to multiple interpretations. The hierarchical view of cultures reflected in Putin's article unwittingly encourages the very ethnicization of economic and social issues which the Kremlin rhetorically opposes. It is our contention that, during Putin's third presidency, the contradictions between different elements in Kremlin-sponsored identity discourse and in the output of broadcasters have become even more apparent than before.

Putin himself has proclaimed 'the preservation' (*sokhranenie*) of 'a new pan-Russian identity' (*rossiiskaia identichnost'*) and 'national consolidation' (*splochennost' natsii*) to be priorities of his new presidency.[11] But what are the defining features of this identity? During the elections and in the post-election period, three incompatible visions have been prominent: the image shaped by isolationist Russian nationalism, which wants to exclude 'ethnic others' from the national body; Russia as the centre of the Eurasian Union, which includes parts of Central Asia and is built on the history of interactions between Orthodox Christianity and Islam; and Russia as a defender of traditional values at a time when they are allegedly being denigrated in the west. The president has himself openly endorsed the last two visions. The first one, however, is promoted by elements within the opposition as well as by influential politicians, and seems to enjoy considerable public support.[12] Its co-optation by official discourse, therefore, has emerged as a primary goal for state-aligned broadcasters. Let us now examine this co-optation process in more detail.

The 'national question' reconsidered

The 'national question' and the issue of migration were virtually ignored by broadcasters at the time of the presidential election campaign. According to the journalists we interviewed, this happened in accordance with instructions to avoid dwelling on contentious and sensitive issues which the state-aligned media received from the authorities.[13] With the lifting of this restriction after the elections, the situation reversed; ethnicity-related issues became a greater focus of attention in television talk shows than ever before and also started to feature more prominently in the news. The fact that Putin identified the 'national question' as a key issue facing Russia can explain the broadcasters' increased attention to it once the elections were over. Yet it does not mean that broadcasters received any clear guidelines from the Kremlin about how to cover this complex issue. As our interviewees indicated, whereas those in charge of the two main channels regularly liaise with the Kremlin regarding broad strategies of the coverage, rank-and-file journalists, in contrast with the case in the Soviet period, do not routinely receive specific instructions on what and how to report. In their own words, they are expected to be guided by their own *adekvatnost*, which one interviewee defined as 'the ability to react appropriately to the conditions in which you find yourself' and another simply as a 'political instinct'.[14] In a situation where broadcasts are expected to reflect both a certain pluralism of opinion (which, even if it is staged, cannot always be fully controlled), as well as the (real or perceived) public mood, the coverage of controversial issues can easily conflict with the Kremlin's pronouncements. As we shall see, in 2012–13, in their interpretation of migration-related issues, state-aligned broadcasters – particularly Rossiia – radically departed from the interpretative framework which was authorized by Putin at the start of his electoral campaign.

The first major programme on inter-ethnic relations to be shown on a federal television channel in the aftermath of the presidential elections was the 29 March 2012 edition of Channel 1's *V kontekste* talk show. While the 1 March edition did not mention the 'national question' as an issue confronting Russia's future new president,[15] the programme's presenter, Maksim Shevchenko, devoted his first post-election broadcast exclusively to that issue. Centring on the fatal shooting in a Jewish school in the French city of Toulouse perpetrated a few days earlier by a terrorist of French Algerian origin, the programme was introduced by Shevchenko as having 'the most direct relevance to Russia, as on the territory of our country people of completely different (*sovershenno razlichnykh*) religions, cultures and ways of life have been brought together by historical fate'.[16] The drawing of such direct parallels between Europe's 'crisis of multiculturalism' and the deteriorating state of inter-ethnic relations in Russia was in itself unusual for the channel. As we saw in Chapter 4, parallels of this sort were drawn only briefly in the immediate aftermath of the Manezhnaia riots. On the whole, and as the report on the French Roma situation with which we opened this book illustrated, the two main federal channels tended to follow the line articulated in Putin's *Nezavisimaia gazeta* article of contrasting the failure of west European multiculturalism with Russia's positive experiences of managing multi-ethnicity.[17]

Shevchenko's use of the expression 'completely different' in describing Russia's cultural diversity is not accidental. As we argued in Chapter 4, it underscores an understanding, widespread among Russian elites, of a so-called 'clash of civilizations' or 'conflict of cultures' as an important interpretative lens through which current relations between peoples of different ethnic origin should be understood. Shevchenko's overall interpretation of the 'national question' through the 'culture conflict' lens was apparent in his relatively carefully managed treatment of Putin's article in the 26 January 2012 broadcast of *V kontekste* analysed in Chapter 7. There it was the Kremlin's position, not pseudo-academic and demotic understandings of ethnicity and race, which framed the show, since it was being broadcast during the election campaign. The 29 March broadcast no longer operated under such constraints.

All participants in the programme could be described as inter-ethnic relations 'experts' representing different political positions, from the isolationist Russian nationalist to the proponent of western-style liberal tolerance. Given that the perpetrator of the Toulouse shooting was a radical Islamist, the inclusion of a mufti as a participant appeared at first glance justified. Yet, as most of the programme was actually about Russia, the choice of a Muslim cleric rather than a representative of any of Russia's other 'traditional religions' implicitly framed Islam and the ethnic groups among which it was the dominant religion as the most closely associated with ethnic conflict. As we will further show in the present chapter, this perception continued to explicitly influence the coverage of ethnicity-related issues in the post-election period.

The diversity of the participants' positions was no barrier to the attainment of consensus. This centred on two issues. Echoing Putin's *Nezavisimaia gazeta* article, all argued that, to some extent, European multiculturalist policy was in crisis.[18] As one might expect in a television discussion, no account of this complex, polysemic concept was offered. Instead, following the line taken in Putin's article, it was used as a self-evident term describing the post-immigration composition of Europe's urban population.[19] However, in contrast to Putin's largely optimistic interpretation of Russia's ethnic diversity, the participants agreed that the condition of inter-ethnic relations in Russia was alarming.

Significantly, the position closest to that of the Kremlin was articulated by the leading proponent of the *rossiiskaia* civic nation concept, the liberal ethnographer Valerii Tishkov, who emphasized that social exclusion and inequalities often underpinned conflicts that were popularly perceived as ethnic in nature. Tishkov also agreed with Putin that it was imperative for migrants to integrate by adhering to the dominant societal norms. He exhibited more optimism than the others regarding the potential success of the pan-Russian (*rossiiskii*) nation-building project which the Kremlin has been championing for over a decade.[20] However, when Tishkov rejected Putin's hierarchical prioritization of Russian culture, his criticism was cut short by Shevchenko.[21]

Despite censoring direct criticism of the leader, Shevchenko himself adopted a position that deviated from the line taken by the Kremlin during the election period. He forcefully projected his own views, exploiting his role as presenter to ensure his voice remained dominant, and at times reducing the polysemy of the discussion to a singular meaning of his own preference. From the frameworks available for discussing inter-ethnic relations he prioritized not those which were being promoted by the Kremlin (corruption, social exclusion and social inequalities) but the popular notion of 'the clash of civilizations',[22] going far further than he did in the 26 January edition.

Shevchenko argued that 'the unifying pan-Russian (*rossiiskaia*) civilization' (a concept endorsed by the Kremlin) existed only at a rhetorical level, claiming: 'But young people say: "On our streets war is going on." Young people are organized according to their ethnicity...'[23] He thus offered a reading of inter-ethnic relations which contradicted the rhetoric of political leaders and that of a certain 'expert' discourse which, it was argued, had failed to correspond to how ordinary people interpreted their social experiences. The notion of street conflicts involving people of different nationalities as a 'war' had long been evoked on the Internet and in alternative media outlets. As we saw in Chapter 4, REN TV's *Nedelia* and *Reporterskaia istoriia* were cases in point.[24] In contrast, on the two main state-aligned television channels, references to 'war' began appearing only after the 2012 presidential elections, in the context of the overall dramatization of threats allegedly facing Russia.

As if to underscore the heightened importance of the 'national question' for the Kremlin's agenda, on the very day that *V kontekste* was broadcast, Putin proposed the setting up of a Presidential Council for Inter-ethnic Relations (*mezhnatsional'nykh otnoshenii*).[25] The proposal raised disturbing questions, implying that inter-ethnic relations had deteriorated to such an extent that close monitoring from the top governmental level was essential. Coverage on *Vremia* and *Vesti* severely constrained the discursive space available for this interpretation, refusing to permit the Kremlin's position to be challenged to the extent that had occurred on Shevchenko's talk show. At the same time, a lengthy report on *Vesti* and a much shorter, but similarly framed, report on *Vremia* about the first major meeting of the Council (on 24 August 2012) indicated that news-writers were struggling to reconcile the popular perception that Russia's inter-ethnic relations were in crisis with the Kremlin-endorsed ideal of multi-ethnicity as the country's major strength.[26]

Russia's European migrant 'problem'

Between the spring of 2012 and the autumn of 2013, Rossiia made a concerted attempt to resolve the discursive struggle in favour of one interpretation. Media and political discourse on migration often reflects broader understandings of national identity held by a society's dominant group.[27] Thus, in his pre-election manifesto article Putin linked his discussion of migration to the alleged ability of the Russians, as a nation, to accommodate and integrate the ethnic 'other'. In 2010 and 2011, the Kremlin's view of the importance of migration for the Russian economy overtly influenced the stance of both channels, which devoted only a very small percentage of news time to the issue.[28] *Vremia* refrained from discussing migration in relation to Russia altogether, only occasionally referring to problems arising from migration in western Europe and blaming them on the ineffective policies of European governments.[29] *Vesti* paid more attention to the issue. There were occasional, though rare, reports on issues involving migrants from Central Asia in Moscow. In contrast to coverage of west European affairs, the Russian government was never held responsible for the situations that arose; *Vesti* seemed to imply the problems were attributable to the migrants themselves. Yet the appalling conditions in which many migrants lived were also mentioned as a root cause of migrant crime, and reports displayed some sympathy for them.[30] Furthermore, at a time when Russia's multi-ethnicity, with its roots in imperial history, was often celebrated as a major strength, influential broadcasters were able to reiterate the pronouncements of Russia's leaders, arguing that Central Asian migrants were 'our former countrymen' and 'today's compatriots' (*sootechestvenniki*); following the collapse of the USSR they became foreigners in the legal, but not the cultural, sense.[31] Until May 2012, both channels generally refrained from representing migration as a serious challenge to Russia's national identity and security.

In the post-election period, however, the relationship between migration and Russian national identity, as well as the assessment of state policies on migration, was redefined by state broadcasters. For Rossiia in particular, migration became intimately linked with corruption and legal inertia – an issue which the Kremlin appropriated from the opposition and turned into a weapon to be used against it. Migration also began to be represented as a major threat to the country's sovereignty, security and identity, as well as to the culture and well-being of the Russian people. The Russian print media and Internet had deployed this frame in their coverage of migration-related issues previously,[32] but it was only during the first year of Putin's third presidency that Rossiia began to follow this line.

State-aligned broadcasters seem to have been responding to real and perceived public concerns. In this regard they can be directly compared to media elsewhere. In the context of the growing politicization of migration across Europe, the myths about migrants propagated by the Russian media are similar to those disseminated by their counterparts in the EU regarding asylum seekers, refugees and labour migrants. In both contexts, these people are presented as a threat to European cultures and the European way of life, and references to a 'clash of civilizations' frame many media reports. Migrants and asylum seekers are portrayed as a burden on local welfare systems and a threat to security, given their allegedly disproportionate association with crime. Both Russian media and those of other European countries suggest that governments are ineffective in confronting the problems posed by migrants and asylum seekers.[33] In common with their Russian counterparts, broadcasters across Europe indiscriminately use the problematic term 'illegal migrant'. They share, too, the practice of referring to people born in European countries as migrants simply because their parents or grandparents happened to have emigrated to Europe.[34]

Nor is Russia alone in seeing itself as having been historically tolerant towards minorities. This self-perception fosters the sense that if complaints about the behaviour of ethnic 'others' are expressed, there must be good reason.[35]

At the same time, within the EU, journalistic practices differ between countries, as well as across media outlets within the same countries. For instance, in Greece and Italy it is not uncommon for public service broadcasters to uncritically reproduce extreme xenophobic and anti-immigrant views; in contrast, in Britain and the Netherlands, PSB news coverage diverges sharply from that of the tabloid newspapers. Broadcasters such as the BBC are (sometimes justifiably) criticized for bias and insensitivity, but they are regulated by a detailed set of editorial guidelines aimed at ensuring balanced and impartial reporting.[36] The BBC is therefore far less likely to uncritically reproduce crude myths and stereotypes than tabloid newspapers are.[37] While it is hard to identify parallels between

Russia's federal broadcasters and leading PSBs in countries like Britain and the Netherlands, their approach certainly resembles that of their Greek and Italian counterparts and, particularly in the case of Rossiia, of the tabloid press across the EU. Moreover, the specificities of official discourse during Putin's third presidency have also prompted Russian federal channels to develop peculiar interpretations of a range of issues which we find in the media coverage of migration in other parts of the world. For instance, in their paranoid adaptation of the western media tendency to associate immigration with security threats, they intimate that western intelligence services use migration to destabilize Russia.

Russian broadcasters also have a particular perception of the kind of coverage the public expects, and their view of what constitutes balanced and impartial reporting seems diametrically opposed to that of, say, the BBC. Where ethnicity or religion (Islam) is perceived to be a factor, it is common for the British broadcaster to deploy reporters of non-white background. In contrast, one of our more liberally minded interviewees informed us that the policy of his independent news channel was never to assign the reporting of ethnicity-related issues to a journalist with an ethnic minority background. In his words: 'Only journalists with a Slavic appearance (*slavianskoi naruzhnosti*) can cover inter-ethnic relations.' He explained that this policy had been developed to accord with his channel's understanding of viewer expectations.[38] That this position had been internalized was intimated by the reply given by one of our interviewees from Channel 1; though positioning himself as a liberal, he suggested that sensitive issues related to Islam could not be covered by a colleague from an ethnic community in which Islam was the predominant religion. The interviewee saw no need to exercise such caution in coverage of issues related to Orthodox Christianity, which was routinely assigned to journalists who self-identified as Orthodox.[39]

Finally, while attributing Russia's 'tolerance of ethnic diversity' to the historically rooted multi-ethnicity of the Russian and Soviet states, Russian broadcasters are less inclined than some of their west European counterparts to acknowledge the imperial and, in relation to Central Asia and the Caucasus, colonial nature of that state and the negative legacy of the hierarchical relations between ethnic groups which it fostered. As we argued in Chapter 6, the Eurasianist ideology, which provides the main interpretative framework for Russia's elites when analysing Russian ethnic diversity, directs its post-colonial critique at west European states rather than Russia.

In the immediate aftermath of the elections, what appears to have been the Kremlin's line on migration was implicitly contradicted in several editions of Rossiia's *Spetsial'nyi korrespondent* fronted by Mamontov. On the one hand, these editions tended to focus on issues which the Kremlin considered particularly important. As was evident from his trilogy of programmes on Pussy Riot, Mamontov's aim appears to be to prepare the public to accept controversial action about to be undertaken by the authorities.[40] The sudden heightening of his interest in migration is unlikely merely to reflect his personal preferences. On the other hand, the radicalism of his views and the crudeness of his reporting methods mean that his output sometimes attracts public criticism even from those loyal to the regime.[41]

In May 2012, Mamontov's programme marked the start of what turned into a full-fledged anti-immigration campaign on Rossiia. He returned to the same issue in October 2012, and then again a year later. The very titles of the three broadcasts – 'Aliens?'(*Chuzhie?*), 'Aliens-2' (*Chuzhie-2*) and 'Aliens-3' – indicate the presenter's propensity for 'othering' migrants.[42] Significantly, Mamontov first selected the title 'Aliens' for an edition of his programme broadcast in August 2010. There, the focus was on foreign 'mercenaries' fighting in Chechnia.[43] These were non-Russian aliens in every sense. In 2012, the same label was applied to citizens of former Soviet states who came to work in Russia. Mamontov's use of the German word *Gastarbeiter* to describe labour migrants further underscores their 'alien' status. In the Russian context, this term's foreign origin facilitates the discursive construction of migrants as not belonging in Russia.

The 'Aliens' series broaches the corruption of law enforcement organs and the immigration authorities, as well as the greed of Russian businessmen. This is even singled out as a core problem by a minority of participants. Nonetheless, ethnicity is placed at the very centre of the discussion, and it is ethnicity which is accorded primary explanatory power in the interpretation of various instances of violence and other crimes reportedly committed by migrants.[44] On this point the positions of the presenter, members of the audience who are invited to speak and the majority of the participating state officials and experts coincide. The visual imagery accompanying the discussion is likewise geared towards underscoring the primacy of ethnicity in all types of societal interactions.

Following the standard format of Mamontov's show, the three programmes are each introduced by films about labour migrants in Russia, with scenes from these films running as a backdrop during the subsequent discussions. The films are presented as carefully researched documentaries, encouraging viewers to regard the programmes as belonging to the macro-genre of informational programming rather than that of light entertainment. The three films convey somewhat different messages. The first highlights the terrible treatment of migrants as modern-day slaves (that is, it makes space for their representation as victims); in contrast, the second and third films depict them entirely as violent criminals. (The third film is dedicated to the issue of 'ethnic criminality', in which migrants are the main protagonists.)

Despite the difference, all the films portray labour migrants from Central Asia as the Russians' main ethnic 'other', emphasizing their distinct racial features through close-up filming of the degrading conditions in which they live and their 'barbaric customs' (e.g. eating food with their bare hands), and thus dehumanizing them.[45] While not giving migrants a voice,[46] the producer describes them as an unstoppable and therefore threatening force: 'They are treated as cattle, they are beaten up and humiliated, they are burned alive, and yet still they come to Russia.'[47]

The selection of officials and experts is such that only a minority of the speakers support the Kremlin-endorsed position – namely, that migration is essential for the country's economy and that inter-ethnic harmony can be achieved through state-sponsored integration schemes. The 'experts' from Moscow-based research institutes and think tanks all regularly challenge this position, interpreting the situation primarily through the 'conflict of culture' and 'ethnic criminality' lenses. Even the first deputy chairman of the State Duma Committee for Nationalities Affairs, who stresses the problem of corruption, agrees that most migrants are of 'the lowest cultural level', which engenders anti-social behaviour and determines their inability to integrate. Throughout the programmes, the participants echo the description of the relationship between the local population and migrants as a 'war'. The rise of isolationist ethnic Russian nationalism that is criticized in Putin's article is justified as the local population's understandable reaction to migrant behaviour.[48]

Mamontov is clearly supportive of what comes across as the majority view and, like Shevchenko, calls on participants to 'take as a starting point the experience of ordinary citizens' when addressing speakers who attempt to deny or downplay ethnic factors in the issues surrounding migration. Indeed, the carefully selected 'ordinary citizens' in the audience invited to comment all happen to be victims of crimes committed by Central Asians and are convinced that their assailants' nationality is directly linked to their behaviour. Their accounts, as well as references to some unspecified statistical data, brazenly challenge the validity of the Kremlin-endorsed view that 'crime has no nationality', despite being reiterated in the discussion by the chairman of the Moscow City Duma.[49]

Like Shevchenko's *V kontekste* of 29 March 2012, Mamontov's shows depart from federal broadcasters' own previous tendency to contrast Russia's experiences of managing multiculturalism with those of western Europe. Now a direct parallel between Russia and western Europe is drawn. Mamontov gives voice to politicians and activists from the west in order to validate the highly negative representation of migrants his shows promote.[50] *Aliens-2* uses the cases of France and Switzerland to stress the inability of migrants from non-European societies ever to integrate. Reflecting the highly tendentious selection of speakers, the French and Swiss 'official' positions are articulated by Marine le Pen, the president of France's right-wing Front National, and by a campaigner against the construction of mosque minarets in Switzerland. West Europeans, Russian viewers are told, 'are horrified by the customs migrants bring with them'.[51] Here, Russia is depicted as a European nation. Rather than identifying with traditional Islam in opposition to degenerate western culture, Mamontov aligns himself with the west against the absolute alterity of the Muslim migrant.

Mamontov's programmes give voice to a new form of Russian nationalism, one which is no longer interested in the state's territorial expansion, but is instead isolationist and inward-looking. It is particularly hostile towards labour migrants, unless they are Slavs, depicting them as a threat to the very essence of 'Russianness'. This kind of nationalism is occasionally criticized by the Kremlin. It openly clashes with the project of creating an integrated Eurasian Union bringing Russia and Kazakhstan closer together, which is personally endorsed by Putin;[52] yet Russia's leaders seem increasingly to perceive this brand of Russian nationalism as an important societal force worthy of being co-opted.[53] The co-optation process is fraught with danger, particularly as it encourages open questioning of the credibility of some of the government's own pronouncements.

Indeed, what Mamontov's programmes do is to treat performative utterances in official discourse, such as claims that Russia's multi-ethnicity is its major strength, as if they were constative utterances capable of subjection to true/false assessments. Numerous pieces of 'evidence' are then cited in order to expose these utterances as empirically false. In Kremlin-sponsored discourse, pronouncements of unity in diversity are intended as performative rebuttals of actions which foster disunity. But in the talk shows, the 'truthfulness' of these pronouncements is instead tested against examples of street conflict and violence.

Significantly, *Aliens-2* was broadcast around the time at which, on more than one occasion, the Kremlin made gestures that could have been interpreted as suggesting a concern with the shift taking place in the discourse about 'absolute otherness'. Developed during the Pussy Riot affair, this attitude was, as we suggested in Chapter 8, now being transposed onto ethnic and other minority groups. We pointed out in that chapter that Putin himself articulated his position specifically in the context of an attempt to justify the harsh treatment meted out to the punk group. In the autumn of 2012, positioning himself as a defender of values of liberal tolerance and wilfully misrepresenting the *Voina* group's earlier satirical 'Decembrists' performance, Putin argued that the Pussy Riot collective consisted of racists and homophobes and, for that reason, should have attracted the attention of the authorities earlier.[54]

In another gesture seemingly critical of isolationist Russian nationalism, the government publicized a draft of the new strategy document on inter-ethnic relations, which depicted Russia as a multi-ethnic state of and for all its citizens. Significantly, the draft made no reference to ethnic Russians as a 'state-forming people'. The enshrining of such a reference in the constitution is a key demand of Russian nationalist leaders.[55] As noted in Chapter 8, the Kremlin simultaneously suggested that the controversial post-Pussy Riot draft law on offending the feelings of religious believers might be excessive.[56] This carefully staged distancing from heated public debates and controversies is a typical Putin ploy (we saw it in action in relation to the *Shkola* serial in Chapter 5), and the stance should not be interpreted as suggesting that any of these developments take place in defiance of the Kremlin's wishes.[57] This approach permits isolationist Russian ethnonationalism to be mainstreamed and co-opted into official discourse without the direct involvement of the Kremlin, which is then able to portray itself as adopting moderate, balanced positions on the 'national question'.

Meanwhile, the overall strategy of maintaining the regime's legitimacy with the help of crude discourse about Russia's multiple 'enemies' remained in place. This discourse only continued to radicalize, and television coverage remained within the broad boundaries of the Kremlin's agenda. Within this context, the anti-immigration campaign also continued to escalate, and it became linked to an important element of the nation-building ideology – that of Russia as a European nation which experiences similar global pressures to other societies, yet is superior to 'the west' in upholding traditional values. On its own terms this message has a degree of coherence, but the campaign against migration on state-aligned television channels unwittingly accentuated the contradictory relationship between this perception of Russian identity and the representation of Russia as the centre of a unique Eurasian civilization.

Given the emphasis on the European nature of Russian identity in media discussions of migration, it is unsurprising that the campaign received a boost in the spring of 2013 from two almost simultaneous events in western Europe – rioting in Stockholm and the murder of a British soldier in London by two young British Islamists of Nigerian descent. These offered both Channel 1 and Rossiia ample opportunities for raising migration-related issues. *Vesti* led the Russia-specific aspect of the campaign. Channel 1 continued its pre-election approach of refraining from covering the migration issue in relation to Russia in its main news programme, *Vremia*, dealing with the topic only in talk shows.[58]

The events in London and Stockholm were exploited by both *Vesti* and *Vremia* to advance more forcefully than ever before the claim that migrants from non-European, particularly Muslim, societies were in principle incapable of integrating into Europe because of their cultural differences. Exhibiting an unashamedly racializing approach, state-aligned broadcasters began arguing that even 'second-generation migrants' (*immigrant vtorogo pokoleniia*) were invariably unintegrated, posing a major threat to the well-being of their host societies.[59] A relentless 'othering' strategy was deployed, with people of Middle Eastern and African origin, all indiscriminately labelled as migrants, presented as violent, unwilling to work or otherwise positively contribute to European societies, yet demanding ever greater benefits from European governments. 'Indigenous' (*korennye*) Europeans were forced to 'adapt' to the rules set by 'migrants', a *Vremia* reporter argued. 'Ethnic Norwegians' (*etnicheskie norverzhtsy*), Russian viewers were told, were now 'refugees in their own countries'.[60]

In this coverage, what appears to be particularly under threat from the influx of migrants is Europe's Christian identity. Western societies began to be represented as divided along religious lines, Christianity versus Islam. According to the presenter of *Vesti nedeli*, 'a cultural transformation (*pererozhdenie*)' of Europe was taking place under the influence of migrants from Muslim countries.[61] Islam was then depicted as core in the identity of all non-European residents in Europe who failed to integrate in their host societies. According to *Vremia*, while Muslim migrants have no desire to integrate, it is 'the indigenous population that is expected to adapt... In London a so-called "Muslim patrol" has appeared... "Radically inclined Muslim youths have no fear of European laws" and "they live the life of parasites on social security benefits".'[62]

Stories covered in the western media were adapted by Russian broadcasters to emphasize the connection between the perceived 'parasitic' behaviour of unintegrated migrants and Islam. Thus in January 2013, *Vesti nedeli* recalled a controversial statement by Baroness Shreela Flather who, back in 2011, criticized the British Bangladeshi and Pakistani communities for having large numbers of children, allegedly in order to claim more benefits.[63] In contrast to the BBC, which made no reference to the religious beliefs of members of these communities when reporting the comment, *Vesti nedeli*'s favourable assessment of Baroness Flather's statement argued that Muslims practised polygamy and that many Muslim mothers were therefore registered as single, allowing them further to exploit British social services. Similarly, the Islamic dimension permeated Russian television coverage of the Stockholm disturbances. Minority communities whose members were involved in the disturbances were identified by *Vremia* as Muslims (something the BBC pointedly avoided doing).[64]

The blame for what is described as 'the failure of multiculturalism' was laid at the door of west European governments, which, having been guided by 'wrongly understood democracy', demonstrated 'excessive tolerance' and political correctness; it was this, broadcasters argued, that had led to the suppression of the rights of 'the Christian majority' in European states.[65] The argument indicated a reframing of the on-going discussion of multiculturalism in Western Europe and Russia. In earlier interpretations west European governments were taken to task by Russian broadcasters for their indifference towards migrants and the lack of will to integrate them.[66] Now western multiculturalist policy was being criticized above all for being too lenient towards the deviant behaviour of non-European minorities and migrants and for its alleged unwillingness to defend European Christian values. These dramatic representations of the 'clash of civilizations' purportedly destroying western Europe seem to serve a double purpose. They were used to bolster demands for tougher measures against migration in Russia, 'before it is too late';[67] they also implied that, in contrast to western Europe, the upper echelons of the Russian leadership did not suffer from misplaced tolerance and were therefore able to govern effectively. Finally, they endorsed a newly invented interpretation of Russian identity as conservative European.

The events in Stockholm and London proved convenient for the organizers of a campaign against Central Asian labour migrants in Russia. They were used to add credibility to anti-immigration measures which some politicians, now backed by state-aligned television, began to advocate. The most sustained campaign against migrants from Central Asia ever seen began on state-aligned television in April 2013. It was marked by two new developments. One was that leading politicians, such as Moscow mayor Sergei Sobianin, began to be depicted by broadcasters as rejecting the ability of migrants with cultures 'completely different' from (ethnic) Russian norms ever to integrate.[68] Significantly, the broadcasters at times tendentiously twisted the politicians' pronouncements. The second was an attempt by state-aligned television directly to link migration to the alleged subversive activities of foreign intelligence services against Russia.[69]

The staged nature of the new turn in the anti-immigration campaign became particularly apparent when, on 14 April, Deputy Prime Minister Dmitrii Rogozin, known for his nationalist views, organized the sudden inspection of a train which followed a route between Dushanbe, the capital of Tajikistan, the origin of most of Russia's labour migrants, and Moscow. The presenter reporting the story on the Rossiia-24 news channel urged the rapid closure of the border with Tajikistan until 'they [citizens of Tajikistan] grow up' (*poka oni ne podrastut*).[70] Further discussion of the issue, with emphasis on migration from Tajikistan, continued on talk shows such as *Poedinok*, one of Rossiia's most widely viewed programmes,[71] on which Zhirinovskii called for the deportation of all migrants from Russia, as they 'bring crime, illness, and reduce the cultural level of our society'.[72] Mamontov also intervened in the campaign with a *Spetsial'nyi korrespondent* edition entitled *Shaitan-train*, which opened with an explanation that the word 'shaitan' means devil in Tajik.[73] Protest against the campaign and its Russian media coverage on the part of Tajikistan's foreign ministry received no mention on either of the two federal channels.

During the campaign, *Vesti* alleged that Putin had promised to introduce a visa regime for the Central Asian states as of 2015.[74] In fact, this seems to have been an inaccurate interpretation of a brief mention, in the presidential address to the Federal Assembly in December 2012, of the possibility of introducing the requirement that citizens of the CIS states other than Belarus and Kazakhstan enter Russia on foreign passports, rather than domestic ID cards.[75] It is unclear whether Rossiia's report merely demonstrated sloppiness in handling information or was a deliberate attempt to add further legitimacy to its anti-migration campaign and to represent the president as responsive to popular concerns.

Vesti also publicized a statement made by the deputy head of the Federal Security Service to the effect that 'foreign intelligence services' used illegal immigration to organize 'subversive activity to the detriment of Russian security', including terrorism.[76] FSB officials have in the past attempted to represent migrants as a major security threat to the country, but while their statements were dramatically reported in 2006 and 2008 by Russian print and online newspapers, the two main federal television channels refrained from covering them.[77]

On the same day, *Vesti* also tendentiously covered the pronouncements on migration made by Sobianin in an interview with the newspaper *Moskovskie novosti*, given during his campaign for re-election as mayor. By focusing exclusively on one statement regarding migration in Sobianin's interview, which in fact covered a wide range of topics, and omitting his qualifying remarks aimed at softening the effect of his apparent attack on migrants, *Vesti* depicted Sobianin's position as more at odds with the Kremlin-sponsored line than was actually the case. From a 4,000-word interview, *Vesti* chose a single provocative remark to the effect that 'people who speak Russian poorly and who have cultures completely different [from ours] had better live in their own country'.[78] In fact, most of Sobianin's pronouncements on migration in this interview corresponded to those of the Kremlin leaders. Russia's multi-ethnicity was represented by the mayor as its strength, the supposedly welcoming attitude of Russia's citizens towards 'guests' was noted, and migrants from Central Asia who knew the Russian language were defined as 'compatriots'.[79] All this was omitted by *Vesti*.

Internal migration' (*Vnutrenniaia migratsiia*)

The labels of migrant and even illegal migrant, both loaded with negative connotations, are applied in the Russian media discourse not only to those who come from abroad, but also to Russia's own citizens. These tend to be almost exclusively residents of the North Caucasian republics who move into areas of the country with a predominantly ethnic Russian majority. In the post-election period, Rossiia has used increasingly dramatic language in representing the so-called 'internal migration' of North Caucasians as a threat to the very existence of ethnic Russians as a national group.

The labelling as 'migrants' of *bona fide* Russian citizens who belong to certain ethnic minority groups is a distinct feature of the Russian discourse of migration. As we argued in Chapters 1 and 4, the origins of current understandings of 'internal migration' can be traced to the specificity of the Russian imperial state and Soviet nationalities policies. Those who call North Caucasians 'migrants' (often unconsciously) evoke the (former) colonial status of the North Caucasus, which has long existed in an unequal power relation with the ethnic Russian core of the country.

The occasional application of the term 'illegal migrant' to Russia's own citizens is rooted in the Soviet *propiska* (residence permit) system, which severely restricted people's freedom to choose their place of residence. In the post-Soviet period, the Soviet-type *propiska* was abolished, but registration of one's place of residence, even if it may now be freely chosen, is still required. The labelling of Russia's citizens as 'illegal migrants' then becomes justified by reference to their failure to register their residency with the appropriate authorities.

The fact that the North Caucasians are citizens of the Russian Federation had until recently placed restraints on how state-aligned television reported population movement from North Caucasus to central Russia, which has intensified in the post-communist period. Indeed, Putin's *Nezavisimaia gazeta* article drew a distinction between 'external' and 'internal migration', stressing the freedom of mobility around the country for all its citizens, although Putin's critics noted a certain vagueness in his statement. In response to this criticism, Putin explicitly clarified that the term 'migrant', which by 2012 had acquired a negative connotation in popular usage, should not be applied to any citizen of the Russian Federation.[80]

The most prominent example of past 'restraint' in the news coverage of 'internal migration' by the two main state-aligned channels can be seen in their reporting of the Manezhnaia riots. As we demonstrated in Chapter 4, after initially allocating partial blame for the riots to North Caucasians themselves, and after evoking the concepts of the 'conflict of cultures' and 'ethnic criminality' in discussion of the causes of the riots during the week following these events, the final annual bulletins of *Vremia* and *Vesti* in December 2010 had purged from their reports all reference to 'internal migration' and threats to Russia's multi-ethnic ideal arising from cultural differences.[81]

By late 2012 such 'restraint' had disappeared from most news reports on Rossiia. *Vesti nedeli* had become especially alarmist. Its editor, Dmitrii Kiselev, proudly noted in an interview with our team that, under his leadership, in December 2012 his programme 'practically staged' (*instsinirovali fakticheski*) the topic of migration of North Caucasians into the Stavropol' region in southeast Russia.[82] In particular, the broadcast of 9 December included a series of dramatic representations of life in Stavropol'. The programme's presenter spoke about an 'invasion' (*ekspansiia*) of North Caucasians into this predominantly ethnic Russian region.

The report presented ethnic Russian residents as being systematically victimized by North Caucasians, who used violence and even murder to take over their land and property. According to the broadcast, the consequence was a mass exodus of ethnic Russians from Stavropol'. It was claimed that in the preceding ten years the situation had deteriorated to a 'shocking' (*oshelomliaiushchaia*) extent – yet the presenter's reference to the 2002 and 2010 censuses, indicating that the proportion of ethnic Russians in the region had decreased by only 1.5 per cent, offered no support for this dramatic assessment. Echoing the Kremlin's line about enemies of Russia wanting to bring about its disintegration and evoking the western conspiracy frame, the report hinted that western intelligence services had a hand in the developments in Stavropol', as western governments were allegedly interested in the separation of Dagestan from Russia.[83]

As in the coverage of migration-related issues in Europe, (radical) Islam emerged as a prominent feature in *Vesti nedeli*'s depiction of 'internal migration' in Stavropol'. A local imam was quoted as making an emotional statement: 'I am a lonely warrior [against radical Islam] on a battlefield which is open to enemy fire from all sides.'[84] Significantly for our understanding of how the media perceive the public mood and of journalists' own views, in April 2013, REN TV offered a similarly alarming depiction of 'an expansion of Islam and its values' in southern regions of Russia. Arguing that most ethnic Russians in the region were understandably 'very frightened', the reporter on REN TV's *Reporterskaia istoriia* was even less inclined to draw any distinctions between the 'moderate, traditional Islam' of Russia and the 'bad' radical Islam imported from abroad than was the presenter of *Vesti nedeli*.[85] From late 2012 onwards, media outlets of varying political allegiances began highlighting, and in some cases fostering, anti-Muslim sentiment in Russia's metropolitan centres too. A mass campaign against plans to build a large mosque in Moscow's Mitino district received extensive and sympathetic coverage from federal channels in the autumn of 2012.[86] In February 2013, *Vesti nedeli* reported on the establishment of 'Muslim patrols' in St Petersburg under the provocative heading 'Islamic Expansion'.[87]

The perceived threat to Russia's large cities posed by Muslims converged with that associated with ethnic minorities more broadly. Prior to the 'staged' coverage of the situation in Stavropol', in October 2012, *Vesti nedeli* had also transformed a minor incident involving North Caucasians in Moscow into an issue of political significance directly relevant to the state of inter-ethnic relations: the firing of shotguns at a wedding celebration. Given that no harm seemed to have been intended and no one was hurt, the incident could have been categorized simply as a human interest story. Those involved were Dagestanis, described in the bulletin as 'guests' in Moscow, despite their Russian citizenship. Rejecting the Kremlin's unity-in-diversity mantra and the Soviet metaphor of the 'friendship of the peoples', the reporter drew on the theory of ethnos developed by late neo-Eurasianist ideologist Lev Gumilev, offering viewers a detailed account of Gumilev's highly contentious arguments. Viewers were informed that alongside instances of the harmonious co-existence of different ethnic groups there could also be highly destructive patterns of co-habitation, in which one ethnic group lived as a parasite (*glist*) or a cancerous metastasis in the body of another. In the report, in which all examples of conflicts involved Caucasians attacking ethnic Russians, there was little doubt about who the parasites were. When describing certain people as 'parasites', however, Gumilev most likely had Jews in mind. The transposition of his anti-Semitic construct onto Caucasians reflects a new trend in Russian ethnonationalism of the past decade. Notwithstanding the lingering presence of anti-Semitism identified in our earlier analysis of the Pussy Riot scandal, the Caucasian has generally replaced the Jew as the main target of vilification.

In the report it was not Putin but the Russian nationalist film director Nikita Mikhalkov who appeared as the 'master figure', positioned outside the debate yet possessing the superior authority sanctioning him to comment on the matter. Mikhalkov's authority derived from the cinematic warning he had issued in 2007 about the rise of highly xenophobic Russian nationalism in response to problems caused by migrants.[88] In his interview with *Vesti nedeli*, Mikhalkov unreflectively promoted a colonial view of power relations between ethnic Russians and North Caucasians, attributing current problems to the breakdown of what he called 'the special language of communication between Russia and the Caucasus'. In his view, this historically shaped 'culture of communication' used to combine 'the languages of force (*sila*) and of mutual respect'. In a gesture familiar from our analysis of Putin's discourse in Chapter 7, with its seemingly universal application, the statement was quickly turned into a reference to an ethnically coloured particular. It soon became clear that Mikhalkov accorded the privilege of using 'the language of force' to Russians alone. Although respect was supposed to be mutual, in the single example offered by the director, respect flowed in one direction only: 'When a Caucasian understands that the one to whom he could be subordinate can be easily bribed... and will do what you demand of him, what respect can be commanded [by the Russians]?'[89]

Channel 1 took a markedly different line. Featuring no reports at all on the Stavropol' region, it filtered the Moscow wedding incident through the prism of the Kremlin's dominant line. A long quote from Putin, claiming that such incidents had nothing to do with ethnicity but were a symptom of social problems which arose as a result of the collapse of the Soviet Union, framed the coverage.[90]

When covering 'staged' stories about migration, whether in Russia or in Europe, television journalists have a great deal of licence and may offer their own interpretations, based on personal beliefs and prejudices. In contrast, the eruption of sudden dramatic conflicts inside Russia continues to elicit a more controlled approach. This was evident for instance when, in July 2013, a major public protest erupted in the town of Pugachev in central Russia following a murderous fight between two youngsters, one of whom happened to be a Chechen. The fight had nothing to do with ethnicity, but led to mass demonstrations demanding that all North Caucasians be deported from the city. Alarmed by the eruption of major public disorder, yet unable to satisfy the unconstitutional demand, the authorities were keen to calm the situation quickly. Under these circumstances, while designating Russian citizens from the North Caucasus as 'migrants' in its first reaction to the protest, *Vesti* then readopted its pre-2012 'restraint' in relation to

Pugachev, omitting all references to 'conflict of cultures' and 'ethnic criminality' and repeatedly insisting that local residents had misunderstood the situation by introducing the ethnic factor into an ordinary, albeit unfortunate, incident induced by excessive drinking.[91] A similarly 'restrained' approach was again apparent in coverage of anti-Caucasian riots in Moscow's Biriulevo district four months later.[92] Following these events, the broad anti-migrant campaign on the two main federal channels subsided. In addition, after the end of the election campaign for the office of Moscow mayor, during which both Sobianin and his opponent Navalnyi turned migration into an important electoral issue, its public discussion notably declined.

***

We have shown that, in the post-election period, the Kremlin's relationship with federal broadcasters, particularly the state-owned Rossiia channel, became more complex than was previously the case. The distinctive role played by broadcasters in shaping (rather than merely reflecting) public discourse paradoxically increased under the growing constraints imposed by Putin's third presidency. Rather than becoming more uniform, as the Kremlin's tightening political control would appear to dictate and as many observers have assumed to be the case, public discourse in 2012 and 2013 became increasingly fragmented and contradictory. Pronouncements by Rossiia journalists and reporters on the 'national question', migration and (radical) Islam in the post-election period were now more rather than less likely to contradict the Kremlin's public pronouncements. Leading media figures such as Kiselev and Mamontov promoted an isolationist version of Russian nationalism manifesting overt hostility to certain ethnic minorities. This nationalism differs from the vision of Russia endorsed by the Kremlin, which persists in seeing Russia's imperial legacy as a major strength and regards the continuation of Russia's rule over the North Caucasus and its close relationship with the Central Asian states as essential for the country's status as a great power.

One of the liberally inclined television reporters whom we interviewed argued that in promoting certain official 'celebrity' journalists, the Kremlin created 'puppets' whom it no longer knew how to control.[93] We have argued that this interpretation, too, oversimplifies the situation. It is true that the increased contradictions within the official discourse and inconsistencies in the mediation of important political issues in the post-election period might be an indication of a particular uncertainty of the regime at the time, as it was searching for new approaches through which to control society in the aftermath of the mass public protests. But even so, the activities of these 'puppets' should not be read as attempts to question the regime's legitimacy. In fact, the anti-immigration campaign on state-aligned television, with its anti-Islamic undertones, broadly followed the logic of the Kremlin-sponsored post-election strategy of controlling society through polarization and the identification of internal enemies alleged to be collaborating with western intelligence services. The representation of these 'enemies' as the Russian nation's 'absolute other' remained at the forefront of the federal broadcasting agenda and the broad, Kremlin-endorsed ideological framework to which it continued to adhere accounted for the specificities in its treatment of migration.

By articulating ideas which directly address real and perceived public concerns, broadcasters likewise reinforced the Kremlin's governing strategy, helping to foster the impression that national leaders were attentive to the concerns of the country's majority without them having to make pronouncements which other citizens and residents might have found offensive. In multi-ethnic, multi-confessional societies such as Russia, the upper echelons of political leadership have to be seen as occupying a moderate position on issues such as the 'national question'.

In several ways, post-election reporting of the 'national question' and of migration, represented in its most distilled form in *Vesti nedeli*, was more internally coherent than the coverage we observed in 2010 and 2011, which struggled to reconcile the tensions between Russia as both a multi-ethnic state and the national homeland of ethnic Russians. Yet it produced a new set of acute tensions, laying bare the contradictions in the official discourse of the Russian nation as both European and uniquely Eurasian and of the west as an evil, conspiring 'other', yet also an occasional ally.

To the extent that the contradictions tended to be resolved, it happened through reporting which fostered, rather than ameliorated, ethnic tensions. In a multi-ethnic, post-imperial society this is a dangerous strategy with the potential to undermine the country's stability and territorial integrity, whose maintenance is paramount. The new virulence with which demands to expel 'the ethnic other' have been expressed throughout Russia's public sphere, particularly in the summer of 2013, bear testament to this danger. In this context, the broadcasters decided to return to a more controlled style of reporting. The fact that the highly alarmist coverage of migration both started and stopped rather abruptly suggests a lack of a coherent government strategy for achieving national cohesion, which the Kremlin insists is its main priority.[94]

The partial convergence of these demands to expel the ethnic 'other' with the intensified campaign against alterity of other forms, such as that represented by Russia's LGBT community and, as we saw in Chapter 8, by other degenerate products of the 'liberal conspiracy', complicates the situation in two ways. First, by subsuming such campaigns into the wider agenda of 'traditional, conservative values', responsibility for the ideological direction of government policy was partially ceded to an uncompromisingly anti-western Orthodox Church hierarchy and to media demagogues such as Mamontov, Kiselev and Shevchenko. Here, again, the triangular relationship between the Kremlin, the public and state-aligned television revealed itself to be less than wholly unilinear. Second, and perhaps more crucially, by blurring the boundaries between 'internal' and 'external' 'others' it risked transforming the 'war on our streets' into an autogenous conflict in which Russia imploded into itself – a prospect of which the rigidly imposed uniformity that was to characterize coverage of the Ukraine conflict is a further harbinger, rather than a refutation.

Notes

1 See, for instance, 'Vozrat k proshlomu', 29 May 2012. Available online at www.kasparov.ru/material.php?id=4FC49C5C35B59 (accessed 20 June 2013).
2 Between March and September 2012, Pussy Riot supporters were represented as opposed to traditional Russian values; this role was then ascribed to critics of the controversial law banning the adoption of children from Russia by US citizens and gay activists. In the spring of 2013 the role was ceded to migrants.
3 Surkov was fired from the post of deputy chairman of the government in May 2013. For an analysis of Surkov's discursive strategies see Lara Ryazanova-Clarke, 'The "West" in the Linguistic Construction of Russianness in Contemporary Public Discourse', in Risto Alapuro, Arto Mustajoki and Pekka Pesonen (eds.), *Understanding Russianness*, London: Routledge, 2012, pp. 3–18.
4 Elisabeth Schimpfossl and Ilya Yablokov, 'Coercion or Conformism? Perceptions of Censorship and Self-Censorship among Russian Television Reporters in the 2010s', *Demokratizatsia*, 22/2, 2014, 295–311.
5 Interview with a Rossiia journalist, 29 March 2013.
6 Interview with a Channel 1 journalist, 3 April 2013.
7 For detailed guidelines which are available for the BBC journalists, see www.bbc.co.uk/bbctrust/assets/files/pdf/review_report_research/impartiality_21century/report.pdf (accessed 15 August 2013).
8 See Catherine Fitzpatrick, 'Naval'nyi Answers the Hard Questions with Hard Answers', 27 July 2013. Available online at www.interpretermag.com/navalny-answers-the-hard-questions-with-hard-answers. See also statements from Naval'nyi's LiveJournal. Available online at http://navalny.livejournal.com/274456.html, and his expression of support for anti-Caucasian rioters in the town of Pugachev in July 2013, available online at http://grani.ru/Society/Xenophobia/d.216686.html (all accessed 2 December 2013).
9 Among affluent 'western' societies the most negative attitudes towards migration have been recorded in Britain, where the 2009–10 citizenship survey indicated that over 75 per cent of the respondents favoured a reduction in immigration. 'UK Public Opinion toward Immigration: Overall Attitudes and Level of Concern', 3 July 2014. Available online at http://migrationobservatory.ox.ac.uk/briefings/uk-public-opinion-toward-immigration-overall-attitudes-and-level-concern (accessed 20 May 2013).
10 'Natsionali'naia politika i otnosheniie k migrantam', 28 November 2012. Available online at www.levada.ru/28-11-2012/natsionalnaya-politika-i-otnoshenie-k-migrantam (accessed 20 May 2013).
11 Putin's speech at the Valdai Club, 19 September 2013. Available online at www.kremlin.ru/news/19243 (accessed 27 November 2013).
12 According to a Levada Centre study of October 2013, over 50 per cent of those polled supported the slogan 'Stop Feeding the Caucasus'. The study concluded that a third of the respondents did not regard North Caucasus as part of Russia. Available online at www.kavkaz-uzel.ru/articles/232217/ (accessed 20 January 2014).
13 Interview with a Rossiia journalist, 29 March 2013.
14 Ibid., and interview with Channel 1 journalist, 5 August 2013. See also Schimpfoessl and Yablokov, 'Coercion or Conformism? Censorship and Self-Censorship among Russian Media Personalities and Reporters in the 2010s'.
15 *V kontekste* (Channel 1), 1 March 2012. Available online at www.1tv.ru/sprojects_edition/si5851/fi14097 (accessed 19 December 2012).
16 *V kontekste* (Channel 1), 29 March 2012. Available online at www.1tv.ru/videoarchive/45226 (accessed 15 December 2013).

17 Our project database suggests that between September 2010 and May 2012 Channel 1 avoided discussing problems related to inter-ethnic relations in Russia. It discussed the 'crisis of multiculturalism' in Europe most intensively when covering the story about the arrival on the Italian island of Lampedusa of refugees fleeing violence and upheaval in North Africa in the spring of 2011. This coverage did not draw any parallels with Russia.

18 *V kontekste* (Channel 1), 29 March 2012.

19 On the complexity of the intellectual tradition associated with the term multiculturalism, see Nasar Meer and Tariq Modood, 'How Does Interculturalism Contrast with Multiculturalism?', *Journal of Intercultural Studies*, 33/2, 2012, 175–96, as well as a discussion this article generated: 197–209.

20 In the aftermath of the 2012 elections, the notion of the multi-ethnic *rossiiskaia* nation received another endorsement in a draft strategy on nationalities' relations, developed by the Presidential Council on Inter-Ethnic Relations in October 2012. Available online at http://rt.com/politics/russia-drafts-nation-report-689/ (accessed 14 December 2012).

21 *V kontekste* (Channel 1), 29 March 2012. Tishkov analysed the utilization of the ethnic nationalist card by politicians during the 2011 and 2012 election campaigns in his article 'Nationalizm i vybornye kampanii', *Vestnik rossiiskoi natsii*, 2/3, 2012, 72–90.

22 Interestingly, Shevchenko adopts a different position in our interview, where he rejects cultural determinism, focusing instead on social and economic reasons for inter-ethnic tensions (interview, 3 April 2013).

23 Ibid.

24 See, for instance, REN TV's *Reporterskaia istoriia* entitled 'Natsional'nyi otvet', 12 September 2010. Available online at www.youtube.com/watch?v=NLdI6Zb3z9c (accessed 19 December 2012).

25 The proposal to set up such a Council was articulated by Putin on 29 March and the relevant decree was issued on 7 June 2012. Available online at www.rg.ru/2012/06/07/sovet-anons.html (accessed 19 December 2012).

26 *Vesti*, 24 August 2012; *Vremia*, 24 August 2012. Available online at http://www.1tv.ru/news/polit/214060 (accessed 19 December 2012).

27 See Harald Bauber, 'Immigration Dialectic in the Media and Crisis as Transformative Moment', in Kerry Moore, Bernhard Gross and Terry Threadgold (eds.), *Migrations and the Media*, New York: Peter Lang, 2012, pp. 34–5, 44–5; Robin Cohen, 'The Incredible Vagueness of Being British/English', *International Affairs*, 76/3, 2000, 575–82; 576; Oxana Shevel, *Migration, Refugee Policy, and State Building in Post-communist Europe*, Cambridge: Cambridge University Press, 2011.

28 See Chapter 2 for details. On the introduction of a 'more liberal approach to migration management' by the Russian government in 2005, see Irina Ivakhnyuk, *Russian Migration Policy and Its Impact on Human Development*, Human Development Reports Research Paper, New York: United Nations Development Programme, 2009. See also Putin's opening remarks at a Security Council meeting on migration policy in March 2005. Available online at http://archive.kremlin.ru/text/appears/2005/03/85300.shtml (accessed 6 March 2014).

29 See, for instance, *Vesti* of 22 September 2011, 28 June 2011 and 2 August 2011.

30 *Vesti* of 28 December 2011 and of 26 August 2011.

31 See, for instance, the debate between Mikhail Leont'ev (a conservative) and Iulii Gusman (a liberal) on *Poedinok* (Rossiia), 25 November 2010. Available online at http://poedinok.net/mixail-leontev-i-yulij-gusman/ (accessed 30 November 2010).

32 V. I. Mukomel', 'Rossiiskie diskursy o migratsii', *Vestnik obshchestvennogo mneniia*, 1, 2005, 48–58; and Vladimir Malakhov, *Ponaekhali tut... Ocherki o natsionalizme, rasizme i kul'turnom pliuralizme*, Moscow: Novoe literaturnoe obozrenie, 2007. See also Marlène Laruelle, 'The Ideological Shift on the Russian Radical Right. From Demonizing the West to Fear of Migrants', *Problems of Post-Communism*, 57/6, 2010, 19–31.

33 Greg Philo, Emma Briant and Pauline Donald, *Bad News for Refugees*, London: Pluto Press, 2013, pp. 32–49; Valeria Alia and Simone Bull, *Media and Ethnic Minorities*, Edinburgh: Edinburgh University Press, 2005.
34 Samuel Bennet, Jessika ter Wal, Artur Lipinski, Malgorzata Fabiszak and Michal Krzyzanowski, 'The Representation of Third-Country Nationals in European News Discourse. Journalistic Perceptions and Practices', *Journalism Practice*, 7/3, 2013: 248–65.
35 Interviews with Rossiia journalists, 27 March 2013 and 3 April 2013. On the situation in Britain, Elizabeth Poole, 'Multiculturalism, Religion and British Identity: The Case of Geert Wilders', paper presented at the conference *Ethnicity, Race and Nationalism in European Media and Film*, University of Manchester, 25–27 May 2013.
36 On the case of Italy, see Giovanna Campani, 'Migrants and Media. The Italian Case', in Russell King and Nancy Wood (eds.), *Media and Migration. Construction of Mobility and Difference*, London: Routledge, 2001, pp. 38–52; Rinella Cere, 'Globalization vs. Localization: Anti-Immigrant and Hate Discourses in Italy', in M. Ardizzoni and C. Ferrari (eds) *Beyond Monopoly: Contemporary Italian Media and Globalization*, Lanham: Lexington Books, 2010, pp. 225–44. On Greece, see Bennet *et al.*, 'The Representation of Third-Country Nationals in European News Discourse', 254. On the Netherlands, see ibid., 255; on Britain, see ibid., 249. See also reports assessing impartiality of the BBC coverage of issues, including migration 'From Seesaw to Wagon Wheel: Safeguarding Impartiality in the 21st Century'. Available online at www.bbc.co.uk/bbctrust/assets/files/pdf/review_report_research/impartiality_21century/report.pdf (accessed 18 October 2013); and 'A BBC Trust Review of the Breadth of Opinion Reflected in the BBC's Output', July 2013. Available online at http://downloads.bbc.co.uk/bbctrust/assets/files/pdf/our_work/breadth_opinion/breadth_opinion.pdf (accessed 18 October 2013).
37 Numerous examples of highly biased, factually inaccurate and otherwise crude reporting on asylum seekers and migrants in the UK by the tabloid press are given in Philo *et al.*, *Bad News for Refugees.*
38 Interview with a REN TV journalist, 5 August 2013.
39 Interview with a Channel 1 journalist, 29 January 2013.
40 See Chapter 4.
41 Interview with Mamontov, 28 March 2013.
42 'Spetsial'nyi Korrespondent. Chuzhie?'. Available online at http://russia.tv/video/show/brand_id/3957/video_id/130159; 'Spetsial'nyi Korrespondent. Chuzhie-2', available online at http://russia.tv/video/show/brand_id/3957/video_id/165399 (accessed 19 December 2012); and 'Spetsial'nyi Korrespondent. Chuzhie-3', available online at http://russia.tv/video/show/brand_id/3957/episode_id/648564/video_id/648564/viewtype/picture (accessed 2 December 2013).
43 'Obratnaia storona. Chuzhie (Rossiia)', 31 August 2010. Available online at http://video.yandex.ru/users/warchechnya-ru/view/148/?cauthor=warchechnya-ru&cid=1 (accessed 27 January 2014).
44 The ethnicization of migration in the programme is particularly noteworthy, given that in an interview with our team, Mamontov prioritized corruption, not inter-ethnic relations, as the key problem facing Russia (interview, 28 March 2013).
45 'Chuzhie?'.
46 It should be noted that migrants, particularly if their residence and work permit in Russia can be questioned, are reportedly reluctant to be interviewed by the media (interview with a journalist from an independent television channel, 15 January 2013).
47 'Chuzhie?'.
48 'Chuzhie-2'. At the same time, the argument is made that no such issues arise with people coming from Ukraine, Belarus and the Transdniester region of Moldova, whose Slavic origin and cultural similarity to the Russians is emphasized.
49 'Chuzhie?'.

50 Similarly, during a public controversy around the wearing of *hidjab* in schools in October 2012, participants – including Putin – referred to west European experiences, in particular a ban on the wearing of any 'ostentatious' religious articles in public places which was introduced in France in 2004.
51 'Chuzhie-2'.
52 See, for instance, Putin's speech of 19 September 2013. Available online at www.kremlin.ru/news/19243 (accessed 13 November 2013).
53 Nicu Popescu, 'The Strange Alliance of Democrats and Nationalists', *Journal of Democracy*, 23/2, 2012, 46–54.
54 See http://rusk.ru/st.php?idar=56640 (accessed 20 September 2012) for Putin's interview with Russia Today on 5 September 2012, when he first made his comments. In November 2012, Putin reiterated his argument during a meeting with the German Chancellor Angela Merkel: see Miriam Elder, 'Putin Reacts Angrily to Angela Merkel's Pussy Riot Comments', *The Guardian*, 16 November 2012. On 28 September 2012, Russia's representative in Brussels was reported by *Vesti* as making a similar statement.
55 'Russia Drafts New Ethnic Policy of 'Multi-people Nation', 18 October 2012. Available online at http://rt.com/politics/russia-drafts-nation-report-689/ (accessed 15 December 2013). It is noteworthy that the 1996 strategy document on the nationalities relations adopted under Yeltsin did contain a reference to ethnic Russians as 'the state-bearing people'.
56 On Putin's concern over the proposed new law on the offence to religious believers, see Gabriele Baczynska, 'Russia May Soften Religion Law over Putin Concern', 2 December 2012. Available online at http://news.yahoo.com/russia-may-soften-religion-law-over-putin-concerns-200345093.html (accessed 15 December 2012).
57 For instance, despite expressing earlier reservations, on 30 June 2013 Putin signed the controversial law on offences to the feelings of religious believers. 'Putin podpisal zakon o nakazanii za oskorblenie chuvstv veruiushchikh'. Available online at www.forbes.ru/news/241477-putin-podpisal-zakon-o-nakazanii-za-oskorblenie-chuvstv-veruyushchih (accessed 4 July 2013).
58 See, for instance, the contribution to the anti-migration campaign in relation to Russia itself made by the channel's talk show 'Svoboda i spravedlivost'', 29 May 2013. Available online at www.1tv.ru/sprojects_edition/si5833/fi23666 (accessed 20 June 2013).
59 *Vremia*, 26 May 2013. Available online at www.1tv.ru/news/world/233811 (accessed 20 June 2013).
60 Ibid. For the development of similar arguments on *Vesti* since the autumn 2012, see, for instance, *Vesti nedeli*, 2 December 2012.
61 *Vesti nedeli*, 2 December 2012.
62 *Vremia*, 26 May 2013, www.1tv.ru/news/world/233811 (accessed 30 May 2013).
63 *Vesti nedeli*, 27 January 2013. See a report by the BBC of 14 September 2011 on Baroness Flather's statement. Available online at www.bbc.co.uk/news/uk-politics-14909062 (accessed 1 June 2013).
64 *Vremia*, 26 May 2013. Available online at www.1tv.ru/news/world/233811 (accessed 27 May 2013).
65 *Vremia*, 26 May 2013. Available online at www.1tv.ru/news/world/233811 (accessed 15 December 2013).
66 See Elena Chebankova, 'Contemporary Russian Multiculturalism', *Post-Soviet Affairs*, 28/3, 2012, 336–7.
67 This argument is made, for instance, in Mamontov's 'Chuzhie' programmes.
68 It should be noted that Aleksei Naval'nyi, who attempted to register as a candidate in the Moscow mayoral election in 2013, also flagged up migration as a key issue of his electoral platform (*Segodnia* [NTV], 1 July 2013).

69 In the past, such arguments were presented only in the non-mainstream media. On how marginal discourses about migration started to be mainstreamed in the spring and summer 2013, see 'Dlia predvybornoi kampanii nuzhen vrag', *Kommersant FM*, 19 June 2013. Available online at www.kommersant.ru/doc/2214647 (accessed 20 June 2013). See also Mukomel', 'Rossiiskie diskursy o migratsii'.
70 *Vesti*, 20 May 2013.
71 Viewing figures for Russian television programmes can be found at www.tns-global.ru/rus/index.wbp (accessed 20 May 2013).
72 The show was broadcast on 18 April 2013: http://russia.tv/video/show/brand_id/3963/video_id/349720 (accessed 20 May 2013).
73 *Vesti*, 20 May 2013.
74 *Vesti*, 21 April 2013.
75 'V MID RF rasskazali o vvedenii vizovogo rezhima so Srednei Aziei', http://lenta.ru/news/2013/03/14/azimov/ (accessed 20 May 2013). Putin's address: http://migrationobservatory.ox.ac.uk/briefings/uk-public-opinion-toward-immigration-overall-attitudes-and-level-concern (accessed 20 May 2013).
76 'Illegal Migration "National Security Threat" – FSB', *RIA Novosti*, 30 May 2013. Available online at http://en.rian.ru/military_news/20130530/181423829/Illegal-Migration-National-Security-Threat—FSB.html and 'FSB: sredi nelegal'nykh migrantov deistvuiut diversanty i shpiony', *Vesti*, 30 May, 2013.
77 Mukomel', 'Rossiiskie diskursy o migratsii'.
78 See 'Sobianin: tem, kto plokho govorit po-russki, luchshe uekhat' iz Moskvy', *Vesti*, 30 May 2013.
79 'Sergei Sobianin, "Moskovskim novostiam": o migratsionnoi politike, grazhdanskoi aktivnosti i konkurentakh', *Moskovski novosti*, 30 May 2013. Available online at www.mn.ru/moscow_authority/20130530/347635870.html (accessed 28 April 2014).
80 See *Vesti nedeli*, 29 January 2012.
81 Stephen Hutchings and Vera Tolz, 'Fault Lines in Russia's Discourse of Nation: Television Coverage of the December 2010 Moscow Riots', *Slavic Review*, 71/4, 2012, 887, 889 and Chapter 4.
82 Interview with Kiselev, 27 March 2013.
83 *Vesti nedeli*, 19 February 2012. Mamontov also devoted one of the broadcasts of his *Spetsial'nyi korrespondent* show titled 'Proetk Kavkaz' (5 February 2013) to similar issues: www.youtube.com/watch?v=VeUra1Uw7ps (accessed 31 May 2013).
84 *Vesti nedeli*, 9 December 2012. Even though in some parts of the North Caucasus the emergence of militant Islamism goes back to the late 1980s, in 2012 and 2013 *Vesti* and *Vremia* represented it as a new phenomenon. See also *Vremia*, 26 May 2013. Available online at www.1tv.ru/news/social/233810 (accessed 30 May 2013). On radical Islamism in Russia and other FSU states see Galina Yemelianova (ed.), *Radical Islam in the Former Soviet Union*, London: Routledge, 2010. On how the current media, political and academic discourses in western Europe and North America about 'radical Islam' draw on a long European tradition of negative stereotyping of Muslims see Elisabeth Poole and John Richardson, *Muslims and the News Media*, London: I.B. Tauris, 2006; Kai Hafez (ed.), *Islam and the West in the Mass Media: Fragmented Images in a Globalizing World*, Cresskill: Hampton Press, 2000 and Richard Jackson, 'Constructing Enemies: "Islamic Terrorism" in Political and Academic Discourse', *Government and Opposition*, 42/3, 2007, 394–426.
85 *Reporterskaia istoriia* (REN TV), 28 April 2013. *Reporterskaia istoriia* is produced by the same team as REN TV's *Nedelia*.
86 *Vesti*, 20 September 2012.
87 *Vesti Nedeli*, 24 February 2013.

88 *Vesti nedeli*, 7 October 2012. Mikhalkov's film '12', released in 2007, was singled out by Rossiia as a contribution to the understanding of Russia's inter-ethnic relations under Putin. Available online at www.trite.ru/news_in.mhtml?PubID=171 (accessed 19 December 2012).

89 Ibid. For distinctly different coverage by Channel 1, see *Vremia*, 7 October 2012, www.1tv.ru/news/social/217112; *Vremia*, 24 August 2012 www.1tv.ru/news/polit/214060 (both accessed 19 December 2012).

90 *Vremia*, 7 October 2012. Available online at www.1tv.ru/newsvideoarchive/pd=07.10.2012 (accessed 19 December 2012).

91 'Chechenskaia diaspora vyvezla chast' svoei molodezhi iz goroda Pugacheva', *Vremia*, 10 July 2013, www.1tv.ru/news/crime/237117; 'Pivolzhskii polpred Prezidenta RF provel soveshchanie v sviazi s sobytiiami v Pugacheve', *Vremia*, 10 July 2013, www.1tv.ru/news/crime/237099 (both accessed 12 July 2013); *Vesti*, 8–10 July 2013.

92 *Vremia*, 13 October 2013, www.1tv.ru/news/crime/243862; *Vremia*, 15 October 2013, www.1tv.ru/news/crime/243976 (both accessed 18 October 2013); *Vesti*, 17 October 2013.

93 Interview with a Channel 1 journalist, 26 March 2013.

94 See, for instance, Putin's Valdai speech of September 2013.

Conclusion

Difference in the balance

From Boston to Simpferopol

One of the perils of researching the contemporary world is that new events have a constant habit of intervening and demanding perpetual re-evaluations of one's analysis. The final year of working on this book was bookended by two world events that provided a jolt to our thinking: the Boston Marathon bombing of April 2013 and the Crimean crisis of spring 2014. However, rather than undermining our conclusions, these events – each of which deeply, but rather differently, entwines Russia with its various others – enabled us to appreciate their wider significance. The Boston bombing coincided with significant new developments on the domestic scene in Russia: the trial for corruption of Aleksei Naval'nyi, finally 'brought to account' for having pressed his anti-Kremlin agenda a step too far; an explosion of anti-migrant rhetoric; and an increase in concern about the threat of radical Islamism in the North Caucasus. Ten months later the Ukrainian conflict, which administered a dramatic shock to the post-Cold War settlement, followed on the heels of Russia's surreal effort to showcase itself to the world in Sochi, location of the 2014 Winter Olympics, a site partly chosen to enable Putin to flaunt Russia's mastery over its 'Caucasian problem'. Recently re-apprehended for protesting at the sentences given to the Bolotnaia protestors, Naval'nyi, meanwhile, spent the Crimean crisis under house arrest.

We will organize our final concluding thoughts around these baneful concatenations of circumstances, centring them on a tripartite notion of difference: (i) the non-sameness (perceived or otherwise) of Russia and the west, brought to a head following Russia's actions in Ukraine, but also of the multiple 'Russias'; (ii) the double ethnic difference represented by the two Boston bombers whose parents came from Russia's North Caucasus and who themselves lived briefly in that region, but whose identities were shaped in no small part by their American upbringing;[1] (iii) difference as the political plurality which, spurred by the 'new cold war' mood now gripping much of Russia's media elite, the Putin regime seems ever more desperate to suppress. We will begin by briefly re-contextualizing the Boston bombing in terms of the concerns addressed in this book, before reassessing their significance in light of Russia's intervention in Ukraine and considering the wider implications for European broadcasters.

One of the two news reports which provided the basis for our observations in the Introduction concerned a Moscow gypsy baron writing to President Sarkozy to protest the treatment of Roma in France. It was deployed by way of distinguishing news coverage of domestic issues of inter-ethnic relations from reporting on similar issues abroad. We noted that the sharp distinctions between these two categories were undermined by the subtext of the gypsy story. The Boston bombing completes the circle, with the domestic and the international converging. As 'warriors' in the global jihad against western infidels, the Tsarnaev brothers who perpetrated the bombing seemed to be willing their insertion into Huntington's notorious 'clash of civilisations'. But as Chechen Muslims whose anti-Russian sentiment provides part of the rationale for their brutal act, they contributed to the domestic 'conflict of cultures' touched on in the talk show discussion which featured in the first story analysed in the Introduction.

The Boston bombing was, as Eliot Borenstein has suggested, troubling for the American media, since it disrupted the standard 'war on terror' narrative;[2] not only did the Tsarnaevs fail to conform to the stereotypical image of the crazed, brown, US-hating Arab terrorist (the younger Tsarnaev was a well-adjusted, pale-faced, gifted young American college student; the elder brother had until very recently dreamed of representing the US in boxing competitions), but they hailed from the same community of brave Chechen independence fighters crushed so brutally by their (former) Cold War nemesis. But despite the obvious propaganda potential that the story of the bombing offered the Russian state media, its own reaction was neither immediate nor entirely consistent. Prominent in initial *Vesti* coverage of the event were statements from the Chechen president Ramzan Kadyrov disowning the Tsarnaevs, denying that they had anything to do with Chechnia (they had, indeed, never lived there). Several *Vesti* reports followed in which the brothers' radicalization was attributed to their upbringing in a US environment that fostered alienation and degeneracy.[3] It was not long before such stories were supplanted by what came to be the dominant line on the bombing: the connection between Russia's struggle against terrorism in its backyard and the global war on terror, and the need to prioritize collaboration over mutual suspicion in the Russia–US relationship.[4] Bolstering the implied criticism of America's current priorities in that relationship, the 'war on terror' angle was supplemented by copious reminders of warnings about the Tsarnaevs that the FSB had given to the CIA a year before the bombing,[5] and by an emphasis on the older brother's admiration for the Syrian opposition movement – support shared by the US, of course – as revealed by postings on his *V kontakte* site.[6]

Inasmuch as it is possible to trace clear lines of causation for the baffling and sudden radicalization of the two privileged young immigrants (though the older brother, Tamerlan, had begun to express his contempt for and alienation from American society some time previously), the global *ummah* facilitated by the virtual world of the Internet clearly played a prominent role; Tamerlan was reported to have been influenced by Sheikh Feiz Mohammed, who was born in Australia, and who regularly issues provocative calls to arms on his website.[7] What is clear is that the multiple and contradictory pathways which led the two brothers to commit their heinous act can be traced as effectively in the Russian mediasphere as in its American equivalent. The labyrinthine combinations, however, are different. For example, the emotional defence mounted by the Tsarnaevs' parents, and their accusations of a CIA 'set-up', were treated more seriously in the Russian media than in either the UK or the US.[8] The CIA angle accorded with the culture of anti-American conspiracy theory encountered in earlier chapters. In the multi-faceted Boston bombing narrative, several conflicting axes converge: the US as a convenient ally in the common struggle against radical Islam; radical Islam as partly a product of western degeneracy; the US as a devious and hostile 'cold warrior'. The identification and counter-identification triad (Russia/Islam/the west) is, as we have seen, subject to perpetual reconfiguration – never more so than in the bizarre tale of the two brothers.

There was for a moment the glimmer of a chance that the Boston bombing might initiate a realignment of Russian and US interests; Russia's role in brokering an end to the Syrian chemical weapons crisis which threatened to unleash another Middle Eastern war in summer 2013 and the convergence of western and Russian views on the challenges posed by *jihadism* at the heart of the Syrian resistance to Assad pointed fleetingly in that direction, but prospects for such realignment rapidly faded and the aftermath of Kiev's Euromaidan revolution dealt a mortal blow to that faint prospect.

There is thus a multi-level correlation between on one hand the narrowing of difference between Russia and its western nemesis and between Russia and its internal 'other' (as we have repeatedly shown, internal and external 'others' are themselves prone to move in and out of alignment), and on the other between the narrowing of these latter differences and the widening of tolerance of (political) heterogeneity within the Russian 'self'.

With inexorable logic, the same correlation, however, operates in reverse. While newly emboldened protestors like Naval'nyi, Tolokonnikova and Alekhina continued to antagonize their Kremlin opponents, anti-migrant discourse was attaining unprecedented levels of vitriol, just as Russia rediscovered expansionist revanchism. Indications are thus that the spiral effect is unlikely to be of the virtuous kind. Of these developments, the sharp escalation in east–west tensions in February–March 2014 was the most dramatic, and we must revisit it. Nonetheless, the multiple threads of the earlier Boston story help establish the basis for our overarching insistence throughout this book on the ties linking the struggle for democracy and respect for the rule of law in Russia, the particularities of Russian public discourse and the broader crisis in diversity management. Let us recapitulate them.

In sum: Sameness as difference

As our frequent cross-national comparisons indicated, Russia both follows common European modes of response to the continent-wide crisis in multiculturalism and deviates from them, but not in the received sense (that of a state-compliant media imposing mono-ethnic order on a heterogeneous population). For in Russia, the Soviet myth of societal unity and an imperial legacy distinctive in both Tsarist and Soviet manifestations encounter a complex and in itself contradictory west European heritage, which includes both post-imperial theories of community cohesion and practices of social exclusion. Simultaneously, historically familiar methods of Russian state control of the media are transformed under the impact of global media culture. Thus, the legacies of the Russian and Soviet past are not reproduced, but rather reinvented in the context of the transnational circulation of concepts and practices of popular and state-sponsored visions of community-building and exclusion. It is striking how, in this globalized context, post-colonial west European and 'post-socialist' representations of the ethnic 'other' both diverge and resemble each other.[9]

Such resemblances challenge the validity of the very concepts of 'post-colonialism' and 'post-socialism' as analytical tools for understanding current processes in Europe east and west. In fact, the current critique of multiculturalism in western European societies and public panics about migration appear to suggest a backlash against the burden of post-colonial guilt that those societies have self-critically borne for decades. In Russia's case, and indeed that of other former Soviet societies, whatever post-socialist guilt had existed was already fading fast by the end of the El'tsin presidency. Instead, former communist nations now indulge in representations of the self as prime victim of past (or current) injustices. In post-Soviet Russia, the public sense of victimhood has acquired historically unprecedented proportions and the ruling elite has actively and intentionally sought to draw benefit from it in the new millennium.[10] The perception of self as victim is particularly strongly manifested in representations of ethnic Russians as the most exploited and discriminated-against people both during Soviet times and at present, by a system allegedly favouring ethnic minorities. Confined to extreme nationalist fringes in the past, such perceptions have been systematically mainstreamed by intellectuals, media personalities and politicians in the past 15 years. In this new context, references to the legacies of the Soviet past – a time when public manifestations of Russian ethno-nationalism were carefully managed and mostly constrained[11] – are of limited help for the purposes of understanding present discourses of nationhood.

Official discourse and state policy in Russia is neither coherent nor univocal. Given that there is no societal consensus regarding common values, that the regime has no clear ideological basis and that, at the same time, it must remain sensitive to public opinion and to the global media environment to a much greater degree than did its Soviet predecessor, the inconsistencies and splits are hardly surprising.[12] In common with other scholars, we note a particularly strong contradiction pitting the official rhetoric of a civic pan-Russian nation, embracing members of all ethnic groups and nationalities as equal citizens, against the discursive representation of Russia as above all the nation state of the ethnic Russian majority.[13] But our analysis looks beyond this contradiction for, after all, the disjunction between the civic and ethnic conceptions of Russian nationhood is acknowledged by Russia's leaders who, as our analysis indicates, collaborate with state-aligned media outlets in deliberately fostering the ambiguity which the disjunction creates in order to render Kremlin-sponsored discourse appealing simultaneously to different societal groups and to different television audiences. The coexistence within Channel 1's output of liberally inclined film-makers like Valeriia Gai-Germanika on one hand and arch-conservative Kremlin loyalists such as Leont'ev on the other testifies to this strategy.[14]

A potentially more complex fault-line is that between the new rhetoric of Russian national unity and community cohesion on one hand, and two reinvented narratives from the past on the other. The first of these is the highly hierarchical account of Russia's diverse cultures which has been shaped and reshaped in turn by imperial, Soviet and European New Right legacies. For despite the vision of new civic Russian nationhood promoted by state-aligned media in the past decade, the rigidity of the hierarchies and of the boundaries between communities defined by ethno-cultural markers has paradoxically increased by comparison with Soviet times and the 1990s; indeed, it has become progressively more dominant during the period of our analysis. The second, related, narrative from the past is that of the non-Russian nationalities as belonging in their own, clearly designated national homelands and sub-state administrative autonomies. This narrative limits the propensity for representatives of ethnic minorities to identify with the Russian Federation as a whole. Equally detrimental to national cohesion, both narratives have been internalized by contemporary elites, including political leaders and media personalities, and by the broader public (Russian majority and minorities alike). As our analysis of the reporting of the Manezhnaia disturbances suggests, these narratives tend to supplant the discourse of Russia as a unified nation in furnishing the default analytical framework within which to interpret tensions and conflicts when they erupt. The significance of this second fault-line lies in its under-appreciation both by the Kremlin and by analysts of Russian societal and political developments.

A troubling parallel appears here with the late Soviet period, when the importance of the ethnicity factor was equally underestimated by Soviet leaders and external observers. The Gorbachev leadership appeared to be completely unprepared for the upsurge of nationalist activism during *perestroika*, just as, on the very eve of the dramatic eruption of ethno-national activism in the late 1980s, influential scholars of Soviet politics argued that the Soviet state was sufficiently successful in managing 'the national question' to make ethno-national political mobilization unlikely.[15] In effect, both parties became hostages to a discourse which was capable merely of simulating the effective management of inter-ethnic problems. Today, the Kremlin is likewise the hostage to a discursive version of reality it has co-constructed with a state television colossus which, despite the tight control to which it remains subject, has acquired its own momentum and its own internal logic of development. Meanwhile many external observers, academic and non-academic, continue unwittingly to be influenced by the transition paradigm, which prioritizes the electoral and party politics of the federal centre and measures the progress made by that centre against the normative model of a mature representative democracy. Adherents to the paradigm thus understate complex societal processes arising from the ethnic diversity in a country for which representative democracy of the strictly western type may not be the ultimate, or even the most appropriate, destiny. It is the same blind spot which causes them to oversimplify the relationship between the Kremlin and the state-aligned media, and to privilege transitive meaning (that conveyed from the state to the media) over transactional meaning (that arising from multi-directional interactions involving state, media, popular and other discourses). Despite the interrogation and criticism to which the paradigm has been subjected for more than a decade, it retains an influential position within scholarship across multiple disciplines.[16]

The compelling significance of the ethnic diversity factor lies in the fact that it introduces into post-Soviet official discourse three related sets of tensions (divergent interpretations of the notions of 'tolerance' and community cohesion; difficulties in establishing the statuses of ethnic Russians, as well as other nationalities, within the multi-ethnic Russian Federation; the absence of a broader set of consensual values). Television, the main source of information for the Russian public, is the primary site of enactment of these tensions. In managing them, television must simultaneously implement a nation-building strategy, performatively instate the boundaries of the 'imagined community' implied by that strategy and provide a gateway through which more extreme variants might enter the public arena and voices hostile to it be expunged. Our account of the abject failure of the recently instituted Day of National Unity as a 'media event' offered one vivid illustration of how acute the difficulties that the nation-building strategy must overcome are. Nor, as the inconsistently framed interventions of both liberal oppositionists and representatives of nationalist fringes into the pre- and post-electoral political fray indicates, does state television fulfil its filtering function in a convincing way (Chapter 3).

We traced the problematic nature of television's discursive struggles not only in news bulletins but across a wide range of genres, and via a broadcasting sphere in which centre (most typically represented by Channel 1's *Vremia*) and periphery are sharply differentiated and the geo-cultural margins are at times the locus for bold challenges to the orthodoxies of the centre (fictional genres aimed at youth audiences on mainstream and specialist channels), and at others an embodiment so quintessential that it, paradoxically, reinforces their peripheral status (coverage of ethnic cohesion on Buriat state television). In another paradox, as we saw in Chapter 9, the more constraints are placed on freedom of speech, the further the prospect of official discourse speaking from the centre point of a stable consensus recedes and the weaker that discourse becomes. In such a situation, television's role in influencing the discourse can become greater.

Alternatives for resolving the tensions depend on the appropriation of new sources of legitimacy. One is represented by a powerful populism centring on widespread revulsion at corruption. A struggle to claim 'ownership' of the anti-corruption platform and establish it as the core of a 'chain of equivalences' pits Putin and the Kremlin against the liberal opposition on one hand and the nationalist right on the other. The bitterness of the struggle leads to the subliminal – and at times overt – ethnicization of corruption and of Russian political discourse more generally. In Chapter 7 we observed in this context a bi-directional inversion of relations between the ethnic and particular and the universal and general (Russo-centric particularism is imposed as the universalist glue binding the Russian Federation; the universal evil of corruption acquires the ethnically particular tint of anti-migrant xenophobia). The second source of legitimacy derives from a revivalist religious traditionalism founded on a rejection of liberal values and a belief in the unifying force of Orthodox Christianity. The influence of this source grew significantly during the course of the period we analysed, as indicated by the shift in the Church's status from a centre of power competing with that of the Kremlin (early celebrations of Day of National Unity; the Church's stance on the Manezhnaia riots) to that of the Kremlin's preeminent discursive tool (later celebrations of the Day of National Unity; the response to the Pussy Riot affair).

We demonstrated that the growing prominence of both sources of legitimacy intersected with a burgeoning state media focus on two related dangers. These are Islamist radicalism (bolstered by world events such as the Syrian crisis, the Boston Marathon bombing and the transnational influence of European 'media panics' about the building of mosques and the wearing of *hidjabs* in public places) and North Caucasian separatism (news stories concerning the constant drip of low-level terrorist activities in Dagestan and Ingushetia). The dangers are themselves increasingly subsumed into two 'master threats' which converge in an inflammatory dynamic: that of the (illegal) migrant incapable of conforming to appropriate behavioural norms, and that of an ever-present, ever-scheming, hostile west. Russian nationalist tendencies of all hues, from the extremist, far-right anti-immigration movement (DPNI) to the democratic opposition leader Aleksei Naval'nyi, have been revitalized by this explosive mix.[17] For this reason, both sources of legitimacy represent a potent challenge to the position of ethnic minorities. The incipient and fluid status of post-Soviet Russian nationhood – highlighted in television's intensive mediation of the Pussy Riot crisis – only adds to the challenge.

The most striking development we observed was the ratcheting up of anti-migrant rhetoric in 2012–13, accompanied by an increase in the coverage of migration-related issues across different television genres, with the inflammatory conflation of 'internal' and 'external' migrant contributing to the promotion of a deeply ethnocentric version of Russian national identity. The trend was apparent in its early stages in coverage of the Manezhnaia riots, which proved to be a touchstone in the recent history of public discourse on inter-ethnic relations in Russia. The period following the 2012 presidential elections saw that trend intensify in a manner which rendered Putin's carefully balanced, if ultimately contradictory, pre-election article on 'the national question' obsolete. Ironically, in the post-election period, an ever more heavy-handed approach to freedom of expression seems to have been accompanied by the initial uncertainty of the regime over how to re-establish control over society following mass public protests. In this period, the fallacy that state television functions solely as a passive instrument of control in the hands of an all-powerful authoritarian regime was revealed particularly clearly. In fact, at the time critics of the official mantras on ethnic cohesion and multi-faith harmony felt *emboldened* rather than cowed by the changing political environment. Here television served actively to shape official discourse rather than be shaped by it. For example, Vladimir Putin's September 2013 annual address to the Valdai Forum marked an unmistakable shift in the Kremlin's political rhetoric toward adoption of a viewpoint that had long been articulated by public intellectuals and television personalities such as Channel 1's Maksim Shevchenko: that of Russia's leading role in promoting the 'traditional moral values' of family, church and nation over the degradation to which those values were being subjected throughout the west.[18] The appointment of the arch-conservative Kiselev to head Rossiia Segodnia hinted that the building of allegiances with social conservatives in the west may have been intended to supplement the current strategy of international broadcaster Russia Today: that of targeting ethnic minorities and other groups with an ingrained hostility to western political hegemony. Given the crucial role television continues to play in mediating the struggle for popular legitimacy within Russia, its propensity to articulate the struggle in covertly ethnicized terms has profound implications for the condition of domestic public discourse, for prospects for democratic renewal and for Russia's very future as a multi-ethnic federation, not to mention its currently virtually broken relationships with other, western, nations.

Our analysis identified the discursive *production*, witting and unwitting, of inter-ethnic relations through multiple sub-genres of media event, each revealing the circulatory logic simultaneously binding, and dividing, television, state and society. The first sub-genre we explored – the openly 'co-operative' type – revealed that the fragile vehicle intended to enact the state mantra of 'unity in diversity', Russia's new Day of National Unity celebration, collapsed under the pressure of its own contradictions, leaving a hollow shell of unconvincing ritual. In television's coverage of the Manezhnaia riots, an example of the 'crisis management' sub-genre, we witnessed the same process in reverse: divergent news accounts of an event which exposed the profound inter-ethnic tensions destabilizing the media–state–society relationship eventually converge in a narrative whose brittle, congealed nature points to the restricted logic of the conceptual apparatus they share. The *Shkola* and *Nasha Russia* phenomena, by contrast, provided examples of the 'endogenous' variety of media event in which outputs at the margins of television itself (with the state's ambivalent involvement) catalyse both broader societal fractures, and a muted struggle by liberal tolerance discourses to pluralize the public sphere. The strangling at birth of the very notion of a second series of *Shkola*, and the fact that the liberally inclined makers of *Nasha Russia* and representatives of Central Asian minorities were at odds over the programme, indicated the struggle's failure to make headway. Finally, the 'collusive' sub-genre was exemplified in the co-ordinated manufacture of a xenophobic, exclusionary form of unity from the divisions and antagonisms of the Pussy Riot scandal.

The outcome of the developments traced by the media events is difficult to determine. The current pall of 'new cold war' sentiments pervading both Russia and the west do not bode well. On one hand, as we suggested in Chapter 7, the temptations of the opposition's campaign against corruption to appropriate popular xenophobia may prove too great to resist, even assuming that progressive, liberal elements in the campaign are capable of regaining the momentum they acquired in 2012 in the face of the ratcheting up of constraints on free speech and constant state media denigration that followed the Ukraine crisis of 2014. As Emil' Pain, one of the more astute commentators on inter-ethnic relations in Russia, observed in 2013:

> the cumulative mass discontent expressed in political protest a year ago has now taken an ethnic turn ... For many, the [nationalist] slogans are much more satisfying to yell than the opposition's rallying cries, which were obscure and generally unrealistic. The simplicity of today's public demands also has its attraction for the regime: 'throw out' and 'don't let in' are concepts close to the Russian apparatchik's heart – much sweeter to their ears than words like 'modernisation' or 'reform'.[19]

In the same context, and echoing our discussion of the logic of populism in Chapter 8, Pain refers to the possibility of 'a new electoral balance in Russia', a disposition of forces in which certain 'democratic' critics of the Putin regime make common cause with its mainstream nationalist opponents.[20]

The dangers that this scenario holds for the future of the Russian Federation are exacerbated by the fact that even the simulated management of tensions referred to earlier has been performed only sporadically, usually in response to the eruption of conflicts. The central task of television news in the simulation process has been to hide the state's inability to manage such conflicts, either by under-reporting relevant developments or by diverting criticism away from the Kremlin.[21] As a result, public phobias and prejudices are not simply given a voice in the coverage in order to show that the leadership is responsive to the popular mood; they are also actively manufactured, particularly within talk shows, with their highly partial selection of evidence, experts and audiences.

The failure of 'the simulation approach' became apparent when the announcement of Putin's decision to run for a third term as president was met with mass public protest. In this context, state-aligned television's attempt to perform national unity via the Pussy Riot affair in fact exposed the fragmented nature of the community which it was exhorting to unite, the absence of societal consensus and the failure of the Kremlin to facilitate its development. The relentless anti-Pussy Riot campaign combined antipathy to west European liberalism with an endorsement of traditional Russian Orthodox values, which it aligned with those of Islam within a shared, and unique, Eurasianist identity. As we argue in Chapter 9, a by-product of this campaign has been the occasional transposition of the mark of absolute difference from the western 'other' to post-Soviet migrants (particularly, and paradoxically, those from Muslim societies) during Putin's third presidential term. Conversely, the figure of the liberal intellectual has become subject to a covert and indirect racial profiling different from, but allied to, the more overt racism suffered by the North Caucasian migrant or the Central Asian *Gasterbeiter*; the use of anti-Semitic motifs in Arkadii Mamontov's diatribe against the westernizing liberals responsible for the 'punk prayer' 'outrage' (Chapter 8) confirms that anti-Semitism is far from eradicated from Russian public discourse.

National television serves as a key facilitator of the process, both aspects of which were epitomized in Mamontov's *Spetsial'nyi korrespondent* series on Pussy Riot and migrant workers respectively.

We have, however, eschewed portraying Russian television as systematically 'racist', taking care not to conflate racism with other, non-biological, forms of prejudice, despite the fact that in practice it often overlaps with them – particularly in the guise of the 'new' or 'cultural' racism at work in the post-Manezhnaia prominence of the 'conflict of cultures' notion. Moreover, we have distinguished between the presence of prejudicial voices and statements *on* Russian television (they are multiple, and often inadvertent), the adoption of racist or ethnically prejudiced positions *by* Russian television journalists, commentators and newsreaders (more prevalent than in the past but hardly endemic) and ethnic/racial prejudice as an impersonal *effect* of the structures of Russian television news discourse (it is this which has preoccupied us most). It would seem perverse to place within the same category as racism attitudes as diverse as homophobia, ideological hostility to western liberalism, visceral disdain for Central Asian migrants, Islamophobia, antipathy to the cultural 'behaviour' of North Caucasian diasporas, ethnically tinged resentment at the corruption and criminality afflicting the Russian state and subliminal anti-Semitic tropes.

Nor, however, is the broader, generalized term 'xenophobia' adequate to the task of accounting for the dizzying plurality of prejudices. Lev Gudkov's problematic if suggestive notion of a post-Soviet Russian 'negative identity', meanwhile, errs both by equalizing the prejudices and by locating them on one side of an antinomy in which the other side is occupied by a falsely unified sense of Russian selfhood.[22] Fully accounting for the multi-directional articulations between these xenophobias, their mutual contaminations, convergences, fragmentations and realignments – and for the complex negotiation of Russian statehood across the boundaries of the civic, the ethnic, the national and the transnational, the Soviet past and the post-Soviet present – is beyond the scope of the media discourse analysis we have undertaken. It would require us to disentangle the complex network of threads linking ideological conflict (anti-liberalism), cultural antagonism (hostility to the traditions and rituals of Roma communities), the clash of faiths (opposition to mosques or to the *hidjab*), resentment at behavioural traits and practices (state corruption and criminal collusion) and biological difference (of race; of sexual orientation) – a task worthy of an entire career! Suffice it to say that our analysis has vividly illustrated the unusual extent of their intertwinement in Russian media discourse.

To turn full circle and re-situate our discussion in the present context, the intertwinement of domestic themes is, at another level, itself imbricated with the mutual predication of domestic and international media perspectives. Throughout the book we have pointed to ways in which trends in reporting issues of inter-ethnic conflict in western Europe have been transposed, re-interpreted, exploited, cross-fertilized, reversed, intensified and transformed in the very different Russian media environment to which, in the global era, they now invariably migrate. Reporting of the European 'Roma crisis' as alternatively a useful, polemical counterpoint to the relative inter-ethnic harmony prevailing in the homelands and a dire warning of what might await those homelands should they adopt the liberal, European multicultural model purportedly responsible for the problems provides one of several examples of the acute and anxious 'dialogism' pervading Russian news discourse. There is no better illustration of that dialogism than the confused *Schadenfreude* with which the Boston Marathon bombing was met by the journalists and commentators of Russian state television.

The interlocking of domestic and international themes and media perspectives attained its apotheosis, however, a year after the Boston bombing, in the events of early spring 2014. The Sochi Winter Olympics provided a symbolic foretaste of what was to unfold within days when, in one of the most bizarre scandals surrounding the controversy-ridden event, a group of Kuban Cossacks brutally horse-whipped Nadezhda Tolokonnikova and Mariia Alekhina as they performed their latest anti-Putin protest song on the outskirts of Sochi, watched and filmed by a gaggle of gawping cameramen. As the metaphorical Cossack beating with which the women were threatened during the 'punk prayer' affair (see Chapter 8) was realized in the form of actual bodily violence, the category confusion besetting Russia's official media suddenly became rather more than a discursive aberration. Violence against the internal 'otherness' represented by Tolokonnikova and Alekhina was about to mutate into the deployment of physical force in Russia's encounter with 'otherness' in an external context: that of a western-backed, 'illegitimate' Ukraine.

The obsessive manner in which the Russian media narrative of the Sochi games centred on paranoid rebuttals of western 'denigration' of the host nation provided a further insight into the context in which the Crimean intervention took place. The Ukrainian crisis thus returns us to the heart of the same dynamic of multiple alterities that the Boston bombing brought to light. In so doing, it illuminates the inextricable ties linking the issues we have discussed in this book to Russia's dysfunctional political culture, and to its increasingly aberrant international posture.

The pretext for Russia's actions in Crimea, and later for its support both tacit and explicit for the separatist rebels in Eastern Ukraine, focused on the protection of its 'compatriots' (*sootechestvenniki*), a term whose arbitrary conflation with 'ethnic Russians' (*etnicheskie russkie*) and 'Russian speakers' (*russkoiazychnye*) was replicated uncritically by many western commentators, who also failed to distinguish the latter terms from the quite distinct notion of 'Russian citizens' (*rossiiskie grazhdane*). There can be no more graphic illustration of the consequences of the confused ethnicization of national identity that our book has traced.

Nor would the bemused alienation expressed in western media outlets at the sight of jubilant crowds welcoming Putin's Crimean annexation have surprised readers of a book which has charted the progressive subjugation of state-aligned Russian broadcasters less to the Kremlin itself than to a Kremlin-endorsed ideology of Russian national pride that has at times threatened to break free from the control of its instigators. The fact that it is an empty, short-circuited ideology whose lack of viable content means that it ultimately has nothing to fill its hollow shell other than an intensified version of itself makes it no less dangerous. The core ideological concepts with which broadcasters frame their news and current affairs programmes are in a state of permanent flux, including, in the period under review in this book, such disparate and often incompatible ideas as Russia's unity in diversity; Russia as a defender of traditional values, interpreted either as exclusively European and Christian or as encompassing Islam and other 'traditional religions'; Orthodox Christianity as the primary pillar of Russian nationhood; and the 'Muslim migrant' as a threat to Russian identity. Against the backdrop of this ideological eclecticism, two currents, both dating back to the 1990s, have been constantly present in the public discourse – Russia as a protector of its 'compatriots' abroad and the west as Russia's perennial foe.[23] Since 2012, the likes of Kiselev have ensured that such ideological frames have been utilized in a particularly confrontational and crude manner by state-aligned broadcasters.

Indeed, Kiselev was at the centre of a frenzy of anti-western Cold War rhetoric that gripped Russian state broadcasters following the imposition of punitive sanctions on Russia. Kiselev used the platform of his *Vesti nedeli* programme to point out that Russia alone among nations has the capacity to turn the US into 'radioactive dust'.[24] He was echoed by Aleksandr Prokhanov, like Kiselev a frequent presence in this book, who announced that his 15-year-long dream of a return to the Cold War had been fulfilled.[25] The two commentators, both close to Putin's inner circle, offer a sobering demonstration of the dependency of Russian national pride, in its distortive, Putinesque manifestation, on the role of the 'treacherous, conspiratorial west' that is Russia's nemesis.

The third corner of the familiar triadic dynamic, Russia's internal ethnic other, was supplied by the Crimean Tatars, coverage of whose predicament by Russian federal television contained its own set of contradictions. A *Vesti nedeli* bulletin of 2 March, for example, acknowledged the Tatar community's unease about the possibility of a Russian takeover. The 9 March *Vesti* broadcast developed this theme further and included an open admission that many Crimean Tatars were not pro-Russian. Other reports, however, echoed Putin's triumphal annexation speech, which insisted (against all the evidence) that most Crimean Tatars supported reunification with Russia.[26] In this representation, the Crimean Tatars were used as a symbol of Crimea's and Russia's unity in diversity.[27] This ambivalent recognition and simultaneous denial of the 'Crimean Tatar problem' exposes the influence of another theme touched upon in earlier chapters: the tension between Putin's neo-imperialist/Eurasianist variant on Russian patriotism (one which, like its nineteenth and twentieth-century predecessors, aspires to square the need for inclusivity and inter-ethnic harmony with the imperative to maintain the dominant ethnic group's power) and the isolationist nationalism of media figures like Kiselev, for whom 'Muslim minorities' constitute a problem.

Euromaidan, Euromedia and the mutuality of responsibility

The symbiosis of national self and antithetical western 'other' in the Russian patriotic imagination, and the fragile condition in which free expression – and with it the future of Russian civil society, and of the Federation – finds itself place great responsibility on European broadcasters when reporting events in Russia. In a globalized news environment, that responsibility extends to their coverage of inter-ethnic encounters in their own, domestic contexts. On one hand, there is a propensity for selective vision in relation to Russia on the part of outlets which, like the BBC, are admirably perspicacious in their treatment of other regions and issues: their blindness to the significance of the Manezhnaia riots contrasted with their blanket coverage of the anti-Kremlin protests of 2011–12; their reluctance to acknowledge the ethnocentrism of some of those protests' leading figures or the degree of societal support that the Orthodox position on the Pussy Riot episode elicited; their denigration of genuine, if exaggerated, Russian anxieties over the xenophobic extremism prevalent in elements within Ukraine's Euromaidan movement. Such omissions evidence the stubborn endurance of Cold War visions of Russian elites as defined entirely, and all too simplistically, by their paranoid aversion to democracy.[28] As it feeds the fires of the anti-western conspiracy theories ravaging large sections of the Russian public sphere, this reductionism is eventually revealed as a self-fulfilling prophecy. The Cold War hysteria of Kiselev and Prokhanov found its match in some of the more extreme pronouncements issuing from the pens of certain western commentators.

On the other hand, Russian broadcasters have much to learn from the sensitivity with which public service broadcasters like the BBC approach controversies surrounding race, ethnicity, public order and security. Only weeks after the Boston Marathon bombing, the UK was shocked when a British soldier's brutal slaying in cold blood on the streets of London's Woolwich district turned out to be the work of two radical Islamists of Nigerian Christian heritage. During the tense period that followed their arrest, there was a fear that a far-right backlash would plunge the country into an abyss of vengeance-seeking crime. That this fear proved largely unfounded but for an initial wave of anti-Islamic protests (and isolated cases of violence) owes something at least to the scrupulous efforts of broadcasters to distinguish the distorted, radical version of Islam pursued by the Woolwich murderers from the mainstream beliefs of the overwhelming majority of British Muslims, to follow strict codes of impartiality and balance and to avoid inflammatory conflations of criminality and ethno-cultural identity.

Russian broadcasters displayed no such restraint. The reporting of the Woolwich incident was cynical and shocking in its inaccuracies, in the unmistakable ethnic profiling it applied to accounts of the two suspects and in its multiple conflations (of foreign worker with British citizen, of ethnicity and nationality with faith, of migration with crime and terrorism, of misplaced resentment against British involvement in Muslim countries with socio-economic unrest in Stockholm). Ironically, Channel 1's coverage of the Woolwich incident proved to be a milestone in the long retreat of the Kremlin-sponsored discourse from its earlier, albeit partial, adherence to the norms and practices of European 'tolerance' (Chapter 9).[29]

At one level, the divergence in Channel 1 and BBC coverage of the Woolwich murder reflects the distinction between reporting on events in one's own vicinity (when remaining alert to community sensitivities is crucial for societal cohesion) and reporting on events that occur at a 'safe' distance (when those events can be conveniently re-construed for domestic consumption). On another level, Russian state television's implicit willingness to equate the crazed actions of two British citizens of Nigerian descent with the purported misdemeanours of Central Asian and North Caucasian migrants in Moscow represents a misperception of similarity between two radically different situations.

On both levels we are dealing with the interplay of sameness and difference which provides our central trope. For what is the dilemma faced by contemporary states everywhere – indeed, by the nation state per se – if not that of achieving the impossible balance of recognizing the reality of increasing rates of global population movement and the value of the 'otherness' that it brings while instilling a shared sense of national sameness across host and migrant communities, and doing so without detriment to the political and ideological pluralism which is both the corollary of that task and its mode of accomplishment? The challenge is that of overcoming the paradox entailed in imagining a cosmopolitan model of citizenship applicable to multiple national contexts. Nowhere is that challenge greater than in Russia, which now experiences rates of migration second only to the United States and which, as a result of imperial Russian and Soviet legacies, is imagined as a society with close to 200 different indigenous ethnic groups across which national 'sameness' is still to be realized.[30] If it has not now abandoned diversity management altogether, Russia still oscillates hesitantly in turn between a model designed to conserve the 'sameness' of its Soviet residue (itself the ideological product of an Enlightenment project shared across many nationalities, cross-fertilized with the specificities of a Russian historical tradition) and one embracing the 'difference' embodied in the principles of European tolerance.

The two key centres of focus of our project converge in the trope of what Roger Silverstone terms the 'dialectic of sameness and difference': where television representation (of the distant other to the intimate self) meets inter-ethnic cohesion (of ethnic self and ethnic other).[31] Silverstone's reference to television's 'capacity to bring people together while simultaneously keeping them apart' overlaps with what Lilia Chouliaraki calls 'the ambivalence of mediation', of 'constantly hovering between closeness and difference', and together they provide the basis for an ethics of mediation.[32] For in bringing dramatic, live images of distant places and people into the proximity of the domestic setting, the television screen retains its status as a privileged and supremely paradoxical tool for experiencing the ethnic 'other' as intimately similar, yet unique and distinct.

It is for this reason that US viewers were so disoriented by the images of the Tsarnaev brothers and their naturalized American relatives which flooded their screens in the aftermath of the Boston Marathon bombing. For here the dialectic is short-circuited: rather than the 'other' being made 'same' via the magical logic of the television screen, that 'other' is disconcertingly revealed to be nothing other than the 'same' from its inception.

Whether Russian television viewers experienced the same disorientation in reverse, of course, is doubtful: their North Caucasian names, origins and demeanours, combined with their American upbringing, meant that the Tsarnaevs would hardly have elicited any sense of proximity among most Russians. Rather, they offered a means of equalizing ('making same') two radically different 'others' (US and Chechen), thereby also reinforcing, rather than disrupting or counter-balancing, the distancing effect. By juxtaposing American and Russian perceptions of the Tsarnaev brothers, we arrive at the outermost layer of the sameness–difference trope: that of the very purpose of this book. For in studying media representations of ethnic difference in Russia from a perspective which reflects the western academic context in which it was developed, we are willy-nilly entering the sameness–difference dynamic with all its twists and turns: attempting to make sense of the peculiarly Russian variant of the crisis in national diversity management strategy that has been induced by the gathering pace of globalization. In doing so, we are making comparisons, both implicit and explicit: drawing parallels with situations in the UK and elsewhere, and noting sharp contrasts. We are providing the grounds for readers to assimilate phenomena to their own experience, or to attribute them to recognizable, specifically Russian and conveniently alien, paradigms. In a paradoxical twist all of its own, readers are in fact grasping at two forms of sameness in which one is the inverse of the other. To identify similarities in the ways that the BBC and Channel 1 discuss multiculturalism, for example, is, in the context of common assumptions about the constraints on the Russian media, to experience the surprise that comes when reality is proven to shatter or break free from its prior conception. Equally, to point to differences between the nuances and balance in BBC reporting of ethnic violence and the crude, univocal prejudices that emerge in NTV coverage of the same issues is ultimately (and, no doubt, correctly in this instance) to see 'sameness': to assimilate Russian television to 'safe', established templates relating to media systems that have yet to shed the shackles of their repressive totalitarian pasts.

As the two final examples illustrate, achieving the appropriate balance of difference and sameness in such contexts is difficult and carries a heavy burden of responsibility. Overstating the parallels between media practices in the UK and Russia – or 'excusing' particularly reprehensible Russian variants, such as the overwhelming tide of distorted half-truths peddled by federal broadcasters while Russia marched roughshod over international law into a sovereign Ukraine, as unavoidable Soviet residues – is tantamount to apologia for a paranoid regime in denial of its desperate need of transformation and renewal;[33] during this period programmes like *Vesti nedeli* and *Voskresnoe Vremia* took the conspiratorial, anti-western bias to hitherto unprecedented levels of intensity.[34] However, to err in the opposite direction of failing to acknowledge western equivalents to Russian media biases, of indulging in reductive, complexity-blind critiques or of refusing to accept the role played by cultural difference in shaping current practice, is to indulge in a form of cultural imperialism, thereby reinforcing anti-western phobias and contributing to the New Cold War dialectic we are currently witnessing.

Above all, it is surely vital to avoid becoming trapped in a hall of mirrors in which difference is merely the other side of sameness: the alien. The appreciation of true difference lies in the delicate balance between accepting alterity in all its strangeness and acknowledging the common humanity that impels us to embrace it. In a spiritual formulation seemingly far removed from his later Marxist sensibility, the great aesthetician Georg Lukacs argued:

> The world is wide, and yet like a home, for the fire that burns in the soul is of the same essential nature as the stars; the world and the self, the light and the fire are sharply distinct, yet they never become permanent strangers to one another, for fire is the soul of all light and all fire clothes itself in light.[35]

Failure to recognize the truth of Lukacs's soaring insight may intensify the post-Ukraine spiral of mutual mistrust and misunderstanding, unleashing a new wave of domestic xenophobia and political repression and transmitting further shock-waves across the external geopolitical landscape.[36] In Russia too, difference is in the balance.

Notes

1 Although one of the brothers was indeed a Russian citizen, the other was born in Kyrgyzstan and never acquired Russian citizenship.

2 Eliot Borenstein, 'Off-White Like Dzhokar', Jordan Center for the Advanced Study of Russia, May 2013. Available online at http://jordanrussiacenter.org/news/off-white-like-dzhokhar/ (accessed 15 December 2013).

3 The title of a report by Evgenii Popov included the quote: 'Americans don't want to admit that they were murdered by their own people' (*Vesti nedeli*, 21 April 2013).

4 One *Vesti* report was entitled 'Tsarnaev Turned Out To Be a Supporter of the Jihad, and a Fan of Basaev' (28 April 2013).

5 See, for example, 'Agent TsRU ne raspoznal v Tsarnaeve terrorista', *Vesti*, 20 April 2013.

6 Several reports on *Vremia* quoted the words of a Damascus-based political scientist, Terry Mason, who drew attention to the number of Chechen mercenaries fighting on the side of the rebels in Syria. See *Voskresnoe Vremia*, 28 April 2013. Available online at www.1tv.ru/news/world/231887 (accessed 15 December 2013).

7 Ted Jeory, 'Britain Warned About Cleric Who May Have Influenced Boston Bombers', *Daily Express*, 20 April 2013. Available online at www.express.co.uk/news/uk/393407/Britain-warned-about-cleric-who-may-have-influenced-Boston-bombers (accessed 15 December 2013).

8 See, for example, an NTV report of 19 April 2013.

9 For the relationship between post-colonialism and post-socialism, see Douglas Rogers, 'Postsocialisms Unbound: Connections, Critiques, Comparisons', *Slavic Review*, 69/1, 2010: 1–15.

10 Serguei Oushakine, *Patriotism of Despair: Nation, War, and Loss in Russia*, Ithaca: Cornell University Press, 2009.

11 On the Russocentrism of Stalin's policies see David Brandenberger, *National Bolshevism: Stalinist Mass Culture and the Formation of Modern Russian National Identity (1931–1956)*, Cambridge: Harvard University Press, 2002 and on intellectual Russian ethnonationalism in the post-Stalin period see Yitzhakh M. Brudny, *Reinventing Russia. Russian Nationalism and the Soviet State 1953–1991*, Cambridge: Harvard University Press, 1998.

12 Even the most repressive top-down Soviet campaigns reflected societal moods and tensions. And despite the rigid Marxist–Leninist ideological framework, Soviet political discourses, including those of nationhood, were ridden with contradiction. Sheila

Fitzpatrick, *The Cultural Front: Power and Culture in Revolutionary Russia*, Ithaca: Cornell University Press, 1992.

13 Oxana Shevel, 'Russian Nation-building from Yel'tsin to Medevdev: Ethnic, Civic or Purposefully Ambiguous?', *Europe-Asia Studies*, 63/2, 2011, 179–202; Vera Tolz, 'Conflicting "Homeland Myths" and Nation-State Building in Post-Communist Russia', *Slavic Review*, 57/2, 1998, 267–94.

14 Shevel, 'Russian Nation-building from Yel'tsin to Medevdev'.

15 Gail Warhofsky Lapidus, 'Ethnonationalism and Political Stability. The Soviet Case', *World Politics*, 36/4, 1984, 555–80.

16 Thomas Carothers 'The End of the Transition Paradigm', *Journal of Democracy*, 13/1, 2002, 5–21; For a nuanced discussion of understanding Russia as a transition society see Richard Sakwa, *The Crisis of Russian Democracy: The Dual State, Factionalism and the Medvedev Succession*, Cambridge: Cambridge University Press, 2010.

17 Emil Pain, Sergei Prostakov, Sergei Mokhov and Egor Poliakov claim that xenophobia is an attribute shared across the political spectrum on the Russian internet: see 'The Runet as the Mirror of Political and National Problems in Russia', 22 November 2012. Available online at www.liberal.ru/articles/5903 (accessed 12 December 2012).

18 For a transcript of Putin's speech, see www.kremlin.ru/news/19243 (accessed 15 December 2013).

19 Emil Pain, 'From Protests to Pogroms', *openDemocracy*, 27 August 2013. Available online at www.opendemocracy.net/od-russia/emil-pain/from-protests-to-pogroms (accessed 15 December 2013). The expression 'a new electoral balance' comes from a report by the Moscow-based think tank the Centre for Strategic Research, which Pain quotes.

20 Ibid.

21 Ivan Krastev and Stephen Holmes, 'An Autopsy of Managed Democracy', *Journal of Democracy*, 23/3, 2012, 33–45, which convincingly argues that 'incapacity-hiding, not capacity building, is at the heart of Putin's statecraft' (40).

22 Lev Gudkov, *Negativnaia identichnost'. Stat'i 1997-2002 godov*, Moscow: Literaturnoe obozrenie, 2004.

23 On the invention and instrumental use of the *sootechestvenniki* concept in the early 1990s, see Neil Melvin, *Russians beyond Russia. The Politics of National Identity*, London: Royal Institute of International Affairs, 1995. On the rise of strong anti-westernism in Russia in the second half of the 1990s, see Michael Urban, 'Remythologising the Russian State', *Europe-Asia Studies*, 50/6, 1998, 969–92.

24 *Vesti nedeli*, 16 March 2014.

25 Reported in Ellen Barry, 'Foes of America in Russia Crave Rupture in Ties', *New York Times*, 15 March 2014. Available online at www.nytimes.com/2014/03/16/world/europe/foes-of-america-in-russia-crave-rupture-in-ties.html?smid=tw-share&_r=0 (accessed 19 March 2014).

26 Full text of Putin's speech at the official website of the Russian President: http://eng.kremlin.ru/news/6889 (accessed 19 March 2014).

27 *Vesti*, 17 March 2014.

28 See for example Edward Lucas, *Deception: Spies, Lies and How Russia Dupes the West*, London: Bloomsbury Publishing, 2013.

29 Scorning 'British tolerance', a *Vremia* reporter remarks that 'It is very British not to realize that a tolerant society can produce absolute evil with radical ideology in its head and a meat cleaver in its hand'. See *Vremia*, 26 May 2013. Available online at www.1tv.ru/news/world/233809 (accessed 15 January 2014).

30 Russian imperial and Soviet ethnographers who outlined the criteria for ethnic group membership named those groups. The number of groups (nationalities) was determined from above. In the 1920s many hundreds were identified by ethnographers who were given free reign, but in the mid-1930s Stalin ordered them to reduce the

number; appropriate policies followed. See Yuri Slezkine, 'The USSR as a Communal Apartment, or How a Socialist State Promoted Ethnic Particularism', *Slavic Review*, 53/2, 1994, 414–52.

31 Roger Silverstone, *Media and Morality: On the Rise of the Mediapolis*, Cambridge: Polity Press, 2007, p. 16.

32 Lilie Chouliaraki, *The Spectatorship of Suffering*, London: Sage, 2006, p. 41.

33 Many journalists we interviewed justified their own malpractices by claiming that their western counterparts acted identically.

34 While REN TV reported multiple views on Russia's actions in Ukraine, it refrained from criticism of the government. The staged pluralism we observed in earlier coverage had by now evaporated. Meanwhile, financial pressure placed on Dozhd' TV brought it to the brink of closure.

35 Georg Lucacs, *Theory of the Novel*, London: Merlin Press, 1971, p. 34.

36 Putin's Crimean annexation speech included a reference to 'a fifth column … a disparate bunch of "national traitors"' with which the West is now purportedly threatening Russia'. . Available online at http://eng.kremlin.ru/news/6889

Bibliography

Agamben, Giorgio, *State of Exception*, Chicago: University of Chicago Press, 2008.

Akturk, Sener, *Regimes of Ethnicity and Nationhood in Germany, Russia and Turkey*, Cambridge: Cambridge University Press, 2012.

Alia, Valeria and Bull, Simone, *Media and Ethnic Minorities*, Edinburgh: Edinburgh University Press, 2005.

Alvarez, Robert R. and Collier, George A., 'The Long Haul in Mexican Trucking: Traversing the Borderlands of the North and the South', *American Ethnologist*, 21/3, 1994, 606–27.

Anderson, Benedict, *Imagined Communities: Reflection on the Origins and Spread of Nationalism*, London: Verso, 1991.

Anderson, Bridget and Blinder, Scott, 'Who Counts as a Migrant? Definitions and Their Consequences', *Migration Observatory Briefing*, COMPAS, University of Oxford, UK, September 2013.

Appadurai, Arjun, *Fear of Small Numbers: An Essay on the Geography of Anger*, Durham: Duke University Press, 2006.

Arnold, Richard, 'Visions of Hate. Explaining Neo-Nazi Violence in the Russian Federation', *Problems of Post-Communism,* 57/2, 2010, 37–49.

Austin, J. L., *How to Do Things With Words*, Cambridge: Harvard University Press, 1962.

Badmatsyrenov, Timur B., 'Sangkha i politika: politicheskie aspekty funktsionirovaniia buddiiskogo dukhovenstva Mongolii i Buriatii', *Vestnik Buriatskogo gosudarstvennogo universiteta*, 14, 2012, 137–43.

Bagdasarian, V. E., *Teoriia zagovora v otechestvennoii istoriografii vtoroi polovini 19–20 vv.*, Moscow: MSU, 1999.

Bakhtin, Mikhail, *The Dialogic Imagination: Four Essays*, Austin: University of Texas Press, 1981.

Bancroft, Angus, *Roma and Gypsy Travellers in Europe: Modernity, Race, Space and Exclusion*, Aldershot: Ashgate Publishing, 2005.

Barker, Martin, *New Racism: Conservatives and the Ideology of the Tribe*, New York: Junction Books, 1981.

Bazarov, B. V. and Zhabaeva, B. V., *Buriatskie natsional'nye demokraty i obshchestvenno-politicheskaia mysl mongolskikh narodov v pervoi treti XX veka*, Ulan-Ude: Izdatelstvo BNTs SO RAN, 2008.

Beers, Daniel, *Renovating Russia. The Human Sciences and the Fate of Liberal Modernity, 1880–1930*, Ithaca: Cornell University Press, 2008.

Beissinger, Mark R., 'The Persisting Ambiguity of Empire', *Post-Soviet Affairs*, 11/2, 1995, 149–84.

Benhabib, Seyla, *The Claims of Culture: Equality and Diversity in the Global Era,* Princeton: Princeton University Press, 2002.

Bennet, Samuel, Ter Wal, Jessika, Lipinski, Artur, Fabiszak, Malgorzata and Krzyzanowski, Michal, 'The Representation of Third-Country Nationals in European News Discourse. Journalistic Perceptions and Practices', *Journalism Practice*, 7/3, 2013, 248–65.

Bernstein, Anya, 'Remapping Sacred Landscapes: Shamanic Tourism and Cultural Production on Olkhon Island', *Sibirica: Journal of Siberian Studies*, 7/2, 2008, 23–46.

Bernstein, Anya, 'More Alive than All the Living: Sovereign Bodies and Cosmic Politics in Buddhist Siberia', *Cultural Anthropology*, 27/2, 2012, 261–85.

Billig, Michael, *Banal Nationalism*, London: Sage, 1995.

Brandenberger, David, *National Bolshevism: Stalinist Mass Culture and the Formation of Modern Russian National Identity, 1931–1956*, Cambridge: Harvard University Press, 2002.

Brubaker, Rogers, *Nationalism Reframed: Nationhood and the National Question in the New Europe*, Cambridge: Cambridge University Press, 1997.

Brubaker, Rogers, 'Ethnicity Without Groups', *Archives Européennes de Sociologie,* XLIII, 2, 2002, 163–89.

Brubaker, Rogers, *Ethnicity Without Groups*, Cambridge: Harvard University Press, 2004.

Brudny, Yitzhakh M., *Reinventing Russia. Russian Nationalism and the Soviet State 1953–1991*, Cambridge: Harvard University Press, 1998.

Burlakov, V. N., *Kriminologiia. Uchebnik dlia vuzov*, St. Petersburg: Izdatel'skii dom Piter, 2004.

Burrett, Tina, *Television and Presidential Power in Putin's Russia*, London: Routledge, 2010.

Bushkovitch, Paul, 'The Formation of a National Consciousness in Early Modern Russia', *Harvard Ukrainian Studies*, 10, 3/4, 1986, 355–76.

Campani, Giovanna, 'Migrants and Media. The Italian Case', in King, Russell and Wood, Nancy (eds.), *Media and Migration. Construction of Mobility and Difference*, London: Routledge, 2001, pp. 38–52.

Carlson, Maria, *'No Religion Higher Than Truth': A History of the Theosophical Movement in Russia, 1875–1922*, Princeton: Princeton University Press, 1993.

Carothers, Thomas, 'The End of the Transition Paradigm', *Journal of Democracy*, 13/1, 2002, 5–21.

Cere, Rinella, 'Globalization vs. Localization: Anti-Immigrant and Hate Discourses in Italy', in Ardizzoni, M. and Ferrari, C. (eds.), *Beyond Monopoly: Contemporary Italian Media and Globalization*, Lanham: Lexington Books, 2010, pp. 225–44.

Cere, Rinella and Brunt, Rosalind (eds.), *Postcolonial Media Culture in Britain*, Basingstoke: Palgrave Macmillan, 2011.

Chebankova, Elena, 'Contemporary Russian Multiculturalism', *Post-Soviet Affairs*, 28/3, 2012, 336–7.

Chouliaraki, Lilie, *The Spectatorship of Suffering*, London: Sage, 2006.

Cohen, Robin, 'The Incredible Vagueness of Being British/English', *International Affairs*, 76/3, 2000, 575–82.

Condee, Nancy, *The Imperial Trace. Recent Russian Cinema*, Oxford: Oxford University Press, 2009.

Cottle, Simon, *Ethnic Minorities and the Media: Changing Cultural Boundaries*, Buckingham: Open University Press, 2000.

Couldry, Nick, *Media Rituals. A Critical Approach*, London: Routledge, 2003.

Couldry, Nick, Hepp, Andreas and Krotz, Friedrich (eds.), *Media Events in a Global Age*, London: Routledge, 2010.

Dagbaev, Erdem D., *Agvan Dorzhiev v istorii Rossiiskogo gosudarstva*, Ulan-Ude: Belig, 2005.

Dagbaev, Erdem D., *Institutsional'nyi dizain politicheskikh protsessov v rossiiskikh regionakh Vnutrennei Azii i Mongolii*, Uland-Ude: Izdatel'stvo BNTs SO RAN, 2011.

Daniliuk, A. Iu., Kondakov, A.M. and Tishkov, V. A., *Kontseptsii dukhovno-nravstvennogo razvitiia i vospitaniia lichnosti grazhdanina Rossii*, Moscow: Prosveshchenie, 2009.

Dayan, Daniel and Katz, Elihu, *Media Events: The Live Broadcasting of History*, Cambridge: Harvard University Press, 1992.

De Benoist, Alain and Champetier, Charles, 'The French New Right in the Year 2000', *Telos*, 115, 1999, 117–44.

Dijk, Teun van, *Racism and the Press*, London: Routledge, 1991.

Douglas, Mary, *Purity and Danger*, London: Routledge, 1984.

Downing, John and Husband, Charles, *Representing 'Race': Racisms, Ethnicities and the Media*, London: Sage, 2005.

Ducrot, Oswald and Todorov, Tzvetan, *Encyclopaedic Dictionary of the Science of Language*, Baltimore: John Hopkins University Press, 1979.

Dunn, John, 'Humour and Satire on Post-Soviet Russian Television', in Milne, Lesley (ed.), *Reflective Laughter: Aspects of Humour in Russian Culture*, London: Anthem Press, 2004, pp. 181–92.

Durkheim, Emile, *The Elementary Forms of Religious Life*, New York: The Free Press, 1995 [1912].

Ellinas, Antonis, *The Media and the Far Right in Western Europe: Playing the Nationalist Card*, Cambridge: Cambridge University Press, 2010.

Ellis, John, 'Television as Working-through', in Gripsrud, Jostein (ed.), *Television and Common Knowledge*, London: Routledge, 1999, pp. 55–71.

Erikson, E. V., 'Ob ubiistvakh i razboiakh na Kavkaze', in Avdeev, V. B. (ed.), *Russkaia rasovaia teoriia do 1917 goda*, vyp. 2, Moscow: FERI-V, 2004.

Erjavec, Karen, 'Media Representation of the Discrimination against the Roma in Eastern Europe: The Case of Slovenia', *Discourse Society,* 12/6, 2001, 699–727.

Evans, Alfred, 'Putin's Legacy and Russia's Identity', *Europe-Asia Studies*, 60/6, 2008, 899–912.

Fiske, John and Hartley, John, *Reading Television*, London: Routledge, 1996.

Fitzpatrick, Sheila, *The Cultural Front: Power and Culture in Revolutionary Russia*, Ithaca: Cornell University Press, 1992.

Flood, Christopher, Hutchings, Stephen, Miazhevich, Galina and Nickels, Henri, 'Between Impartiality and Ideology: The BBC's Paradoxical Remit and the Case of Islam-related Television News', *Journalism Studies*, 12/2, 2011, 221–38.

Flood, Christopher, Hutchings, Stephen, Miazhevich, Galina and Nickels, Henri, *Islam, Security and Television News: A Comparative Study*, Basingstoke: Palgrave Macmillan, 2012.

Foxall, Andrew, 'Discourses of Demonisation: Chechens, Russians, and the Stavropol' Riots of 2007', *Geopolitics,* 15/4, 2010, 684–704.

Fredrickson, George, *Racism. A Short History*, Princeton: Princeton University Press, 2002.

Genette, Gerard, *Narrative Discourse: An Essay in Method*, Ithaca: Cornell University Press, 1983.

Gerasimov, Ilya, Glebov, Sergey and Mogilner, Marina, 'The Postimperial Meets the Postcolonial: Russian Historical Experience and the Postcolonial Moment', *Ab Imperio*, 2, 2013, 97–135.

Gigliotti, Simone, 'Unspeakable Pasts as Limit Events: The Holocaust, Genocide, and the Stolen Generations', *Australian Journal of Politics and History*, 49/2, 2003, 164–81.

Gramsci, Antonio, *Selections from the Prison Notebooks*, New York: International Publishers, 1971.

Gray, Herman, *Watching Race: Television and the Struggle for Blackness*, Minneapolis: University of Minnesota Press, 1995.

Gudkov, Lev, *Negativnaia identichnost'. Stat'i 1997–2002 godov*, Moscow: Novoe literaturnoe obozrenie, 2004.

Gudkov, Lev (ed.), *Obraz vraga*, Moscow: OGI, 2005.

Hafez, Kai (ed.), *Islam and the West in the Mass Media: Fragmented Images in a Globalizing World*, Cresskill: Hampton Press, 2000.

Hall, John A., 'Nationalisms: Classified and Explained', *Daedalus*, 122/3, 1993, 1–28.

Haywood, Susan, *Cinema Studies: The Key Concepts*, London: Routledge, 2000.

Hechter, Michael, *Internal Colonialism: The Celtic Fringe in British National Development, 1536–1966*, Berkley: University of California Press, 1975.

Herman, Edward and Chomsky, Noam, *Manufacturing Consent: The Political Economy of the Mass Media*, New York: Pantheon Books, 1988.

Hirsch, Francine, *Empire of Nations. Ethnographic Knowledge and the Making of the Soviet Union*, Ithaca: Cornell University Press, 2005.

Hobsbawm, Eric and Ranger, Terence (eds.), *The Invention of Tradition*, Cambridge: Cambridge University Press, 1993.

Horowitz, David, *The Deadly Ethnic Riots*, Berkeley: University of California Press, 2001.

Humphrey, Caroline, *Marx Went Away but Karl Stayed Behind*, Ann Arbor: University of Michigan Press, 1998.

Humphrey, Caroline, '"Eurasia". Ideology and the Political Imagination in Provincial Russia', in Hann, Christopher M. (ed.), *Postsocialism. Ideals, Ideologies and Practices in Eurasia*, London: Routledge, 2002, pp. 258–76.

Hunter, Shireen, *Islam in Russia: The Politics of Identity and Security*, New York: M.E. Sharpe, 2004.

Hutchings, Stephen, *Screening Intercultural Dialogue: Russia and its Other(s) on Film*, Basingstoke: Palgrave Macmillan, 2008.

Hutchings, Stephen, 'The Gypsy as Vanishing Mediator in Russian Television Coverage of Interethnic Tension', *Nationalities Papers*, 41/5, 2013, 1–17.

Hutchings, Stephen, Miazhevich, Galina, Flood, Christopher and Nickels, Henri, 'The Impact of "Islamic Extremism" on TV News Representations of Multiculturalism, *Intégration* and *Mnogonarodnost*: A Comparative Analysis', *Russian Journal of Communication*, 1/1, 2008, 43–70.

Hutchings, Stephen and Rulyova, Natalia, *Television and Culture in Putin's Russia: Remote Control*, London: Routledge, 2009.

Hutchings, Stephen and Rulyova, Natalia, 'Commemorating the Past/Performing the Present: Television Coverage of the Second World War Victory Celebrations and the (De)construction of Russian Nationhood', in Beumers, Birgit, Hutchings, Stephen and Rulyova, Natalia (eds.), *The Post-Soviet Media: Conflicting Signals*, London: Routledge, 2009, pp. 137–57.

Hutchings, Stephen and Tolz, Vera, 'Fault Lines in Russia's Discourse of Nation: Television Coverage of the December 2010 Moscow Riots', *Slavic Review*, 71/4, 2012, 873–99.

Ivakhnyuk, Irina, *Russian Migration Policy and Its Impact on Human Development*, Human Development Reports Research Paper, New York: United Nations Development Programme, 2009.

Jackson, Richard, *Writing the War on Terrorism: Language, Politics and Counter-terrorism*, Manchester: Manchester University Press 2005.

Jackson, Richard, 'Constructing Enemies: "Islamic Terrorism" in Political and Academic Discourse', *Government and Opposition*, 42/3, 2007, 394–426.

Jacobs, Ronald and Townsley, Eleanor, *The Space of Opinion: Media Intellectuals and the Public Sphere*, New York: Oxford University Press, 2011.

Jameson, Fredric, 'Imaginary and Symbolic in Lacan: Marxism, Psychoanalytic Criticism, and the Problem of the Subject', *Yale French Studies*, 55/56, 1977, 338–95.

Johnson, Douglas, 'The New Right in France', in Cheles, Luciano, Ferguson, Ronnie and Vaughan, Michalina (eds.), *Neo-Fascism in Europe*, London: Longman, 1991, pp. 234–44.

Johnson-Cartee, Karen S., *News Narratives and News Framing: Constructing Political Reality*, Oxford: Rowman and Littlefield, 2005.

Jusic, Tarik, 'Media Discourse and the Politics of Ethnic Conflict: The Case of Yugoslavia', in Kolstø, Pål (ed.), *Media Discourse and the Yugoslav Conflicts. Representations of Self and Other*, Farnham: Ashgate, 2009, pp. 24–38.

Kelly, Catriona, *Children's World: Growing Up in Russia, 1890–1891*, London and New Haven: Yale University Press, 2007.

Kelly, Raymond, 'Witchcraft and Sexual Relations: An Exploration in the Social and Semantic Implications of the Structure of Belief', in Brown, Paula and Buchbinder, Georgeda (eds.), *Man and Woman in the New Guinea Highlands*, Washington DC: American Anthropological Association, 1976, pp. 36–53.

Khalid, Adeeb, 'The Soviet Union as an Imperial Formation: A View from Central Asia', in Stoler, Ann, McGranahan, Carole and Perdue, Peter (eds.), *Imperial Formations*, Santa Fe: SAR Press, 2007, pp. 123–51.

Khamutaev, V. A., *Natsional'nyi vopros v Buriatii. 1980–2000-e gody*, Ula-Ude: Izdatel'stvo BNTs SO RAN, 2008.

King, Russell and Wood, Nancy (eds.), *Media and Migration. Construction of Mobility and Difference*, London: Routledge, 2001.

Kirill, the Patriarch of Moscow and All Russia, *Freedom and Responsibility: A Search for Harmony*, London: Darton, Longman and Todd, 2011.

Klandermans, Bert and Mayer, Nonna, *Extreme Right Activists in Europe: Through the Magnifying Glass*, London: Routledge, 2006.

Knight, Peter, *Conspiracy Nation: The Politics of Paranoia in Postwar America*, New York: New York University Press, 2002.

Knox, Zoe, *Russian Society and the Orthodox Church: Religion in Russia after Communism*, New York: Routledge, 2005.

Knox, Zoe, 'The Watch Tower Society and the End of the Cold War: Interpretations of the End-Times, Superpower Conflict, and the Changing Geo-Political Order', *Journal of the American Academy of Religion,* 79/4, 2011, 1018–49.

Knox, Zoe, 'Preaching the Kingdom Message: The Jehovah's Witnesses and Soviet Secularization', in Wanner, Catherine (ed.), *State Secularism and Lived Religion in Russia and Ukraine*, Oxford: Oxford University Press, 2012, pp. 244–71.

Knox, Zoe, 'Church, State, and Belief in Post-Soviet Russia', *The Russian Review,* 71/1, 2012, 122–7.

Kolstø, Pål (ed.), *Media Discourse and the Yugoslav Conflicts. Representations of Self and Other*, Farnham: Ashgate, 2009.

Koltsova, Olessia, *New Media and Power in Russia*, London: Routledge, 2006.

Kornblatt, Judith, *The Cossack Hero in Russian Literature: A Study in Cultural Mythology*, Madison: University of Wisconsin Press, 1992.

Krastev, Ivan and Holmes, Stephen, 'An Autopsy of Managed Democracy', *Journal of Democracy*, 23/3, 2012, 33–45.

Kristeva, Julia, *Powers of Horror: An Essay on Abjection*, trans. Leon Roudiez, New York: Columbia University Press, 1982.

Kroz, Mikhail, 'Ksenofobiia v rossiiskikh SMI kak proiavlenie ekstremizma', in Laruelle, Marlène (ed.), *Russkii natsionalizm. Sotsial'nyi i kul'turnyi kontekst*, Moscow: Novoe literaturnoe obozrenie, 2008, pp. 425–44.

Laclau, Ernesto, *Hegemony and Socialist Strategy: Towards a Radical Democratic Politics*, London: Verso, 1985.

Laclau, Ernesto, *On Populist Reason*, London: Verso, 2005.

Lapidus, Gail Warhofsky, 'Ethnonationalism and Political Stability. The Soviet Case', *World Politics*, 36/4, 1984, 555–80.

Laruelle, Marlène, *Russian Eurasianism: The Ideology of Empire*, Baltimore: Johns Hopkins University Press, 2008.

Laruelle, Marlène, *In the Name of the Nation. Nationalism and Politics in Contemporary Russia*, Basingstoke: Palgrave Macmillan, 2009.

Laruelle, Marlène, 'The Ideological Shift on the Russian Radical Right. From Demonizing the West to Fear of Migrants', *Problems of Post-Communism*, 57/6, 2010, 19–31.

Laruelle, Marlène, 'Alexei Navalny and Challenges in Reconciling "Nationalism" and "Liberalism"', *Post-Soviet Affairs*, 30/4, 2014, 276–97.

Laryš, Martin and Mareš, Miroslav, 'Right-Wing Extremist Violence in the Russian Federation', *Europe-Asia Studies,* 63/1, 2011, 129–54.

Layton, Susan, *Russian Literature and Empire: Conquest of the Caucasus from Pushkin to Tolstoy*, New York: Cambridge University Press, 1994.

Lemon, Alaina, *Between Two Fires: Gypsy Performance and Romani Memory from Pushkin to Post-Socialism*, Durham: Duke University Press, 2000.

Lentin, Alan, 'Imagining the West, Perceiving Race: Social Sciences and Political Imagination', in Bottici, Chiara and Challand, Benoit (eds.), *The Politics of Imagination*, Abingdon: Birkbeck Law Press, 2012, pp. 109–23.

Lentin, Alan and Titley, Gavan, *The Crises of Multiculturalism: Racism in a Neoliberal Age*, New York: Zed Books, 2011.

Lloyd-Jones, Gaynor, 'Design and Control Issues in Qualitative Case Study Research', *International Journal of Qualitative Methods*, 2/2, 2003, 1–20.

Lockyer, Sharon (ed.), *Reading Little Britain: Comedy Matters on British Television*, London: I.B. Tauris, 2010.

Lotman, Yurii, *Universe of the Mind: A Semiotic Theory of Culture*, Indianopolis: Indiana University Press, 2001.

Lucacs, Georg, *Theory of the Novel*, trans. Anna Bostock, London: Merlin Press, 1971.

Lucas, Edward, *Deception: Spies, Lies and How Russia Dupes the West*, London: Bloomsbury Publishing, 2013.

MacFadyen, David, *Russian Television Today: Primetime Drama and Comedy*, London: Routledge, 2007.

Maiorova, Olga, *From the Shadow of Empire: Defining the Russian Nation through Cultural Mythology, 1855–1870*, Madison: The University of Wisconsin Press, 2010.

Malakhov, Vladimir, '"Natsional'naia politika" kak fenomen politicheskoi, rechi', in Malakhov, V. S., *Ponaekhali tut: Ocherki o natsionalizme, rasizme i kul'turnom pliuralizme*, Moscow: Novoe literaturnoe obozrenie, 2007.

Malakhov, V. S., *Ponaekhali tut... Ocherki o natsionalizme, rasizme i kul'turnom pliuralizme*, Moscow: Novoe literaturnoe obozrenie, 2007.

Martin, Terry, 'The Origins of Soviet Ethnic Cleansing', *The Journal of Modern History*, 70/4, 1998, 813–61.

Martin, Terry, *The Affirmative Action Empire. Nations and Nationalism in the Soviet Union, 1923–1939,* Ithaca: Cornell University Press, 2001.

Martin-Barbero, Jesus, *Communication, Culture and Hegemony: From the Media to Mediations*, London: Sage, 1993.

McCombs, Maxwell, 'Building Consensus: The News Media's Agenda-Setting Roles', *Political Communication*, 14/4, 1997, 433–43.

McCrone, David and McPherson, Gayle (eds.), *National Days. Constructing and Mobilising National Identity*, Basingstoke: Palgrave Macmillan, 2009.

Meer, Nasar and Modood, Tariq, 'How Does Interculturalism Contrast with Multiculturalism?', *Journal of Intercultural Studies*, 33/2, 2012, 175–96.

Melvin, Neil, *Russians beyond Russia. The Politics of National Identity*, London: Royal Institute of International Affairs, 1995.

Mickiewicz, Ellen, *Split Signals: Television and Politics in the Soviet Union*, New York: Oxford University Press, 1988.

Mickiewicz, Ellen, *Changing Channels: Television and the Struggle for Power*, Durham: Duke University Press, 1999.

Mickiewicz, Ellen, *Television, Power, and the Public in Russia*, Cambridge: Cambridge University Press, 2008.

Miller, Aleksei, 'The Empire and the Nation in the Imagination of Russian Nationalism', in Miller, Aleksei and Rieber, Alfred J. (eds.), *Imperial Rule*, Budapest: Central European University Press, 2004, pp. 9–26.

Miller, Aleksei, 'Istoricheskaia politika v Rossii: novyi povorot?', in Miller, A. and Lipman, M. (eds.), *Istoricheskaia politika v XXI veke: sbornik statei*, Moscow: Novoe literaturnoe obozrenie, 2012, pp. 334–7.

Miller, Aleksei, 'Istoriia poniatiia natsiia v Rossii', in Sdvizhkov, D. and Schirle, I. (eds.), *Poniatiia o Rossii*, vol. 2, Moscow: Novoe literaturnoe obozrenie, 2012, pp. 7–49.

Mogilner, Marina, *Homo imperii. Istoriia fizicheskoi antropologii v Rossii (konets XIX-nachalo XX vv.)*, Moscow: Novoe literaturnoe obozrenie, 2008.

Moore, Kerry, Gross, Bernhard and Threadgold, Terry (eds.), *Migrations and the Media*, New York: Peter Lang, 2012.

Morozov, Evgenii, *The Net Delusion: The Dark Side of Internet Freedom*, New York: Public Affairs, 2011.

Mukomel', V. I., 'Rossiiskie diskursy o migratsii', *Vestnik obshchestvennogo mneniia*, 1, 2005, 48–58.

Nora, Pierre, 'Between Memory and History: Les Lieux de Mémoire', *Representations*, 26, 1989, pp. 7–24.

Nora, Pierre and Kritzman, Lawrence D. (eds.), *Realms of Memory: Rethinking the French Past*, vol. 1, New York, Columbia University Press, 1996.

Nordstrom, Carolyn, *Shadows of War: Violence, Power and International Profiteering in the Twenty-First Century*, Los Angeles: University of California Press 2004.

Oates, Sarah, *Television, Democracy and Elections in Russia*, London: Routledge, 2008.

Osipov, Alexander, 'Minority Questions: Ethnicity, Discrimination, and Extremism in Russia', *Problems of Post-Communism,* 57/2, 2010, 50–60.

Oushakine, Serguei, *Patriotism of Despair: Nation, War, and Loss in Russia*, Ithaca: Cornell University Press, 2009.

Oushakine, Serguei, 'V poiskakh mesta mezhdu Stalinym i Gitlerom: o postkolonial'nykh istoriiakh sotsializma', *Ab Imperio*, 1, 2011, 209–33.

Park, Robert E., 'Human Migration and the Marginal Man', *The American Journal of Sociology*, XXXIII/6, 1928, 881–93.

Pavlinskaia, L. R., *Buriaty. Ocherki etnicheskoi istorii*, St. Petersburg: Evropeiskii dom, 2008.

Philo, Greg, Briant, Emma and Donald, Pauline, *Bad News for Refugees*, London: Pluto Press, 2013.

Platonov, Rachel, *Singing the Self: Guitar Poetry, Community and Identity in the Post-Stalin Period*, Evanston: Northwestern University Press, 2012.

Poole, Elisabeth, *Reporting Islam: Media Representations of British Muslims*, London: I.B. Tauris, 2002.

Poole, Elisabeth and Richardson, John, *Muslims and the News Media*, London: I.B. Tauris, 2006.

Popescu, Nicu, 'The Strange Alliance of Democrats and Nationalists', *Journal of Democracy*, 23/3, 2012, 46–54.

Postill, John, *Media and Nation Building. How the Iban Became Malaysian*, New York: Berghahn, 2006.

Prokhorova, Elena, 'Can the Meeting Place Be Changed? Crime and Identity Discourse in Russian TV Series of the 1990s', *Slavic Review,* 62/3, 2003, 512–24.

Rogers, Douglas, 'Postsocialisms Unbound: Connections, Critiques, Comparisons', *Slavic Review*, 69/1, 2010, 1–15.

Roth-Ey, Kristin, *Moscow Prime Time: How the Soviet Union Built the Media Empire That Lost the Cultural Cold War*, Ithaca: Cornell University Press, 2011.

Roudometof, Victor, 'Toward an Archaeology of National Commemoration in the Balkans', in Geisler, Michael E. (ed.), *National Symbols, Fractured Identities: Contesting the National Narrative,* Hanover: University Press of New England, 2005, pp. 35–59.

Rulyova, Natalia and Zagibalov, Taras, 'Blogging the Other: Construction of National Identities in the Blogosphere', *Europe-Asia Studies,* 64/8, 2012, 1524–45.

Ryazanova-Clarke, Lara, 'The "West" in the Linguistic Construction of Russianness in Contemporary Public Discourse', in Alapuro, Risto, Mustajoki, Arto and Pesonen, Pekka (eds.), *Understanding Russianness*, London: Routledge, 2012, pp. 3–18.

Said, Edward, *Covering Islam: How the Media and the Experts Determine How We See the Rest of the World*, London: Vintage, 1997.

Sakwa, Richard, *The Crisis of Russian Democracy: The Dual State, Factionalism and the Medvedev Succession*, Cambridge: Cambridge University Press, 2011.

Sarsenov, Karin, 'Televising Soviet Tropes: Reforging a Supraethnic Cultural Identity', *Slavica Lundensia,* 24, 2009, 257–77.

Schain, Martin, Zolberg, Aristide and Hossay, Patrick (eds.), *Shadows over Europe: the Development and Impact of the Extreme Right in Western Europe*, New York: Palgrave, 2002.

Schenk, Caress, 'Nationalism in the Russian Media: Content Analysis of Newspaper Coverage Surrounding Conflict in Stavropol, 24 May–7 June 2007', *Nationalities Papers,* 40/5, 2012, 783–805.

Schimpfossl, Elisabeth and Yablokov, Ilya, 'Coercion or Conformism? Perceptions of Censorship and Self-Censorship among Russian Television Reporters in the 2010s', *Demokratizatsia. The Journal of Post-Soviet Democratization*, 22/2, 2014, 295–311.

Shevel, Oxana, *Migration, Refugee Policy, and State Building in Postcommunist Europe*, Cambridge: Cambridge University Press, 2011.

Shevel, Oxana, 'Russian Nation-building from Yel'tsin to Medevdev: Ethnic, Civic or Purposefully Ambiguous?', *Europe-Asia Studies*, 63/2, 2011, 179–202.

Shevtsova, Lilia, 'Putinism under Siege. Implosion, Atrophy, or Revolution?', *Journal of Democracy*, 23/2, 2012, 19–32.

Shevtsova, Lilia, 'Russia under Putin: Titanic Looking for Its Iceberg?', *Communist and Post-Communist Studies*, 45, 2012, 209–16.

Shlapentokh, Dmitry (ed.), *Russia between East and West: Scholarly Debates on Eurasianism*, Leiden: Brill, 2007.

Shnirel'man, Viktor A., *'Porog tolerantnosti'. Ideologiia i praktika novogo rasizma*, 2 vols., Moscow: Novoe literaturnoe obozrenie, 2011.

Shohat, Ella and Stam, Robert, *Unthinking Eurocentrism: Multiculturalism and the Media*, London: Routledge, 1994.

Silverstone, Roger, 'Proper Distance: Toward an Ethics for Cyberspace', in Liestøl, Gunnar, Morrison, Andrew and Rasmussen, Terje (eds.), *Digital Media Revisited: Theoretical and Conceptual Innovations in Digital Domains*, Cambridge: MIT Press, 2003, pp. 469–90.

Silverstone, Roger, *Media and Morality: On the Rise of the Mediapolis*, Cambridge: Polity Press, 2007.

Simon, Vera, 'Reunification Day – Day of German Unity?', in McCrone, David and McPherson, Gayle (eds.), *National Days. Constructing and Mobilising National Identity*, Basingstoke: Palgrave Macmillan, 2009, pp. 151–65.

Skrynnikova, Tatyana and Amogolonova, Darima, 'Symbols of Post-Soviet Buriat Consolidation', *Zhurnal Sibirskogo Federalnogo Universiteta*, 4/6, 2011, 797–8.

Slezkine, Yuri, 'The USSR as a Communal Apartment, or How a Socialist State Promoted Ethnic Particularism', *Slavic Review*, 53/2, 1994, 414–52.

Sonneman, Toby, 'Dark Mysterious Wanderers: The Migrating Metaphor of the Gypsy', *The Journal of Popular Culture*, 32/4, 1999, 119–39.

Stroganova, E. N., *Buriatskoe national'no-kulturnoe vozrozhdenie*, Moscow-Irkutsk: Natalis, 2001.

Strukov, Vlad, 'Russia's Internet Media Policies: Open Space and Ideological Closure', in Beumers, Birgit, Hutchings, Stephen and Rulyova, Natalia (eds.), *The Post-Soviet Russian Media: Conflicting Signals*, London: Routledge, 2009, pp. 208–23.

Suny, Ronald G., *The Revenge of the Past. Nationalism, Revolution, and the Collapse of the Soviet Union*, Stanford: Stanford University Press, 1993.

Suny, Ronald G., 'The Empire Strikes Out: Imperial Russia, "National Identity", and Theories of Empire', in Suny, Ronald G. and Martin, Terry (eds.), *A State of Nations. Empire and Nation-Making in the Age of Lenin and Stalin*, New York: Oxford University Press, 2001, pp. 23–66.

Thussu, Dayan, *News as Entertainment: The Rise of Global Infotainment*, London: Sage, 2007.

Tishkov, Valerii, *Ethnicity, Nationalism and Conflict In and After the Soviet Union*, London: Sage, 1997.

Tishkov, Valerii, 'Samoopredelenie rossiiskoi natsii', *Mezhdunarodnye protsessy*, 3/2, 2005, 17–27.

Tishkov, Valerii, 'Nationalism i vybornye kampanii', *Vestnik rossiiskoi natsii*, 2–3, 2012, 72–90.

Todorov, Tzvetan, *On Human Diversity: Nationalism, Racism and Exoticism in French Thought*, Cambridge: Harvard University Press, 1993, pp. 57–60.

Tolson, Andrew, *Media Talk: Spoken Discourse on TV and Radio*, Edinburgh: Edinburgh University Press, 2006.

Tolz, Vera, 'The Radical Right in Post-Communist Russian Politics', in Merkel, Peter H. and Weinberg, Leonard (eds.), *The Revival of Right-Wing Extremism in the Nineties*, Portland: Frank Cass, 1997, pp. 177–202.

Tolz, Vera, *Russian Academicians and the Revolution: Combining Professionalism and Politics*, Basingstoke: Macmillan, 1997.

Tolz, Vera, 'Conflicting "Homeland Myths" and Nation-State Building in Post-Communist Russia', *Slavic Review*, 57/2, 1998, 267–94.

Tolz, Vera, 'Forging the Nation: National Identity and Nation Building in Post-Communist Russia', *Europe-Asia Studies*, 50/6, 1998, 993–1022.

Tolz, Vera, *Russia: Inventing the Nation*, London: Arnold/Hodder, 2001.

Tolz, Vera, 'A Future Russia: A Nation-state or a Multi-national Federation', in Slater, Wendy and Wilson, Andrew (eds.), *The Legacy of the Soviet Union*, Basingstoke: Palgrave Macmillan, 2004, pp. 17–38.

Tolz, Vera, *'Russia's Own Orient'. The Politics of Identity and Oriental Studies in the Late Imperial and Early Soviet Periods*, Oxford: Oxford University Press, 2011.

Treisman, Daniel, *The Return: Russia's Journey from Gorbachev to Medvedev*, New York: Free Press, 2011.

Trenin, Dmitri, 'No Return to the Past for Russia', *The International Spectator*, 47/3, 2012, 8–12.

Triandafyllidou, Anna, 'Migrants and the Media in the Twenty-First Century. Obstacles and Opportunities for the Media to Reflect Diversity and Promote Integration', *Journalism Practice*, 7/3, 2013, 240–7.

Umland, Andreas (ed.), *Theorizing Post-Soviet Extreme Right: Comparative Political, Historical and Sociological Approaches*, Armonk: M. E. Sharp, 2008.

Umland, Andreas, *The Nature of Russian "Neo-Eurasianism": Approaches to Aleksandr Dugin's Post-Soviet Movement of Radical Anti-Americanism,* Armonk: M. E. Sharpe, 2009.

Urban, Michael, 'Remythologising the Russian State', *Europe-Asia Studies*, 50/6, 1998, 969–92.

Volkov, Denis, 'The Protesters and the Public', *Journal of Democracy*, 23/3, 2012, 55–62.

Voloshinov, Valentin, *Marxism and the Philosophy of Language*, Cambridge: Harvard University Press, 1986.

Weisberger, Adam, 'Marginality and Its Directions', *Sociological Forum*, 7/3, 1992, 425–46.

Weitz, Eric D., Hirsch, Francine, Weiner, Amir and Lemon, Alaina, 'Discussion of Eric D. Weitz's "Racial Politics without the Concept of Race: Reevaluating Ethnic and National Purges"', *Slavic Review*, 61/1, 2002, 1–65.

White, Hayden, 'The Value of Narrativity in the Representation of Reality', in Mitchell, W. J. T. (ed.), *On Narrative*, Chicago: Chicago University Press, 1981, pp. 1–23.

White, Richard, 'National Days and the National Past in Australia', *Australian Cultural History*, 22, 2003, 55–72.

Wikan, Unni, *Generous Betrayal: Politics of Culture in the New Europe*, Chicago: Chicago University Press, 2002.

Wilson, Andrew, *Virtual Politics: Faking Democracy in the Post-Soviet World*, New Haven: Yale University Press, 2005.

Wolchik, Sharon, 'Putinism under Siege: Can There Be a Color Revolution?', *Journal of Democracy*, 23/3, 2012, 63–70.

Wolfsfeld, Gadi, *Media and the Path to Peace,* Cambridge: Cambridge University Press, 2004.

Yablokov, Ilya, 'Conspiracy Discourse in Post-Soviet Russia: Political Strategies of Capture of the Public Sphere', PhD thesis, Russian and East European Studies, University of Manchester, UK, 2014.

Yemelianova, Galina (ed.), *Radical Islam in the Former Soviet Union*, London: Routledge, 2010.

Ytreburg, Espen, 'Moving out of the Inverted Pyramid: Narratives and Descriptions in Television News', *Journalism Studies*, 2/3, 2001, 357–71.

Zaslavsky, Victor and Brym, Robert, *Soviet-Jewish Emigration and Soviet National Policy*, New York: St. Martin's Press, 1983.

Zassoursky, Ivan, *Media and Power in Post-Soviet Russia*, Armonk: M.E. Sharpe, 2004.

Žižek, Slavoj, *The Ticklish Subject*, London: Verso, 1999.

Žižek, Slavoj, *Living in the End Times*, London: Verso, 2010.

Zuev, Dennis, 'The Russian Ultranationalist Movement on the Internet: Actors, Communities and Organization of Joint Actions', *Post-Soviet Affairs*, 27/2, 2011, 121–57.

Zuev, Dennis, 'The Russian March: Investigating the Symbolic Dimension of Political Performance in Modern Russia', *Europe-Asia Studies*, 65/1, 2013, 102–26.

Zvereva, Vera, '"Zakon i kulak": Rossiiskie militseiskie serialy', *Novoe literaturnoe obozrenie*, 78, 2006, 305–25.

Index

For Product Safety Concerns and Information please contact our EU representative GPSR@taylorandfrancis.com
Taylor & Francis Verlag GmbH, Kaufingerstraße 24, 80331 München, Germany

www.ingramcontent.com/pod-product-compliance
Ingram Content Group UK Ltd.
Pitfield, Milton Keynes, MK11 3LW, UK
UKHW020016200926
475702UK00014B/1907
9780815362326